Adversarial Example Detection and Mitigation Using Machine Learning

Ehsan Nowroozi • Rahim Taheri • Lucas Cordeiro
Editors

Adversarial Example Detection and Mitigation Using Machine Learning

Springer

Editors

Ehsan Nowroozi
Centre for Sustainable Cyber Security (CS2)
University of Greenwich
London, UK

Rahim Taheri
School of Computing
University of Portsmouth
Portsmouth, UK

Lucas Cordeiro
Department of Computer Science
University of Manchester (UoM), Systems and Software Security (S3) Research Group
Manchester, UK

ISBN 978-3-031-99446-3 ISBN 978-3-031-99447-0 (eBook)
https://doi.org/10.1007/978-3-031-99447-0

This Springer imprint is published by the registered company Springer Nature Switzerland AG
The registered company address is: Gewerbestrasse 11, 6330 Cham, Switzerland

Preface

In an era where artificial intelligence shapes critical sectors from healthcare to transportation, the security and robustness of AI systems have become paramount. Yet, the rise of adversarial threats—from data poisoning to backdoor attacks and beyond—has exposed significant vulnerabilities within even the most advanced machine learning models. This book brings together leading researchers and practitioners to address these pressing challenges, offering a comprehensive exploration of adversarial attacks and defenses across a wide range of AI applications.

The landscape of adversarial machine learning is rapidly evolving. Attacks targeting federated learning environments, biometric systems, code generation models, and large language models (LLMs) reveal that no domain is immune. Meanwhile, advances in quantum computing and generative AI introduce new complexities and risks. As adversaries innovate, so too must defenders. This book captures this dynamic arms race, covering both theoretical foundations and practical defense strategies with a focus on real-world applicability.

A key strength of this volume is its breadth and depth. From cluster-based defenses against GAN-driven poisoning attacks to explainable AI's role in securing systems, from detecting backdoors in text classification to mitigating adversarial threats in Digital Twin technologies, the chapters present cutting-edge research and actionable insights. Special attention is given to ethical and regulatory considerations, recognizing that technical solutions must align with societal values and legal standards.

The contributors span a global network of experts across academia, industry, and government, each bringing unique perspectives to the conversation. Whether exploring formal verification techniques, anomaly detection methods for IoT networks, or emerging threats to AI security, they provide a rich, multidimensional view of the adversarial AI frontier.

This book is intended for researchers, engineers, cybersecurity professionals, and students interested in the security, privacy, and resilience of AI systems. A basic understanding of machine learning concepts will enhance the reader's experience, but the material is accessible to anyone with a keen interest in understanding the new challenges that adversarial AI poses.

As the capabilities of AI continue to advance, so too must our ability to defend it. We hope this collection serves as both a resource and an inspiration—equipping readers with the knowledge to contribute to a safer, more trustworthy AI-powered future.

London, UK Ehsan Nowroozi
Portsmouth, UK Rahim Taheri
Manchester, UK Lucas Cordeiro

Acknowledgement The editors wish to express their deep appreciation to all those who contributed to the successful completion of this book. Foremost, heartfelt thanks go to the chapter authors, whose expertise, research excellence, and dedication in developing their contributions from the cornerstone of this volume. Their commitment to advancing the field of adversarial machine learning and AI security is reflected throughout the book.

We are equally grateful to the reviewers, whose careful evaluations and insightful feedback strengthened the scholarly quality of the chapters. Their critical perspectives and constructive suggestions were essential in refining and elevating the final work.

Special recognition is due to the publishing team, whose professional guidance and support throughout the editorial and production process were invaluable. Their meticulous attention to detail and encouragement made the journey from manuscript to publication both smooth and rewarding.

Finally, we extend our sincere gratitude to the broader research community and our families, colleagues, and institutions for their encouragement and unwavering support. It is our hope that all those who contributed to this effort will take pride in being part of this collaborative achievement.

1. Ghita Lazrek, Lab LIASSE, Sidi Mohammed Ben Abdellah University, ENSA, Fez, Morocco
2. Manimuthu Arunmozhi, Aston University
3. Raj Mani Shukla, Anglia Ruskin University
4. Muhammad Taimoor Khan, University of Greenwich
5. Madhurima Chakraborty, University of California, Riverside
6. Mohammad Saedi, University of London
7. Mahboobeh Dorafshanian, Université Laval
8. Mohammadreza Mohammadi, RISE Research Institutes of Sweden
9. Sima Sinaei, RISE Research Institutes of Sweden
10. Mohsen Eslamnejad, 5G/6G Innovation Centre, Institute for Communication Systems (ICS), University of Surrey, UK
11. Yousef Emami, University of Porto
12. Maryam Karimi, FM, Department of Innovation and Future of Industry
13. Ali Mehrabi, Western Sydney University
14. Hadiseh Razaei, University of Portsmouth
15. Saman Shojaei Chaeikar, Sydney International School of Technology and Commerce, Sydney, Australia
16. Mo Adda, University of Portsmouth
17. Nasreen Anjum, University of Portsmouth
18. Rosiane de Freitas Rodrigues, Institute of Computing, Federal University of Amazonas
19. Ramin Taheri, University of Portsmouth

20. Tihanyi Norbert, Eötvös Loránd University, Technology Innovation Institute
21. Edjard Mota, Federal University of Amazonas
22. Behrooz Razaghi, School of Engineering and Applied Sciences, Harvard University
23. Alireza Jolfaei, Flinders University

Contents

Part III Defense Techniques and Robustness Strategies

About the Editors

Dr. Ehsan Nowroozi is a Senior Lecturer (Associate Professor) in Cybersecurity at the University of Greenwich, United Kingdom, where he is affiliated with the Centre for Sustainable Cyber Security (CS2), from January 2025 to now, and research assistant at the City St George's, University of London, from May 2025 to July 2025. His academic journey is marked by a PhD in Information Engineering and Mathematics from the University of Siena, Italy, where he conducted pioneering research on machine learning techniques for image forensics in adversarial settings. His research has significantly contributed to areas including adversarial machine learning, multimedia and digital forensics, federated learning security, and resilient AI systems. Prior to his appointment at Greenwich, he held academic and research fellow positions across Europe, including at Ravensbourne University London, Queen's University Belfast (Centre for Secure Information Technologies), Bahçeşehir University in Istanbul, Sabanci University (VERİM), and the University of Padua (SPRITZ research group). His scholarly output includes numerous high-impact journal and conference publications, many of which are published in IEEE Transactions and Springer volumes. He has also contributed to internationally funded research projects such as the DARPA MediFor project and the EU's PREMIER and Lighthouse initiatives. He holds patents on novel deep learning approaches to network security and malicious URL detection. He serves as an Associate Editor for the *IEEE Transactions on Network and Service Management* and reviews for several prestigious journals including *IEEE*

Transactions on Industrial Informatics and *Computers & Security*. An active member of the IEEE (Senior Member) and ACM, he is a frequent presenter at major conferences such as IEEE ICASSP and WIFS and is involved in various technical committees and organizing roles. His teaching experience spans undergraduate and graduate levels, covering modules like Digital Forensics, Secure Programming, and AI for Security. He has supervised numerous MSc theses and has developed new curricula aligned with emerging threats in cybersecurity. Fluent in English, Persian, and Turkish, and with working knowledge of Italian, he brings an international perspective to research collaboration and education. His academic profile is complemented by a strong commitment to advancing the frontiers of secure and explainable AI, with a focus on ensuring the robustness and trustworthiness of machine learning systems in real-world applications.

Dr. Rahim Taheri is a Senior Lecturer at the School of Computing, University of Portsmouth, United Kingdom. He received his PhD in Computer Science and has over ten years of experience teaching computer science and programming at both undergraduate and postgraduate levels. He has supervised several PhD students and remains actively involved in mentoring and academic development. He has held research positions at King's College London, where he was affiliated with the King's Communications, Learning and Information Processing (KCLIP) lab, and at the University of Padua, Italy, where he collaborated with the SPRITZ Security and Privacy Research Group. These experiences have contributed to his international research profile in AI and cybersecurity. His research focuses on secure and privacy-preserving artificial intelligence, including federated learning, adversarial machine learning, and sustainable AI. He is particularly interested in defending AI models against data poisoning and other adversarial threats in applications such as IoT, connected vehicles, and distributed systems. He has published widely in high-impact journals and international conferences. He is a Senior Member of the IEEE and an ACM Professional Member. He is actively engaged in exploring innovative and ethical AI solutions to address modern security challenges.

Dr. Lucas C. Cordeiro is a Full Professor in the Department of Computer Science at the University of Manchester (UoM), where he leads the Systems and Software Security (S3) Research Group. He is also the Business Engagement and Innovation Director and the Arm Centre of Excellence Director at UoM. He is affiliated with the Trusted Digital Systems Cluster, the Formal Methods Group, and postgraduate programs at the Federal University of Amazonas, Brazil. He is the Chief Technology Officer at VeriBee, a UoM spinout focused on transforming the software testing market. He has co-authored over 170 peer-reviewed publications in prestigious venues and earned major international awards, including the Most Influential Paper at ASE'23, Distinguished Paper Awards at ICSE'11 and ASE'24, and 50 competition awards (2012–2025). He has secured over USD 13M in research funding from organizations such as EPSRC, Intel, Motorola, Samsung, Nokia Institute of Technology, CNPq, FAPEAM, the British Council, and the Royal Society.

Part I
Foundations of Adversarial Machine Learning

Chapter 1
A Brief Survey of Emerging Threats to AI Security

Carine Tannous, Hichem Faraoun, and Kassem Kallas

1.1 Introduction

In today's rapidly evolving world, Deep Learning (DL) models have been widely adopted across various fields, driving significant technological advancements due to their ability to process vast amounts of data and solve complex problems. However, with their increasing use in critical applications, they also face a wide spectrum of security threats. This chapter explores these threats, providing an overview of existing attacks targeting deep learning models. As a broad field, Artificial Intelligence (AI) encompasses various algorithms and techniques for developing applications, with machine learning as a key subset. Machine learning utilizes complex statistical methods, following a structured pipeline to generate predictions or insights for specific tasks.

The lifecycle of a machine learning model typically begins with data collection, where relevant data is gathered from various sources. This is followed by data cleaning and pre-processing, which entails correcting errors, handling inconsistencies, and normalizing the dataset to ensure accuracy and reliability. Once prepared, the model enters the training phase, where it learns patterns and complex relationships from the dataset, referred to as the training set. Training occurs over multiple iterations, known as epochs, where each epoch represents a complete pass through

---

These authors contributed equally to this work.

C. Tannous · H. Faraoun
IMT Atlantique, Department of Data Science, Plouzané, France

K. Kallas (✉)
National Institute of Health and Medical Research, Inserm UMR, Brest, France
e-mail: kassem.kallas@inserm.fr,
https://www.inserm.fr/

E. Nowroozi et al. (eds.), *Adversarial Example Detection and Mitigation Using Machine Learning*, https://doi.org/10.1007/978-3-031-99447-0_1

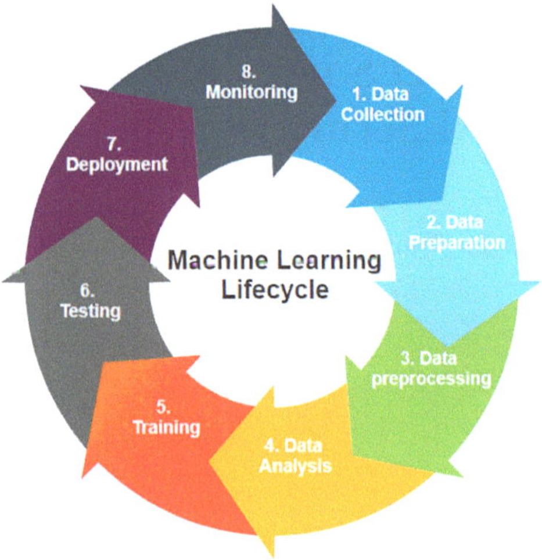

Fig. 1.1 A typical machine learning life cycle

the dataset. During this process, the model continuously refines its parameters using optimization techniques to minimize disparities between predictions and the ground truth values. After training, the model undergoes testing, where its generalization ability is evaluated on a separate dataset of unseen samples. This step ensures the model can correctly learn the function that maps input data to output while avoiding overfitting—where it performs well on training data but poorly on testing data. Based on test results, hyperparameters can be tuned to enhance overall performance. A well-optimized model should maintain high accuracy on testing samples. Finally, once the model meets the required accuracy and reliability, it is deployed to handle real-world, unseen data. This entire process—from data collection to pre-processing, training, testing, and optimization is known as the machine learning lifecycle and forms the foundation of machine learning, as illustrated in Fig. 1.1, and consists of the following steps:

1. **Data preparation:** this step includes data collection, preparation, pre-processing, and analysis.
2. **Training:** train the model to learn the approximation function that maps between the data inputs to the ground truth.
3. **Testing:** test the trained model on unseen data and evaluate its generalization performance.
4. **Deployment:** deploy the model in a production environment for predictions or decisions.
5. **Monitoring:** monitor and evaluate the performance of the deployed model on real-life data.

Each phase of the machine learning lifecycle is vulnerable to various attacks and security threats. For example, during the training phase, risks arise due to the large size of training datasets and the computational complexity of the process. Key threats include outsourcing training to third-party entities, which may introduce security risks; relying on pre-trained models that serve as intellectual property; and using large-scale data from unreliable sources, increasing the risk of adversarial manipulation, especially when proper data inspection and validation are lacking. Additionally, with the widespread adoption of Machine Learning as a Service (MLaaS), cloud-based model deployment has become common, allowing clients to interact with trained models through prediction APIs. However, when a deployed model is trained on a vast amount of sensitive data, its learned parameters become valuable assets, making it a prime target for attackers during the inference phase, where it interacts with real-world, unseen data. Adversaries may attempt to extract model parameters, infer sensitive training data, or manipulate inputs to exploit vulnerabilities.

A discussion in [1] explores the motivations behind attacks, highlighting ongoing debates on why such attacks succeed. In [2], Goodfellow et al. argue that the non-linearity of deep neural networks (DNNs) in high-dimensional spaces makes them vulnerable to adversarial examples. Meanwhile, [3] and [4] suggest that incomplete training data, which fails to cover all regions within the decision boundary, is a major cause of adversarial examples. As a result, these examples often represent low-probability edge cases in the test set, indicating insufficient data coverage during training. Yeom et al. [5] examine the impact of overfitting on training data, identifying it as a key factor that adversaries can exploit. Their findings highlight how overfitting facilitates membership inference attacks, where an attacker attempts to infer or reconstruct training data samples. Due to the opaque nature of deep learning models, fundamental questions about these attacks remain unresolved. Are adversarial examples an inherent property of the model, or do they represent anomalies? Furthermore, why is it possible to recover or reconstruct sensitive training data through seemingly normal inferences?

Machine learning security is built on the foundational CIA Triad: Confidentiality, Integrity, and Availability. While organizations like ANSI, NIST, and CISSP define these principles in slightly different ways, they all emphasize the same core security objectives. Figure 1.2 illustrates various attacks that threaten each axis. Confidentiality risks include model extraction, data reconstruction, and membership inference, which expose sensitive information. Integrity is compromised when adversarial or backdoor attacks manipulate model behavior, leading to unintended or malicious outcomes. Availability threats, such as energy-latency attacks and denial-of-service (DoS) attacks, aim to disrupt normal system operations. The following sections provide a deeper analysis of these threats and emphasize the importance of developing strong defensive measures. For a more detailed discussion on privacy attacks, see [6].

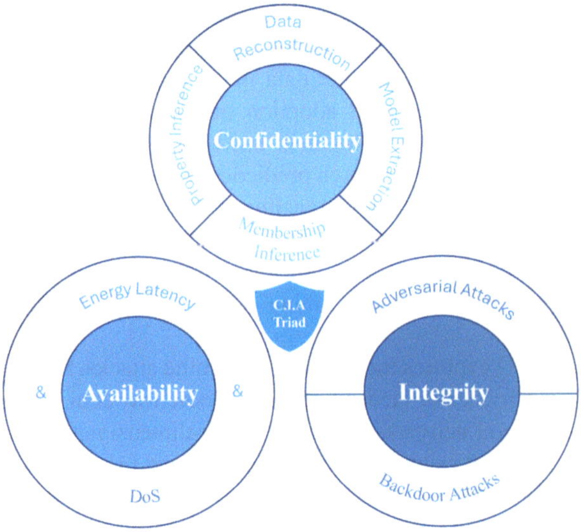

Fig. 1.2 CIA triad

1.2 Adversarial Attacks

Adversarial examples, introduced in [3], represent inputs carefully crafted to mislead DNN models into making wrong predictions or classifications. They exploit flaws in the decision-making mechanisms of the model, typically by injecting unidentifiable noise into the data samples, as shown in Fig. 1.3.

We can categorize adversarial attacks based on their goal and the attacker's level of access to the DL system. They may reduce confidence in the actual class label (**Confidence Reduction**), cause misclassification into any incorrect class (**Untargeted**), or deliberately force a misclassification into a specific incorrect class (**Targeted**). Access levels range from full knowledge of the model's architecture,

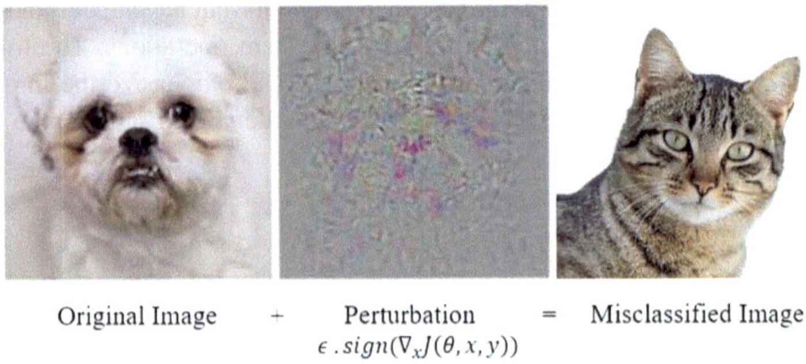

Fig. 1.3 Adversarial attack example

parameters (weights and biases), training algorithm, and potentially the training data (res. **white-box attacks**) to knowing only the model's type and outputs without access to its internal structure or training data (res. **black-box attacks**) [7].

These attacks come in a variety of modalities, ranging from image, text, audio, and video classification. They exhibit a *transferability* property, wherein attacks for a given model can, in general, fool another of similar architecture, or, an attack on a model trained on a given dataset can succeed on another trained on a different dataset of similar characteristics. This property makes it easier for an attacker, as they can develop the adversarial attack on their own local model before executing it on the victim model. However, there can be exceptions, especially in Convolutional Neural Networks (CNN)-based image forensics [8].

There are different ways of crafting adversarial examples, which involve computation of a gradient w.r.t. image pixels, or solving an optimization problem. Among current attacks, is using L-BFGS as a start point, that is computationally expensive, which prompted introduction of the Fast Gradient Sign Method (FGSM) by Goodfellow et al. [2] and which is a simple method in which the attacker attempts adding a minimal perturbation in input image so it steps in the opposite direction of the gradient, beyond a decision boundary towards a lower-confidence class. The process of this method is in three steps: calculation of a loss function post forward propagation $J(\theta, x, y)$, calculation of a gradient w.r.t. input image $\nabla_x J(\theta, x, y)$, and adjustment of image pixels by adding η in it, which is a step size, in a direction so it maximizes a previously estimated loss. Mathematically, it is given as in Eq. 1.1. And there have been stronger attacks since, such as PGD [9], and Shapeshifter [10] as an example of a white-box attack; BIM, ILCM [11] as universal adversarial perturbations—weaker than a white-box attack, though more transferable—, HipSkipJump [12] and others as a black-box attack, and, Auto-Attack [13] which is an ensemble of PGD, FAB and Square attack.

$$\hat{x} = x + \eta, \quad \text{where} \tag{1.1}$$

$$\eta = \epsilon \cdot \text{sign}(\nabla_x J(\theta, x, y)) \tag{1.2}$$

where, x is the input, θ represents the model parameters, and ϵ is a hyperparameter controlling perturbation magnitude.

1.3 Membership Inference

Membership inference attacks (MIAs) seek to find out whether a given data sample was used in a model's training, which make it a severe threat to privacy. Homer et al. [14] originally introduced these attacks, which exploit the property of the model having distinct behavior on training data vs. unseen data. The attacks have a similar connection with model over-fitting, which necessitates well-generalized models on unseen data samples (Fig. 1.4).

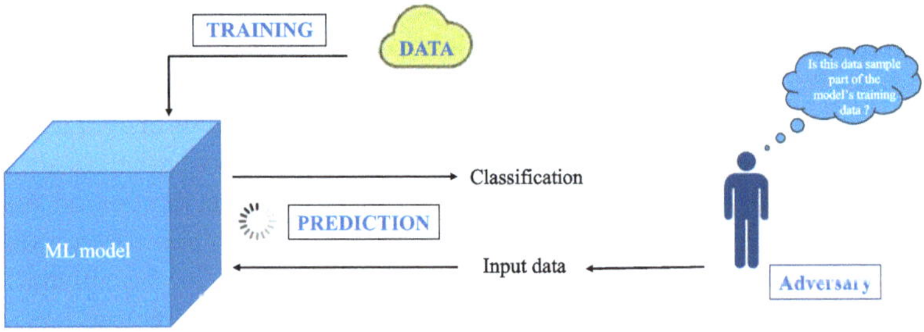

Fig. 1.4 A sketch of membership inference attack

MIAs have various implications: they can invade privacy by exposing sensitive training data; beneficial for data privacy and compliance, such as being employed in confirming a platform has deleted a given data point appropriately following a request for its erasure, under legal requirements of the right of being forgotten; and can serve as a stepping stone for more complicated attacks such as model extraction [15]. While MIAs have potential, success is not guaranteed and is dependent on various factors. The effect of MIAs is different in various fields; for instance, in medicine, there is a large risk of breaching privacy, since exposing an individual in a dataset may expose their medical status [16]. In other fields, such as image categorization, there is a smaller risk.

Several methods have attempted to identify the success of attacks such as:

- **Metric-based attacks**: The loss-based attack of Yeom et al. [17] assumes an adversary who knows a model's loss function and its loss-value distribution on private data. The attack assumes a weaker adversary who knows just a set of training data's expected loss and uses this quantity as a decision boundary. MIA is possible in a scenario of a lack of shadow models, according to Salem et al. [18], by establishing a predictive boundary of confidence. The label-only MIA began in [19] by querying a perturbation of a model's label predictions. Decision-based MI attacks have been proposed by Li et al. [20], which shows label-only instances being attackable too, forming transfer and boundary attacks.
- **Shadow Model attacks**: Shadow Model attacks make use of shadow models in order to mimic the actions of the target model and train a binary classifier for inference of membership [21]. Shokri et al. [21] originally proposed this attack, making use of shadow training in building an attack model of greater capability compared to random guesses. With auxiliary information, an adversary builds a shadow dataset of similar size and shape as the target dataset, trains a shadow model of similar shape as the target model, and accumulates vectors of predictions. The outputs in return train an attack model for inference of membership status. Later works made relaxation of assumptions [18] and generalized for various model architectures [22]. One of the well-known attacks, LiRA [23], provides a large increase in accuracy along with minimal false positives. It excels in terms of the Area Under the Curve (AUC), balancing between efficiency and complexity.

Prior studies have widely discussed membership inference attacks on DL models in a black-box setting, in which an adversary is granted query access to a trained model [17, 21, 23]. The growth in **Machine Learning as a Service (MLaaS)** usage further underlines concerns for privacy breaches as an adversary can craft MIAs based on knowing which particular service is being used [24]. Public privacy tools such as **TensorFlow Privacy** [25] and **ML Privacy Meter** [26] offer implementations of MIAs.

1.4 Data Reconstruction

Data reconstruction attacks, introduced by Dinur and Nissim [27], represent among the strongest threats to privacy, as they allow adversaries to extract personal information from aggregate data that appears to be anonymous. Recovery of data traditionally required an exponentially large number of queries, although subsequent studies have reduced this requirement down to a polynomial number of queries [28].

A related class of attacks, model inversion attacks, have been introduced by Fredrickson et al. [29] in an effort to reconstruct representative class samples of a model. While these attacks do not exactly reproduce training data, they generate outputs that closely resemble actual training instances. In a more recent effort, a reconstructor network has been proposed by Balle et al. [30], which can pull a data sample out of a neural network model under the assumption that the attacker knows all other training samples. Haim et al. [31] further investigated reconstruction attacks based on insights into implicit bias in neural networks [32], demonstrating that training data of a binary classifier can be recovered using its model parameters. One such similar attack is an attribute inference attack, in which an adversary attempts to infer sensitive attributes based on a partial assumption of other attributes in a training dataset [33].

Memorization of training data by neural networks intensifies threats to privacy. Zhang et al. [34] conducted studies on the memorization of randomly selected datasets by a model, and label memorization is required for the best achievable generalization error, as proposed by Feldman [35]. Empirical studies by Feldman and Zhang [36] indicated that memorization, in some cases, can improve generalization performance. However, this memorization also increases the vulnerability of data reconstruction. Recent studies have demonstrated that homogeneous neural networks can reconstruct a substantial portion of their trained data using only the learned model parameters, though a deeper understanding is required to determine the conditions and specific data points susceptible to reconstruction.

Threats to privacy in federated learning have been highlighted in newer data reconstruction works [37]. Shared models and gradient updates can be manipulated by an adversary to recover training data. Lastly, it was illustrated by Haim et al. [38] that Large Language Models (LLMs) are susceptible to data reconstruction attacks, describing a mechanism that not only reconstructs the entire fine-tuned private dataset but also the most informative words. Reconstruction attacks have been taken a step further by Salem et al. [39] using single-gradient updates in complex models. Conversely, another type of attack, which directly retrieves model parameters, enabling efficient reconstruction of data points even in fully retrained simple models, was introduced by Bertran et al. [40]. Furthermore, recent research has focused on gradient-based reconstruction attacks, primarily targeting untrained models [41].

1.5 Model Extraction/Stealing

Model extraction attacks pose a major threat for MLaaS providers, wherein it is of utmost priority to protect model architectures, parameters, and intellectual property. In this attack, an attacker attempts to obtain secret information about models, steal parameters, or clone a target model in an attempt to recreate a similar-functioning model compromising the model's confidentiality. From literature, it is in [42] that we have evidence of MLaaS systems' vulnerability to such attacks, capable of extracting models ranging from traditional ML algorithms such as decision trees and logistic regression all the way to deep networks. While it is theoretically unachievable for exact extraction, as claimed by Jagielski et al., other studies concede that it is achievable for an attack to produce a model with similar functionality to the target, raising concerns about model confidentiality and integrity.

Several strategies have been suggested for model extraction attacks, each using a different approach. Direct attacks make use of the mathematical modeling of deep networks, which allows an adversary to solve for model weights using algebra [42, 43]. Learning-based attacks, such as active learning and reinforcement learning, extract model parameters by optimizing queries in a more efficient manner. Another class includes side-channel attacks, which take advantage of unintended information disclosures, such as electromagnetic radiation or hardware flaws like Rowhammer, in an attempt to infer model parameters.

Model theft is typically accompanied by more powerful white-box or gray-box attacks, which call for more stringent defense measures. These attacks can generally be categorized into model attribute extraction, which focuses on stealing critical intellectual-property-related components such as parameters, hyperparameters, architecture, and decision boundaries, and model functionality extraction, which aims to replicate the target model's behavior. Examples of the former include StealML for parameter extraction [42], Steal-Hyperparameter [44], and StealNN, which utilizes timing side channels for architecture inference. For functionality extraction, attacks like StealClassifier and KnockoffNets target classifiers ranging from traditional algorithms like Naive Bayes and SVMs to deep learning models [45]. As model extraction techniques continue to evolve, understanding their practicality and effectiveness remains crucial in ensuring the security of cloud-based machine learning services. Several survey studies provide comprehensive analyses of existing model extraction attacks and defenses, as discussed in [46].

1.6 Property Inference

Deep learning models, known for their ability to learn high-level features from training data and make decisions, are susceptible to property inference attacks [47]. These attacks exploit the knowledge embedded in a model to extract relevant information about the training data distribution or even to reconstruct approximately training samples. Property inference attacks can target a wide range of models, from traditional machine learning algorithms like Support Vector Machines (SVMs) and Hidden Markov models (HMMs) [48] to deep learning architectures [49, 50]. Moreover, these threats extend to decentralized frameworks like Federated Learning [51] and even Generative Adversarial Networks [52].

Property inference attacks aim to undermine confidentiality by allowing adversaries to extract detailed information about the training data used to develop machine learning models. These attacks rely on analyzing model parameters to infer sensitive properties of the dataset. Furthermore, while watermarking techniques are commonly used to protect intellectual property (IP) and copyright [53–55], such attacks can be leveraged to reveal such watermarks. They can also be used to analyze test environments, enabling attackers to bypass detection models.

However, property inference attacks face significant challenges, particularly when dealing with complex and over-parameterized model architectures. One key difficulty is training meta-classifiers to infer properties, as the increasing number of equivalent representations within the neural networks complicates this process. Moreover, using naive flattened parameter vector as feature representations is not sufficient, highlighting the need for more sophisticated strategies to enhance the effectiveness of such attacks [47].

1.7 Energy Latency and DoS

Recent advances in deep learning have highlighted the growing dependency on acceleration hardware, such as GPUs and TPUs, which have become essential for training energy-intensive models and improving computational efficiency. However, the high computational demands of deep learning models make them vulnerable to security threats, particularly Denial of Service (DoS) such as sponge attacks, which exploit excessive resource consumption. Such attacks induce excessive neuron activations, resulting in excessive power usage and increased prediction latency. As illustrated in Fig. 1.5, a sponge model can be trained to maintain accuracy while significantly raising computational costs by activating more neurons than necessary.

This vulnerability spans various hardware architectures, encompassing CPUs, GPUs [56], TPUs [57], and even ASIC simulators [58]. Additionally, research has shown that sponge attacks are not limited to language models alone; they also affect computer vision models, albeit to a lesser extent [59]. A notable real-world example of such an attack was observed in Microsoft Azure's translation

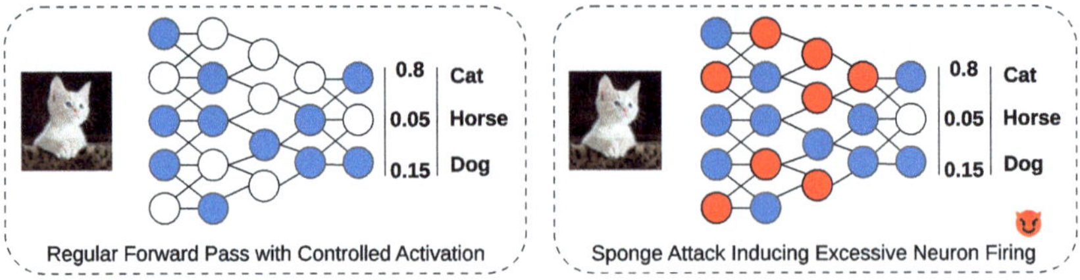

Fig. 1.5 Example of Sponge Attack: While both models correctly classify the cat, the model on the right—affected by a sponge attack—shows a much higher number of activated neurons (highlighted in red and blue) compared to the benign model, which only activates the blue neurons

service, where a sponge attack caused a staggering 6000-fold increase in response time [60]. These threats pose significant concerns, particularly for cloud-based machine learning providers–such as Amazon, Facebook, Microsoft, and Google–as well as for edge devices, where they can lead to excessive battery drain and temporary unavailability [60, 61]. Sponge attacks are generally crafted using two primary techniques: gradient-based methods and genetic algorithms, as outlined in [59], encompassing both black-box and white-box scenarios. Moreover, these attacks are not confined to test-time inputs; some studies, such as [62, 63], have investigated the injection of sponge examples during training, where malicious modifications can severely compromise critical systems and degrade decision-making efficiency [60].

1.8 Backdoors

In recent years, numerous studies have highlighted the susceptibility of deep learning models to backdoor attacks [64–66], demonstrating how these models can simultaneously learn both the main task that represents the primary task of the model and a hidden malicious objective. This presents severe security risks, as backdoor attacks enable adversaries to embed malicious behaviors during training, which can later be exploited at inference time. For example, in a facial recognition system, an attacker could introduce imperceptible modifications to an input image, granting unauthorized access [67]. Likewise, in autonomous driving, a compromised stop sign could mislead the model into misclassifying it, potentially causing the vehicle to disregard the stop command and continue at full speed. Given their potential to compromise model integrity and security, backdoor attacks remain a pressing challenge in deep learning applications (Fig. 1.6).

Backdoor attacks allow adversaries to manipulate training data by embedding a trigger signal into select samples, known as poisoned samples, causing the model to learn the association between the trigger and a specific behavior during training. As a result, at test time, when the trigger is present, the model misclassifies the

Fig. 1.6 Example of Backdoor Attack: The backdoored model on the right misclassify the input image with trigger

input to the attacker predefined target error class, while behaving normally in its absence. These attacks can be non-targeted, aiming to induce unpredictable errors whenever the trigger appears, or targeted, where the model reliably maps poisoned samples to an attacker-specified class while maintaining performance on the main task. Triggers can take various forms, such as pixel patterns, L_2 noise perturbations, or semantic features. Backdoor attacks are categorized into Data Poisoning Attacks [64], which operate in a black-box manner by modifying training data alone, and Model Poisoning Attacks [66], where adversaries directly manipulate model weights–particularly concerning in outsourced model scenarios or Federated Learning. A notable variant, Neural Trojans [68], embeds malicious neurons within the model, enabling backdoor activation under specific conditions.

Several recent studies have highlighted the stealthiness and robustness of backdoor attacks. One notable example is the clean-label backdoor [65], where the attacker must preserve the original label of poisoned samples, making detection more challenging. Latent backdoor attacks, introduced in [69], are designed to remain effective even after the model undergoes fine-tuning with clean data. Dynamic backdoor attacks [70] further enhance stealth by optimizing trigger characteristics, such as position or content, during training. Moreover, backdoor attacks are not restricted to computer vision tasks; they have been extended to other domains, including audio [71], natural language processing [72], and even large language models, where they can manipulate auto-completion or generate inconsistent responses when a trigger word appears. As backdoor threats continue to evolve and expand across different modalities, ongoing research is crucial to developing more robust and resilient deep learning models.

1.9 Defense Strategies and Countermeasures

As deep learning models become more prevalent, the range of potential threats continues to expand, highlighting the need for effective defenses and countermeasures to ensure system security. One widely used approach is adversarial training, where the model is exposed to both normal and adversarial examples during training. This process helps the model develop robust features, making it more resilient to adversarial attacks. Since adversarial examples often exploit gradient information, limiting or obfuscating access to these gradients can significantly hinder an attacker's ability to generate adversarial perturbations [73]. Similarly, membership inference and data reconstruction attacks can be mitigated using differential privacy, which introduces noise into the model's output to obscure sensitive training data. Regularization techniques, such as Dropout, further enhance security by reducing overfitting, improving generalization, and preventing the model from memorizing specific training samples that could be exploited [74]. To counteract model extraction attacks, implementing API interaction rate limits and continuously monitoring suspicious activity are effective strategies. Additionally, obfuscating predictions, as suggested in [75], can disrupt an attacker's ability to construct an accurate surrogate model, reducing the effectiveness of extraction attempts.

Energy-based and denial-of-service (DoS) attacks, such as sponge attacks, can be mitigated through model pruning and quantization, which reduce computational overhead and prevent excessive resource consumption [76]. Additionally, continuous monitoring of resource usage helps detect anomalies before they escalate into serious security threats. For backdoor attacks, anomaly detection is crucial in identifying irregular model behavior caused by hidden triggers, while preprocessing training data can help filter out poisoned samples. Techniques like neural pruning further minimize a model's susceptibility to embedded triggers [67], and model restoration methods, such as backdoor unlearning, work to remove learned backdoor triggers, restoring the model's integrity [77]. Another promising approach involves analyzing the interaction between attackers and defenders within a game-theoretic framework, as explored in [78], to better understand the strategic performance of both sides. As deep learning continues to evolve, ongoing research and the development of innovative defense mechanisms remain essential for strengthening the security and resilience of AI-driven systems.

1.10 Conclusion

While deep learning has driven remarkable technological advancements, it remains vulnerable to a wide range of security threats throughout its lifecycle. This survey explored key attack vectors, including adversarial perturbations that deceive models, membership inference and data reconstruction techniques that compromise privacy, and property inference and model extraction threats that endanger confidentiality. Additionally, we examined energy-latency and denial-of-service (DoS) attacks that disrupt system availability, as well as backdoor attacks that embed hidden vulnerabilities within models.

These challenges highlight the pressing need to integrate security measures at every stage of AI development. As machine learning continues to play a vital role in critical fields such as healthcare, finance, and autonomous systems, ensuring resilience against emerging attack strategies is more important than ever. Future research should prioritize refining defense mechanisms, better development of theoretical insights into attack methodologies, and rigorously stress-testing models under real-world conditions to maintain a balance between security and performance in AI-driven applications.

Acknowledgments This work was partially supported by the industrial chair CYBAILE, partly funded by a European and Brittany region grant (reference FEDER-AIDEN-76568) and SSF-ML-DH.

References

1. Xue, M., Yuan, C., Wu, H., Zhang, Y., Liu, W.: Machine learning security: threats, countermeasures, and evaluations. IEEE Access **8**, 74720–74742 (2020)
2. Goodfellow, I.J., Shlens, J., Szegedy, C.: Explaining and harnessing adversarial examples. Preprint. arXiv:1412.6572 (2014)
3. Szegedy, C., Zaremba, W., Sutskever, I., Bruna, J., Erhan, D., Goodfellow, I.J., Fergus, R.: Intriguing properties of neural networks. CoRR abs/1312.6199 (2013). https://api.semanticscholar.org/CorpusID:604334
4. Zhang, J., Li, C.: Adversarial examples: opportunities and challenges. IEEE Trans. Neural Netw. Learn. Syst. **31**(7), 2578–2593 (2019)
5. Yeom, S., Giacomelli, I., Menaged, A., Fredrikson, M., Jha, S.: Overfitting, robustness, and malicious algorithms: a study of potential causes of privacy risk in machine learning. J. Comput. Secur. **28**(1), 35–70 (2020)
6. Rigaki, M., Garcia, S.: A survey of privacy attacks in machine learning. ACM Comput. Surv. **56**(4), 1–34 (2023)
7. Chakraborty, A., Alam, M., Dey, V., Chattopadhyay, A., Mukhopadhyay, D.: A survey on adversarial attacks and defences. CAAI Trans. Intell. Technol. **6**(1), 25–45 (2021)
8. Barni, M., Kallas, K., Nowroozi, E., Tondi, B.: On the transferability of adversarial examples against CNN-based image forensics. In: ICASSP, pp. 8286–8290. IEEE Press, New York (2019)
9. Madry, A., Makelov, A., Schmidt, L., Tsipras, D., Vladu, A.: Towards deep learning models resistant to adversarial attacks. Preprint. arXiv:1706.06083 (2017)
10. Chen, S.T., Cornelius, C., Martin, J., Chau, D.H. Shapeshifter: Robust physical adversarial attack on faster R-CNN object detector. In: ECML PKDD 2018, pp. 52–68. Springer (2019)
11. Kurakin, A., Goodfellow, I., Bengio, S.: Adversarial examples in the physical world. Preprint. arXiv:1607.02533 (2017)
12. Chen, J., Jordan, M.I., Wainwright, M.J.: Hopskipjumpattack: a query-efficient decision-based attack. In: IEEE S&P. IEEE, Piscataway (2020)
13. Croce, F., Hein, M.: Reliable evaluation of adversarial robustness with an ensemble of diverse parameter-free attacks. In: ICML, pp. 2206–2216. PMLR (2020, November)
14. Homer, N., Szelinger, S., Redman, M., Duggan, D., Tembe, W., Muehling, J., Pearson, J.V., Stephan, D.A., Nelson, S.F., Craig, D.W.: Resolving individuals contributing trace amounts of DNA... PLoS Genet. **4**(8), e1000167 (2008)
15. Bai, L., et al.: Membership inference attacks and defenses in federated learning: a survey. ACM Comput. Surv. **57**(4), 1–35 (2024)
16. Xu, T., Liu, C., Zhang, K., Zhang, J.: Membership inference attacks against medical databases. In: Int. Conf. Neural Inf. Process., pp. 15–25. Springer (2023)
17. Yeom, S., Giacomelli, I., Fredrikson, M., Jha, S.: Privacy risk in machine learning. In: CSF, pp. 268–282. IEEE Press, New York (2018)

18. Salem, A., et al.: ML-leaks: Model and data independent membership inference attacks... arXiv:1806.01246 (2018)
19. Choquette-Choo, C.A., Tramer, F., Carlini, N., Papernot, N.: Label-only membership inference attacks. In: ICML, pp. 1964–1974. PMLR (2021)
20. Li, Z., Zhang, Y. Membership leakage in label-only exposures. In: CCS, pp. 880–895. ACM, New York (2021)
21. Shokri, R., Stronati, M., Song, C., Shmatikov, V.: Membership inference attacks against ML models. In: SP, pp. 3–18. IEEE, Piscataway (2017)
22. Liu, H., et al.: Encodermi: membership inference against pre-trained encoders... In: CCS. ACM, New York (2021)
23. Carlini, N., Chien, S., Nasr, M., Song, S., Terzis, A., Tramer, F.: Membership inference attacks from first principles. In: SP, pp. 1897–1914. IEEE, New York (2022)
24. Bertran, M., Tang, S., Roth, A., Kearns, M., Morgenstern, J.H., Wu, S.Z.: Scalable membership inference attacks via quantile regression. In: Advances in Neural Information Processing Systems (NeurIPS), vol. 36, pp. 314–330 (2023)
25. Song, S., Marn, D.: Introducing a new privacy testing library in TensorFlow. Google AI Blog (2020). https://blog.tensorflow.org/2020/06/introducingnew-privacy-testing-library.html
26. Murakonda, S.K., Shokri, R.: Ml privacy meter.... arXiv:2007.09339 (2020)
27. Dinur, I., Nissim, K.: Revealing information while preserving privacy. In: PODS, pp. 202–210. ACM, New York (2003)
28. Dwork, C., Yekhanin, S.: New efficient attacks on statistical disclosure control. In: Crypto, pp. 469–480. Springer, Heidelberg (2008)
29. Fredrikson, M., Jha, S., Ristenpart, T.: Model inversion attacks. In: CCS, pp. 1322–1333. ACM, New York (2015)
30. Balle, B., Cherubin, G., Hayes, J.: Reconstructing training data with informed adversaries. In: SP (2022)
31. Haim, N., Vardi, G., Yehudai, G., Shamir, O., Irani, M.: Reconstructing training data from trained NNs. In: NeurIPS (2022)
32. Buzaglo, G., et al.: Reconstructing training data from multiclass neural networks. arXiv:2305.03350 (2023)
33. Jayaraman, B., Evans, D.: Are attribute inference attacks just imputation? In: CCS (2022)
34. Zhang, C., Bengio, S., Hardt, M., Recht, B., Vinyals, O.: Understanding deep learning (still) requires rethinking generalization. Commun. ACM **64**(3),107–115 (2021)
35. Feldman, V.: Does learning require memorization? In: STOC (2020)
36. Feldman, V., Zhang, C.: What neural networks memorize and why. In: NeurIPS (2020)
37. Yin, H., et al.: See through gradients: Image batch recovery via gradinversion. In: CVPR (2021)
38. Panchendrarajan, R., Bhoi, S.: Dataset reconstruction attack against LMs. In: CEUR (2021)
39. Salem, A., et al.: Updates-Leak. In: USENIX Sec (2020)
40. Bertran, M., et al.: Reconstruction attacks on machine unlearning. arXiv:2405.20272 (2024)
41. Wang, Z., Lee, J., Lei, Q.: Reconstructing training data from model gradient, provably. In: AISTATS (2023)
42. Tramèr, F., Zhang, F., Juels, A., Reiter, M.K., Ristenpart, T.: Stealing ML models via prediction APIs. In: USENIX Sec (2016)
43. Wang, B., Gong, N.Z.: Stealing hyperparameters in ML. In: SP (2018)
44. Carlini, N., Jagielski, M., Mironov, I.: Cryptanalytic extraction of NNs. In: Crypto (2020)
45. Orekondy, T., Schiele, B., Fritz, M.: Knockoff nets. In: CVPR (2019)
46. Gong, X., et al.: Model extraction attacks and defenses on cloud-based ML models. IEEE Commun. Mag. **58**(12), 83–89 (2020)
47. Ateniese, G., et al.: Hacking smart machines with smarter ones. Int. J. Secur. Netw. **10**(3), 137–150 (2015)
48. Parisot, M., Pejó, B., Spagnuelo, D.: Property inference attacks on CNNs. arXiv:2104.13061 (2021)
49. Mahloujifar, S., et al.: Property inference from poisoning. In: SP (2022)

50. Suri, A., Evans, D.: Formalizing and estimating distribution inference risks. arXiv:2109.06024 (2021)
51. Melis, L., et al.: Exploiting unintended feature leakage in collaborative learning. In: SP (2019)
52. Zhou, J., et al.: Property inference attacks against GANs. arXiv:2111.07608 (2021)
53. Adi, Y., et al.: Turning your weakness into a strength. In: USENIX Sec (2018)
54. Zhang, J., et al.: Protecting intellectual property of DNNs with watermarking. In: AsiaCCS (2018)
55. Wang, T., Kerschbaum, F.: RIGA: covert and robust white-box watermarking. In: WWW (2021)
56. Hazelwood, K.M., et al.: Applied machine learning at Facebook. In: HPCA (2018)
57. Jouppi, N.P., et al.: In-datacenter performance analysis of a TPU. In: ISCA (2017)
58. Parashar, A., et al.: SCNN: an accelerator for CNNs. In: ISCA (2017)
59. Shumailov, I., et al.: Sponge examples. In: EuroS&P (2021)
60. Martin, T., et al.: DoS attacks on battery-powered mobile computers. In: PerCom (2004)
61. Palmieri, F., et al.: Energy-oriented DoS attacks. J. Supercomput. **71**, 1620–1641 (2015)
62. Ciná, A.E., et al.: Energy-latency attacks via sponge poisoning. arXiv:2203.08147 (2022)
63. Meftah, H.FZ., et al.: Energy backdoor attack to deep neural networks. arXiv:2501.08152 (2025)
64. Chen, X., et al.: Targeted backdoor attacks on DL systems using data poisoning. arXiv:1712.05526 (2017)
65. Barni, M., Kallas, K., Tondi, B.: A new backdoor attack in CNNs by training set corruption. In: ICIP (2019)
66. Dumford, J., Scheirer, W.: Backdooring CNNs via targeted weight perturbations. In: IJCB (2020)
67. Le Roux, Q., Bourbao, E., Teglia, Y., Kallas, K.: Survey on backdoor attacks in face recognition. IEEE Access **12**, 47433 (2024)
68. Liu, Y., Xie, Y., Srivastava, A.: Neural trojans. In: ICCD (2017)
69. Yao, Y., et al.: Latent backdoor attacks. In: CCS (2019)
70. Salem, A., et al.: Dynamic backdoor attacks. In: EuroS&P (2022)
71. Shi, C., et al.: Audio-domain backdoor attack. In: MobiCom (2022)
72. Chen, X., et al.: BadNL: backdoor attacks in NLP models. In: ACSAC (2021)
73. Qiu, H., et al.: Mitigating advanced adversarial attacks with gradient obfuscation. arXiv:2005.13712 (2020)
74. Li, Y., et al.: Towards robustness of DNNs via regularization. In: ICCV (2021)
75. Zheng, H., et al.: BDPL: A boundary differentially private layer. In: ESORICS (2019)
76. Wang, Z., et al.: Energy-latency attacks to on-device NNs via sponge poisoning. In: STDS Workshop (2023)
77. Le Roux, Q., Kallas, K., Furon, T.: Restore: black-box defense against DNN Backdoors. In: SaTML (2024)
78. Kallas, K., Le Roux, Q., Hamidouche, W., Furon, T.: Strategic safeguarding: game-theoretic defense. In: EURASIP JIS (2024)

Chapter 2
Ethical Considerations and Regulatory Standards for Adversarial Defense

Iris-Panagiota Efthymiou

2.1 Introduction to Adversarial Defense

Adversarial attacks on machine learning and AI exploit model vulnerabilities by manipulating input data to force wrong decisions or steal sensitive information [1]. Adversarial attacks can be classified as poisoning attacks (training data manipulation), evasion attacks (input manipulation to misleading models), model extraction attacks (reproduction of model behaviour), and inference attacks (stealing confidential information). Attackers deploy adversarial examples, tiny and imperceptible changes that trick models but seem fine to humans [2].

Adversarial defense for machine learning and AI are methods utilized to safeguard models from adversarial attacks, such that slight, imperceptible modifications of input data can modify model predictions [3]. Defenses involve adversarial training, defensive distillation, and gradient masking to promote model robustness and security [4]. They form a crucial part in increasing the resilience, security, and stability of machine learning systems, especially those that involve high-risk uses such as healthcare, finance, and transportation [5]. Adversarial attacks represent pivotal vulnerabilities in these sectors where ML models form the basis of decision-making. The attacks leverage vulnerabilities in ML models through subtly crafted and low-amplitude manipulations in input data, normally imperceptible to human operators, leading to erroneous and in some cases lethal outputs. The consequences of these attacks are not just ordinary system crashes, which result in loss of funds, security, propagation of misinformation, and in severe cases, even harmful consequences [2].

I.-P. Efthymiou (✉)
University of Greenwich, Old Royal Naval College, London, UK,
https://www.gre.ac.uk/

E. Nowroozi et al. (eds.), *Adversarial Example Detection and Mitigation Using Machine Learning*, https://doi.org/10.1007/978-3-031-99447-0_2

In healthcare, adversarial attacks have the potential to mislead diagnostic AI models and offer incorrect medical assessments and treatment [6]. Adversarial attacks on medical imaging equipment, for example, can quietly modify scans to lead to misclassification, making a malignant tumour appear benign [7]. Poisoning attacks during training can also contaminate datasets by adding fabricated patient records, distorting the model's learning process and resulting in systemic misdiagnoses. Moreover, model inversion attacks present a privacy risk where attackers can rebuild sensitive patient information from AI models, thereby potentially revealing confidential medical data [8].

The finance industry is most susceptible to adversarial attacks, especially given the prevalence of using ML models in fraud detection, credit risk evaluations, and algorithmic trading. Attackers can implement adversarial attacks by making minor changes to fraudulent patterns of transactions, which enables them to evade fraud detection systems without being detected [9]. Poisoning attacks can be utilized to manipulate credit risk models by injecting malicious financial records, making banks make erroneous lending decisions [8]. Model extraction attacks also present a great threat, as attackers can systematically query trading models to reverse-engineer proprietary algorithms, resulting in market manipulation and competitive losses for financial institutions [5].

In the transport industry, adversarial attacks can be fatal, particularly in self-driving cars that use ML models for navigation and object recognition. Computer vision systems evasion attacks can deceive traffic signs by making minor changes, which will result in a self-driving vehicle misinterpreting a stop sign for a speed limit sign, resulting in an unsafe situation [10]. Poisoning attacks may impact training data employed for lane detection and obstacle avoidance, introducing vulnerabilities into the model that may result in unstable driving behaviour [8]. Model inversion attacks may also be employed to reconstruct driver profiles from AI-powered fleet management systems, revealing personal data [11].

2.1.1 Ethical Considerations and Regulatory Compliance in the Development of Defense Mechanisms

The creation of AI defense strategies must be carried out with special attention to ethical standards and regulations to guarantee security, equity, and accountability [12]. Privacy is among the main ethical issues since most defense approaches are based on mass data gathering, which has the potential to violate user rights if not effectively regulated [13]. A balance between security and openness is another delicate issue, where openness in defense can promote trust, but excessive exposure can make weaknesses available to adversaries [14]. Additionally, AI defense system prejudice is a critical issue since flawed defense mechanisms may unknowingly discriminate against specific segments of users and enhance inequality. Regulatory compliance frameworks, such as the General Data Protection Regulation (GDPR) and the National Institute of Standards and Technology (NIST) framework, are critical in ensuring that AI defense systems comply with legal and ethical standards [15]. Accountability remains a problem, as it becomes increasingly hard to ascertain liability in the event of an adversarial attack with advanced AI systems [16]. Moreover, the dual-use character of AI security technologies raises ethical issues since the technologies can be employed for malicious offensive cyber operations rather than for defense (Continuing Education and Development, Inc., n.d.). To overcome the above challenges, a multidisciplinary solution that incorporates ethical principles of AI, legal provisions, and technical solutions is required for the proper utilization of AI defense strategies.

2.1.2 Ethical Implications of Adversarial Defense

Adversarial defenses are important in keeping security strong, but they also have the ethical considerations taken seriously. First and most important among these is ensuring that adversarial defenses would not unfairly affect marginalized communities. Algorithmic discrimination in programming code or training data can potentially have disproportionate protection with some groups of people being analysed more thoroughly and exposed to risk [17]. For example, [17] found facial recognition systems exhibited severe racial and gender bias against darker-skinned populations, particularly women, with higher error rates. These disparities highlight the risk that adversarial defenses will reinforce social bias rather than provide equal protection [18]. Ethically developing AI should prioritize fairness by continuously testing and refining adversarial defense methods to prevent discriminatory outcomes. Without such monitoring, security practices risk strengthening the existing social inequalities, further marginalizing, and causing injury [19].

Another ethical challenge in adversarial defense is the balancing of transparency and secrecy. Transparency, on the one hand, is vital to maintaining public trust, allowing external audits, and determining potential defense biases. For instance, when OpenAI initially developed GPT-2, it censored the whole model due to fear of misuse, with debates on whether censorship delayed independent research efforts or protected against dangerous exploitation [20]. Excessive openness, however, can allow hackers to go around or alter security measures, rendering them ineffective. This creates a dilemma in which the need for accountability hinders the need for effective security. Balancing the two requires careful consideration to allow ethical regulation without compromising on protection against possible misuse [21].

Accountability is yet another key ethical issue in adversarial defense. Defensive measures that have unintended effects need to be defined in terms of responsibility and redress mechanisms. For instance, Facebook's automated content moderation system has incorrectly labelled activism-related posts as harmful content, raising censorship issues [22]. In 2021, there were reports that the AI on the Instagram and Facebook inadvertently deleted content associated with Palestinian activism, questioning algorithmic bias and responsibility [23]. If an adversarial defense system mislabels legitimate behaviour as malicious, the individuals or organizations impacted could face reputational or financial damage with no opportunity for appeal. This also creates fundamental questions of responsibility–developers, organizations implementing the technology, or policymakers who implement regulatory requirements. Without clear frameworks of responsibility, adversarial defenses can harm with no suitable oversight and control measures in place [24].

Along with accountability, overcoming biases in defense systems is crucial to providing fair protection to all groups [25]. Most adversarial defenses are based on machine learning models learned from past data, which could have inherent biases. A prominent example is Amazon's recruitment algorithm, which was discovered to prefer male candidates over female candidates based on historical recruitment data biases [26]. While this was not an adversarial defense system, it illustrates the way biased data can result in unjust outcomes. If these biases are not determined and addressed, some groups of people may not receive consistent defense or may find themselves at risk of being improperly targeted. Ethical AI deployment needs to include thorough testing, diverse datasets, and representative design practices to help ensure that adversarial defenses act fairly for all people. Unless there are active measures taken to combat bias, security controls may worsen existing inequalities instead of alleviating them [27].

2.1.3 Global Regulatory Frameworks and Standards

Some of the current regulations offer directions on data privacy and security, which indirectly affect adversarial defense mechanisms. The European Union's General Data Protection Regulation (GDPR) focuses on transparency, responsibility, and the rights of the user and obliges organizations to put in place strong security measures to secure individual data [28]. In the United States, the National Institute of Standards and Technology (NIST) has established standards for cybersecurity that encompass adversary threat specifications, outlining best practices for organizations regarding threat defense [29].

The standards provide a basis for the ethical use of mechanisms of adversarial defense without infringing on the rights of individuals. Despite that, regulatory variations across regions are significant between the EU and the US. The EU over-reliably depends on excellent data protection and privacy law with stringent sanctions against failing GDPR compliance, while the US has sectoral and voluntary compliance methods through guidelines such as the NIST Cybersecurity Framework [15]. This difference impacts adversarial defense applications because international institutions must be in line with varying legal requirements. Besides, China has enacted stringent cybersecurity regulations focusing on state control of data security making it harder to create an overall regulatory framework [30]. These variations show the difficulty of creating global uniform standards for adversarial defense.

International institutions are vital in organizing regulation activity and formulating common standards. The International Organization for Standardization (ISO) and the Institute of Electrical and Electronics Engineers (IEEE) have established guidelines to resist cybersecurity threats, including adversarial attacks against AI systems [31]. ISO/IEC 27001 offers a corporate-level framework for information security management, while IEEE's Ethically Aligned Design has ethical AI development guidelines [32]. These guidelines enable organizations to leverage adversarial defenses in a way that is transparently, securely, and equitably balanced. However, the application of these global standards is voluntary, and therefore there may be disparity of application in different industries and geographies.

Despite the progress in regulatory frameworks, there are still wide gaps in addressing adversarial defense explicitly. The majority of existing regulations are focused on general cybersecurity practices rather than specific threats from adversarial machine learning and AI-based attacks. For instance, GDPR requires data protection practices but does not specify how to defend against adversarial AI attacks [33]. Similarly, NIST guidelines emphasize cybersecurity best practices without mandating standardized means for countering adversarial attacks on AI-based systems. The absence of direct regulatory guidance compels companies to develop their own safeguarding mechanisms, which can lead to dissimilar and disjointed implementations [34]. Closing these gaps will necessitate ongoing cooperation among governments, business leaders, and global organizations to create end-to-end regulations that provide both security and ethical compliance in adversarial defense.

2.1.4 Alignment of Adversarial Defense with Ethical AI Principles

Alignment of the adversarial defense mechanisms with the principles of ethical AI is paramount to guarantee security measures promote fairness, accountability, transparency, and human control. Ethical AI models require defense strategy designs that do not differentially affect a particular group of individuals or have unforeseen negative effects. Adversarial defense must be equitable so that the security systems utilized are unbiased and do not adversely impact particular segments of individuals [35]. Science has established that AI-driven security controls are capable of magnifying existing biases unless carefully designed, as in the case where facial recognition systems expose racial and gender biases by imbalanced training data [36]. To circumvent such threats, adversarial defense techniques need to incorporate fairness-aware machine learning models and undergo comprehensive testing for biases to ensure balanced protection for everyone.

Accountability is also a vital ethical concern in adversarial defense. Where AI systems are making security choices, there should be transparent accountability lines. The European Commission's Ethics Guidelines for Trustworthy AI highlight that AI systems need to have clear frameworks of accountability for dealing with any unwanted harms or mistakes [37]. For example, in cybersecurity applications, if an adversarial defense system mistakenly flags legitimate transactions as fraudulent, businesses and users could suffer financial losses without an effective mechanism for redress. To prevent such issues, organizations must define accountability structures that clarify whether responsibility lies with developers, deploying organizations, or policymakers [38].

Transparency plays a crucial role in ensuring that adversarial defense mechanisms operate in an explainable and trustworthy manner. The OECD AI Principles (1.3) stress that AI actors should ensure transparency by providing clear, context-appropriate information about AI systems, their capabilities, decision-making processes, and avenues for users to challenge adverse outcomes [39, 40]. However, a challenge arises in balancing transparency with security, as revealing too much information about defense mechanisms could enable adversaries to exploit system vulnerabilities. One approach to addressing this dilemma is differential transparency, where information about AI decision-making is made available to regulators and auditors while maintaining operational secrecy from potential attackers [41].

The UNESCO Recommendation about the Ethics of Artificial Intelligence, endorsed by 193 Member States, presents an international basis for the ethical utilization of AI in order to ensure AI systems become normative to the principles of fairness, transparency, accountability, and inclusiveness [42]. Of the core matters to be considered is that AI will further exacerbate social biases, specifically on gender, racial, and socioeconomic bases. For protection against such dangers, UNESCO facilitates robust institutional and legal frameworks facilitating responsible AI regulation while upholding human rights and basic freedoms. The recommendation further highlights better governance of data as well as responsible uses of AI in education, labour markets, health, and the environment. For these values to bear fruit, UNESCO has initiated ambitious projects such as the Global Observatory on AI Ethics, the AI Ethics Experts Without Borders Network, and the Women for AI Ethics Network [43]. They all provide a point of entry for global cooperation that enables policymakers, companies, and civil society to come together to create AI systems that are inclusive, ethical, and beneficial to all.

2.1.5 Practical Challenges in Meeting Ethical and Regulatory Standards

Integrating AI defenses that meet ethical and legal requirements involves several technical and operational challenges. Ensuring transparency and explainability in security systems powered by AI is one of the most critical technical challenges. Most AI models, particularly those based on deep learning for cybersecurity, are "black boxes," and it is hard to audit their decision-making against global ethical and legal requirements [44]. For example, GDPR requires explainability in automated decision-making, but AI-powered threat detection tools tend to be based on black-box algorithms that security teams find difficult to interpret [45]. This challenge is met by improvements in explainable AI (XAI) and standardized frameworks for auditing AI security controls without compromising efficacy against cyber threats [46].

The other important obstacle is the lack of resources for organizations, especially small and medium-sized businesses (SMEs), to comply with international ethical and legal requirements. Implementation of standards such as the UNESCO AI Ethics Recommendation or the OECD AI Principles typically involves huge investments in talent, infrastructure, and ongoing monitoring [47]. For instance, to become compliant with ISO/IEC 42001 (AI Management System Standard) there are training personnel expenses, revising security procedures, and meeting new legal demands arising from ongoing legal developments [48]. Most organizations face these economic and technical challenges to adopt effective and ethically compliant AI defense.

There are also contradictions between national security interests and international ethical standards that complicate AI regulation. For instance, governments usually choose national security over ethics and thus create contradictions in international regulation of AI. For instance, certain nations use AI surveillance technologies for counterterrorism and cybersecurity operations, triggering concerns regarding violations of privacy and human rights [49]. The application of AI-based surveillance by state actors in manners that are contrary to universal human rights principles, including the UNESCO Recommendation's requirement for transparency and accountability, is a case in point [50]. Furthermore, although international frameworks promote data protection, some national security policies require the collection of vast amounts of data, leading to a clash between regulatory compliance and state security needs [51].

2.1.6 Strategies for Ethical and Compliant Adversarial Defense

A formalized risk assessment model is necessary to measure adversarial defense mechanisms against legal and ethical criteria. Models like the National Institute of Standards and Technology (NIST) AI Risk Management Framework and the ISO/IEC 42001 AI Management System Standard assist organizations in assessing risks, bias, and gaps in compliance within their security frameworks systematically [48]. For instance, adversarial defenses in cyber security have to balance security performance with data privacy regulations like GDPR and CCPA. A risk-based approach makes sure that AI-powered threat detection systems do not unintentionally break ethical principles, like over-surveillance or biased profiling [52].

Applying human-centred design (HCD) principles guarantees that adversarial defense systems are transparent, fair, and accessible. Security solutions based on artificial intelligence should be developed with interpretability, so security professionals can comprehend and take over automated decision-making when needed [53]. For example, fraud detection adversarial machine learning defenses should not perpetuate discrimination against particular demographic groups so that security protocols are fair and do not unfairly mark particular groups disproportionately [54]. Adversarial training that is aware of fairness can reduce bias while still being highly effective as a defense.

Ethical adversarial defense mechanisms need constant monitoring and auditing to keep up with changing threats and regulatory updates. Continuous auditing tools, like explainable AI (XAI) dashboards, assist organizations in monitoring AI-based security choices and determining their consistency with ethical standards [55]. Periodic audits can identify latent biases in AI security models, ensuring that adversarial defense solutions remain effective as well as compliant with the law. For instance, banks that employ AI fraud systems perform regular audits to ensure that they comply with anti-discrimination legislation and ensure strong security levels [56].

Closing regulatory gaps in adversarial defense calls for collaboration among governments, private sector players, and academia. Public-private partnerships can enable the sharing of best practices to ensure that AI security solutions are in line with national security priorities as well as international ethical standards. The EU's AI Act and the US-EU Trade and Technology Council are some of the initiatives that promote cooperation on AI governance, reconciling differences in regulatory strategies [57]. Also, interdisciplinary collaboration between ethicists and cybersecurity specialists can guide the creation of defense systems that are both technically solid and ethically defensible.

2.2 Case Studies and Best Practices

The adoption of adversarial defense mechanisms consistent with ethical and legal requirements has been a prime area of focus for leading technology firms. By studying actual case studies, we can determine successful strategies and lessons from failed attempts.

1. **Google's AI Principles in Security and Ethics**
   Google AI Principles, which were created in 2018, inform the ethical design of AI in a wide range of applications, including adversarial defense systems. Google incorporates these principles into cybersecurity through the implementation of AI-powered threat detection and mitigation tools that comply with privacy and fairness standards [58]. Through partnerships with NGOs, industry leaders, and academics, Google refines its adversarial defenses to reduce bias and enhance transparency. One key success is the Secure AI Framework (SAIF), which ensures compliance with data protection laws while strengthening cybersecurity measures [59].
2. **Microsoft's Responsible AI in Cybersecurity**
   Microsoft incorporates ethical AI principles into adversarial defense through a structured Responsible AI Framework. This framework emphasizes fairness,

reliability, and transparency in AI-powered security solutions, such as fraud detection and cyberattack prevention [60]. The company applies stringent oversight to AI models that impact individuals and organizations, ensuring adversarial defenses comply with GDPR and industry standards. Microsoft's AI-driven cybersecurity solutions, such as Defender and Sentinel, utilize explainable AI (XAI) to ensure transparency in threat detection, reducing false positives and enhancing user trust [61].

3. **IBM's Trustworthy AI for Adversarial Threats**
   IBM's commitment to Trustworthy AI ensures that adversarial defense systems undergo continuous validation to maintain ethical and legal compliance [62]. The company's Responsible Use of Technology Framework fosters a culture of accountability, ensuring AI models in cybersecurity remain reliable, unbiased, and privacy-conscious. IBM's AI-driven security solutions, such as Watson for Cybersecurity, use adversarial testing to identify vulnerabilities while complying with industry regulations. The World Economic Forum has recognized IBM's ethical AI governance as a model for other organizations [63].
4. **Mastercard's AI-Driven Fraud Detection System**
   Mastercard successfully implemented an AI-powered fraud detection system that adheres to GDPR and consumer protection laws while defending against adversarial financial fraud. The system incorporates fairness-aware machine learning to prevent bias in fraud detection algorithms [64]. This case demonstrates how adversarial defenses can be designed to align with global regulatory requirements while maintaining ethical AI principles.

2.2.1 Failures and Ethical Oversights in Adversarial Defense

Failures in adversarial defense systems often stem from biases, lack of transparency, and inadequate adaptability to evolving threats.

Facebook's AI-Based Misinformation Detection Failure

Facebook's misinformation detection system built with artificial intelligence has been largely criticized for being unable to combat political extremism, disinformation, and conspiracy theories, which have led to offline violence and communal divisions. While AI was intended to filter harmful content, opaque algorithms have instead amplified misinformation, empowered authoritarian regimes, and weakened democratic processes [65]. In response, Facebook initially strengthened its fact-checking efforts by incorporating third-party human reviewers and refining its AI models. However, acknowledging the biases and censorship concerns tied to fact-checking, the company is now shifting to a Community Notes program, similar to X's approach, where users collectively determine the accuracy of content. Additionally, Facebook is easing content restrictions, limiting over-enforcement, and revising its policies to prioritize free expression. The platform is also refining its approach to political content by personalizing recommendations, ensuring users have more control over the civic content they see [66]. These changes reflect Facebook's broader commitment to balancing misinformation control with freedom of speech while addressing the shortcomings of its previous moderation strategies.

COMPAS Algorithm Bias in Criminal Justice

The COMPAS risk assessment tool was designed to predict recidivism by generating risk scores based on a defendant's survey responses and criminal history, influencing judges' decisions on pretrial release. However, a 2016 ProPublica investigation revealed that COMPAS disproportionately classified Black defendants as high-risk, nearly twice as often as white defendants with similar records [67]. Following these allegations, Northpointe (now Equivant) defended COMPAS by arguing that its algorithm prioritized predictive parity rather than error rate balance and allowed external audits to review its methodology. Despite legal challenges, courts upheld COMPAS as a recommendation tool and not a determinative system, leading to wider debates on fairness in AI-driven sentencing [68].

Google's AI Ethics Board Controversy

In 2019, Google dissolved its AI ethics advisory board only a week after it was established because of public outcry and internal pressure [69]. The company subsequently changed its approach by fortifying in-house AI ethics teams and undertaking a more formal review process. Although the Advanced Technology External Advisory Council (ATEAC) was an idea initially, it was not pursued. Rather, Google prioritized internal governance, publishing AI ethics reports, and including stricter ethical reviews within its AI development pipelines [58]. This case shows how businesses address ethical issues through improved AI policies, greater transparency, and compliance with international ethical standards.

2.3 Conclusion

Adversarial defense in machine learning and AI is a significant research and development area, which involves safeguarding models from malicious attacks that exploit vulnerabilities to manipulate outputs or steal confidential data. As the advancement in adversarial attacks continues, the necessity for efficient defense methods like adversarial training, defensive distillation, and gradient masking has never been more paramount. These defenses are critical to the trustworthiness, safety, and fairness of AI systems, especially for safety-critical domains such as medicine, finance, transport, and cybersecurity, where the consequences of an attack by an adversary would be disastrous [70].

The regulatory and ethical aspects of adversarial defense are also very significant. Defense activities must be supported by a sense of fairness, transparency, responsibility, and inclusiveness to ensure that they don't reinforce social biases or trespass on people's rights. Global standards like GDPR, NIST guidelines, and UNESCO guidelines on AI ethics are central to it, yet there are openings to be completed in dealing with specific adversarial AI challenges still in dispute. Completing those openings involves multi-disciplinary activity, balancing technology innovation with moral guidelines and transnational cooperation to create consistent norms and best practices.

Real-world case studies of MasterCard, IBM, Microsoft, and Google show effective inclusion of ethical issues into adversarial defense mechanisms, while the inability to remove misinformation from Facebook and the COMPAS algorithm shows the possibility of inattention and prejudice. They are proof that monitoring, auditing, and adjustments should periodically be performed for new threats and regulation updates.

In the future, human-centered design, fairness-aware machine learning, and explainable AI will be crucial to building adversarial defense so that security solutions are effective and fair. Public-private partnerships and cross-disciplinary approaches will play a critical role in surmounting the technical, ethical, and regulatory challenges of adversarial defense. By linking adversarial defense strategies with ethical AI principles and international standards, we can develop AI systems that not only exist as a defense against adversarial attacks but are also credible, inclusive, and positive for society in general.

References

1. Ibitoye, O., Abou-Khamis, R., Shehaby, M.E., Matrawy, A., Shafiq, M.O.: The threat of adversarial attacks on machine learning in network security – a survey. Preprint. arXiv:1911.02621 (2019)
2. Birch, L.: AI under attack: Six key adversarial attacks and their consequences. Mindgard. Available at: https://mindgard.ai/blog/ai-under-attack-six-key-adversarial-attacks-and-their-consequences (2025, September 29)
3. Liu, N., Du, M., Guo, R., Liu, H., Hu, X.: Adversarial attacks and defenses: an interpretation perspective. ACM SIGKDD Explor. Newsl. **23**(1), 86–99 (2021)
4. Ren, K., Zheng, T., Qin, Z., Liu, X.: Adversarial attacks and defenses in deep learning. Engineering **6**(3), 346–360 (2020)
5. Lumenova AI: Adversarial attacks in ML: Detection & defense strategies. Lumenova AI. https://lumenova.ai/blog/adversarial-attacks-ml-detection-defense-strategies/ (2024, July 2)
6. Efthymiou-Egleton, I.: Trends in Health Care: A Global Challenge. Xlibris Corporation, New York (2016)
7. Kaviani, S., Han, K.J., Sohn, I.: Adversarial attacks and defenses on AI in medical imaging informatics: A survey. Expert Syst. Appl. **198**(2), 116815 (2022)
8. SentinelOne: What is data poisoning? Types & best practices. SentinelOne (2024, October 21)

9. Malik, P., Chourasia, A., Pandit, R., Bawane, S., Surana, J.: Credit risk assessment and fraud detection in financial transactions using machine learning. J. Electr. Syst. **20**(3s), 2061–2069 (2024)
10. Badjie, B., Cecílio, J., Casimiro, A.: Adversarial attacks and countermeasures on image classification-based deep learning models in autonomous driving systems: a systematic review. ACM Comput. Surv. **57**(1), 1–52 (2024)
11. Khosravy, M., Nakamura, K., Hirose, Y., Nitta, N., Babaguchi, N.: Model inversion attack by integration of deep generative models: privacy-sensitive face generation from a face recognition system. IEEE Trans. Inf. Forens. Secur. **17**(1), 357–372 (2022)
12. Vousınas, G.L., Simitsi, I., Livieri, G., Gkouva, G.C., Efthymiou, I.P.: Mapping the road of the ethical dilemmas behind artificial intelligence. J. Polit. Ethics New Technol. AI **1**(1), e31238 (2022)
13. Kisselburgh, L., Beever, J.: The ethics of privacy in research and design: principles, practices, and potential. In: Modern Socio-Technical Perspectives on Privacy, pp. 395–426. Springer International Publishing, Cham (2022)
14. Jonker, A., Gomstyn, A., McGrath, A.: What is AI transparency? IBM Think. https://www.ibm.com/think/topics/ai-transparency (2024, September 6)
15. Baig, A., Ismail, M., Hasan, A.: A comparative analysis of the NIST Privacy Framework vs. the EU's GDPR. Securiti AI. https://securiti.ai/nist-privacy-framework-vs-eu-gdpr/ (2025, January 26)
16. U.S. Department of Homeland Security: Risks and mitigation strategies for adversarial artificial intelligence threats: A DHS S&T study preparedness series. U.S. Department of Homeland Security (2023, June)
17. Ntoutsi, E., Fafalios, P., Gadiraju, U., Iosifidis, V., Nejdl, W., Vidal, M.E., Staab, S.: Bias in data-driven artificial intelligence systems—an introductory survey. Wiley Interdiscip. Rev. Data Mining Knowl. Discov. **10**(3), e1356 (2020)
18. Buolamwini, J.A.: Gender shades: intersectional phenotypic and demographic evaluation of face datasets and gender classifiers. Doctoral dissertation, Massachusetts Institute of Technology (2017)
19. Jadon, A.: Ethical AI development: mitigating bias in generative models. EasyChair Preprint 14104. https://easychair.org/publications/preprint/1Klv (2024, July)
20. Ray, P.P.: Chatgpt: a comprehensive review on background, applications, key challenges, bias, ethics, limitations and future scope. Internet Things Cyber-Phys. Syst. **3**, 121–154 (2023)
21. Charfeddine, M., Kammoun, H.M., Hamdaoui, B., Guizani, M.: Chatgpt's security risks and benefits: offensive and defensive use-cases, mitigation measures, and future implications. IEEE Access **12**, 30263 (2024)
22. York, J.C., Collings, P., Greene, D.: Meta's new content policy will harm vulnerable users. If it really valued free speech, it would make these changes. Electronic Frontier Foundation (2025, January 9)
23. Human Rights Watch: Meta's broken promises: systemic censorship of palestine content on Instagram and Facebook. Technical report (2023, December 21)
24. Alzaidy, S., Binsalleeh, H.: Adversarial attacks with defense mechanisms on convolutional neural networks and recurrent neural networks for malware classification. Appl. Sci. **14**(4), 1673 (2024)
25. Singhal, A., Neveditsin, N., Tanveer, H., Mago, V.: Toward fairness, accountability, transparency, and ethics in AI for social media and health care: scoping review. JMIR Med. Inform. **12**(1), e50048 (2024)
26. Dastin, J.: Insight - Amazon scraps secret AI recruiting tool that showed bias against women (2018, October 11)
27. Radanliev, P., Santos, O., Brandon-Jones, A., Joinson, A.: Ethics and responsible AI deployment. Front. Artif. Intell. **7**(4), 1377011 (2024)
28. Sartor, G., Lagioia, F.: The impact of the general data protection regulation (GDPR) on artificial intelligence. Technical Report PE 641.530, European Parliamentary Research Service, Scientific Foresight Unit (STOA) (2020)

29. Kohnke, A., Sigler, K., Shoemaker, D.: Implementing Cybersecurity: A Guide to the National Institute of Standards and Technology Risk Management Framework. Auerbach Publications, New York (2017)
30. Austin, G.: Cybersecurity in China: The Next Wave. Springer International Publishing, Cham (2018)
31. Ladkin, P.: Functional safety and oracular subsystems: an observation on ISO/IEC TR 5469: artificial intelligence—functional safety and AI systems. Saf. Crit. Syst. eJ. **3**(2), 25–26 (2024)
32. Shahi, A., Bansal, S., Chawla, S.K., Elngar, A.A.: Navigating compliance for the industrial IoT landscape: frameworks, security, standards, and key ethical considerations. In: Industrial Internet of Things Security, pp. 81–99. CRC Press, Boca Raton (2024)
33. Ijiga, O.M., Idoko, I.P., Ebiega, G.I., Olajide, F.I., Olatunde, T.I., Ukaegbu, C.: Harnessing adversarial machine learning for advanced threat detection: AI-driven strategies in cybersecurity risk assessment and fraud prevention. Open Access Res. J. Sci. Technol. **11**(1), 1–24 (2024)
34. Chimamiwa, G.: Managing cyber risks in the face of AI-and ML-driven adversarial attacks. SBS Swiss Business School (2024)
35. Radanliev, P.: AI ethics: integrating transparency, fairness, and privacy in AI development. Appl. Artif. Intell. **39**(1), 2463722 (2025)
36. Ferrara, E.: Fairness and bias in artificial intelligence: a brief survey of sources, impacts, and mitigation strategies. Preprint. arXiv:2304.07683 (2023)
37. European Commission: Ethics guidelines for trustworthy AI. Technical report (2019)
38. Shan, W.: AI-powered fraud detection in banking: innovations, challenges and preventive strategies. PhD thesis, Doctoral dissertation (2025)
39. Brennen, A.: What do people really want when they say they want" explainable AI?" we asked 60 stakeholders. In: Extended Abstracts of the 2020 CHI Conference on Human Factors in Computing Systems, pp. 1–7 (2020)
40. OECD AI Policy Observatory and GPAI: Transparency and explainability (principle 1.3). OECD AI Dashboard
41. Puplampu, R.: What Everyone Should Know About the Rise of AI: AI Transparency, Privacy, and Ethics Best Practices. Puplampu Books (2024)
42. Asenov, E.: The global framework for the regulation of artificial intelligence. Godishnik na UNSS **1**(1), 7–21 (2024)
43. Para, E.: The UNESCO recommendation on the ethics of artificial intelligence. Soroptimist International (2024)
44. Shafik, W.: Artificial intelligence and machine learning with cyber ethics for the future world. In: Future Communication Systems Using Artificial Intelligence, Internet of Things and Data Science, pp. 110–130. CRC Press, Boca Raton (2024)
45. Farinu, U.: Fairness, accountability, and transparency in AI: Ethical challenges in data-driven decision-making. Technical report, SSRN (2025)
46. Akhtar, M.A.K., Kumar, M., Nayyar, A.: Privacy and security considerations in explainable AI. In: Towards Ethical and Socially Responsible Explainable AI: Challenges and Opportunities, pp. 193–226. Springer Nature Switzerland, Cham (2024)
47. Ashraf, Z.A., Mustafa, N.: AI standards and regulations. In: Intersection of Human Rights and AI in Healthcare, pp. 325–352 (2025)
48. Reivo, J.: Artificial intelligence governance, management and risk management-a look into EU AIA, standards and other frameworks from practical level. Master's thesis, Institution/University Name (2024)
49. Ünver, H.A.: Artificial intelligence (AI) and human rights: Using AI as a weapon of repression and its impact on human rights. In-Depth Analysis requested by the Subcommittee on Human Rights (DROI) of the European Parliament, Policy Department for External Relations, Directorate-General for External Policies of the Union, PE 754.450. Available at: https://www.europarl.europa.eu/RegData/etudes/IDAN/2024/754450/EXPO_IDA(2024)754450_EN.pdf (2024, May)

50. Garrido Rebolledo, V.: Impact of artificial intelligence on international relations: towards a global algorithms governance. Rev. UNISCI **67**, 1–27 (2025)
51. Rehman, Z.: Beyond borders: international law and global governance in the digital age. J. Account. Bus. Arch. Rev. **1**(1), 1–12 (2023)
52. Ijaiya, H., Odumuwagun, O.O.: Advancing artificial intelligence and safeguarding data privacy: a comparative study of EU and US regulatory frameworks amid emerging cyber threats. Int. J. Res. Publ. Rev. **5**(12), 3357–3375 (2025)
53. Capel, T., Brereton, M.: What is human-centered about human-centered AI? A map of the research landscape. In: Proceedings of the 2023 CHI Conference on Human Factors in Computing Systems, pp. 1–23 (2023, April)
54. Olushola, A., Mart, J.: Fraud detection using machine learning. ScienceOpen Preprints (2024)
55. Patil, D.: Explainable artificial intelligence (Xai), enhancing transparency and trust in machine learning models. SSRN (2024)
56. Sharma, R., Mehta, K., Sharma, P.: Role of artificial intelligence and machine learning in fraud detection and prevention. In: Risks and Challenges of AI-Driven Finance: Bias, Ethics, and Security, pp. 90–120. IGI Global, Hershey (2024)
57. Broadbent, M.: What's ahead for a cooperative regulatory agenda on artificial intelligence. Technical report, Center for Strategic and International Studies (CSIS) (2021). Accessed 14 May 2021
58. Google: AI principles: Progress update. Technical report (2023)
59. Hansen, R., Venables, P.: Introducing Google's Secure AI framework. Google Blog. https://blog.google/technology/safety-security/introducing-googles-secure-ai-framework/ (2023, June 8)
60. Microsoft: AI considerations for your cloud strategy. Microsoft Learn (2025, February 3)
61. Jakkal, V.: Microsoft unveils expansion of AI for security and security for AI at Microsoft Ignite. Microsoft Security Blog. https://www.microsoft.com/en-us/security/blog/2023/11/15/microsoft-unveils-expansion-of-ai-for-security-and-security-for-ai-at-microsoft-ignite/ (2023, November 15)
62. Sullivan, M.: Key principles for ethical AI development. Transcend (2023, October 20)
63. World Economic Forum: Responsible use of technology: The IBM Case Study. World Economic Forum. Available at: https://www.weforum.org/publications/responsible-use-of-technology-the-ibm-case-study/ (2021, September)
64. Mastercard: Mastercard accelerates card fraud detection with generative AI technology. Retrieved from https://www.mastercard.com/us/en/news-and-trends/press/2024/may/mastercard-accelerates-card-fraud-detection-with-generative-ai-technology.html (2024, May 22)
65. Lauer, D.: Facebook's ethical failures are not accidental; they are part of the business model. AI Ethics **1**(4), 395–403 (2021)
66. Kaplan, J.: More Speech and Fewer Mistakes. Meta News. Available at: https://about.fb.com/news/2025/01/meta-more-speech-fewer-mistakes/ (2025, January 7)
67. Geiger, R.S., Tandon, U., Gakhokidze, A., Song, L., Irani, L.: Making algorithms public: reimagining auditing from matters of fact to matters of concern. Int. J. Commun. **18**(1), 634–655 (2024)
68. Park, A.L.: Injustice ex machina: Predictive algorithms in criminal sentencing. UCLA Law Review (2019)
69. Piper, K.: Exclusive: Google cancels AI ethics board in response to outcry. Vox (2019)
70. Efthymiou, I.-P., Egleton, T.E.: The impact of deep fakes in markets and economies. In: Deepfakes and Their Impact on Business, pp. 19–50. IGI Global Scientific Publishing, Hershey (2025)

Chapter 3
Vulnerability Detection: From Formal Verification to Large Language Models and Hybrid Approaches: A Comprehensive Overview

Norbert Tihanyi, Tamas Bisztray, Mohamed Amine Ferrag, Bilel Cherif, Richard A. Dubniczky, Ridhi Jain, and Lucas C. Cordeiro

3.1 Introduction

Software testing is critical to software development, ensuring reliability, security, and performance. Various testing techniques are employed to detect bugs, vulnerabilities, and inefficiencies throughout different development stages. In the literature, software testing is categorized into different groups. Figure 3.1 illustrates the six primary high-level categories of software testing. *Static Application Security Testing (SAST)* inspects code without executing it, detecting potential issues early in development [28]. *Dynamic Application Security Testing (DAST)* validates software behavior during runtime, identifying vulnerabilities that may only surface under real-world conditions. *Interactive Application Security Testing (IAST)* blends

N. Tihanyi (✉)
Eötvös Loránd University (ELTE), Budapest, Hungary

Technology Innovation Institute (TII), Abu Dhabi, UAE

T. Bisztray
University of Oslo, Oslo, Norway

Cyentific AS, Oslo, Norway

M. A. Ferrag
Guelma University, Guelma, Algeria

B. Cherif · R. Jain
Technology Innovation Institute (TII), Abu Dhabi, UAE

R. A. Dubniczky
Eötvös Loránd University (ELTE), Budapest, Hungary

L. C. Cordeiro
The University of Manchester, Manchester, UK

Federal University of Amazonas, Manaus, Brazil

E. Nowroozi et al. (eds.), *Adversarial Example Detection and Mitigation Using Machine Learning*, https://doi.org/10.1007/978-3-031-99447-0_3

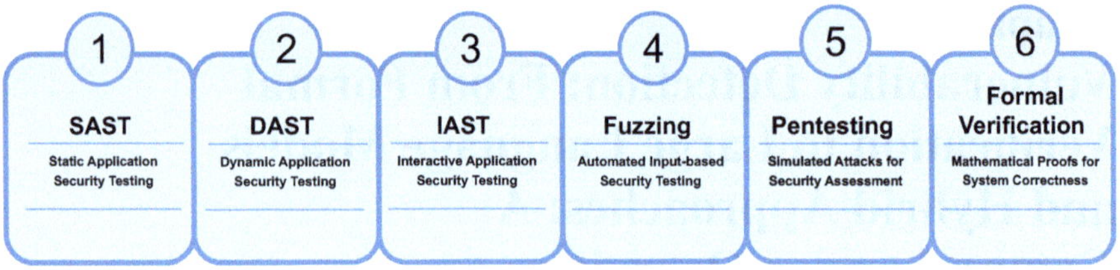

Fig. 3.1 High-level categories of software testing

insights from both static and dynamic analysis by instrumenting the application to observe security issues during execution [38]. *Fuzzing* leverages unexpected or malformed inputs to uncover hidden weaknesses [32]. In contrast, *Penetration Testing* (often a manual or semi-automated process) simulates attacker behaviors to find exploitable security flaws in realistic scenarios [53]. *Formal verification (FV)* is a rigorous mathematical approach to proving that hardware or software systems fulfill their intended requirements under all possible scenarios [10]. Unlike traditional testing that only covers some executions, FV either formally proves a system meets its properties or provides a concrete counterexample, eliminating false positives.

These six categories provide a broad perspective on modern application security strategies. However, some sources may classify them differently, such as grouping fuzzing under dynamic testing or recognizing only three primary categories—SAST, DAST, and IAST—while others introduce new categories like *Run-time Application Security Protection (RASP)* [4]. However, this review will treat these six as the main categories, since each demands specialized expertise and distinct methodologies that naturally separate them based on the required knowledge.

Lately, *Large Language Models (LLMs)* have been introduced and are now used in every part of software testing, from fuzzing [21] to invariant generation [40], and also play an important role in the generation of source code. As LLMs are expected to dominate future software development, the need for robust automated testing has become more urgent, as multiple studies reveal that LLMs can introduce security vulnerabilities in all programming languages examined [31, 44, 50, 55]. In the field of software testing, FV and LLMs present a promising combination for addressing these challenges. Each offers unique strengths and weaknesses, but they can complement and compensate for one another's limitations.

Among the various FV techniques, *Bounded Model Checking* (BMC) [2] has gained particular prominence for its scalability. Rather than exhaustively analyzing the entire state space, BMC confines its search to a predefined bound, employing powerful satisfiability (SAT/SMT) solvers to detect property violations. While this approach excels at uncovering memory safety and concurrency issues, it fails to address various vulnerabilities, such as SQL injection or cross-site scripting (XSS). Moreover, the *state explosion problem* can make verification computationally infeasible for large or highly concurrent systems. Finally, most FV tools focus on a small set of languages—primarily C, C++, and Java, with python being recently introduced [13]—limiting their relevance for modern software stacks.

In parallel, LLMs have gained significant traction in software security research. These models use neural architectures trained on massive source code and natural language corpora to detect vulnerabilities. Unlike traditional static or dynamic analysis, which relies on handcrafted rules or symbolic execution, LLM-based approaches leverage patterns that generalize across different programming languages and codebases. However, current LLMs face challenges: they may overfit training data, generate false positives, and lack the rigorous guarantees formal methods offer. Nonetheless, as LLMs mature and incorporate advanced learning paradigms, they promise to complement FV techniques by providing broader coverage of language-specific and domain-specific vulnerabilities. This review article offers the following key contributions:

1. **Analysis of Formal Verification-Based Methods.** We provide a comprehensive overview of how FV techniques are used to detect vulnerabilities, outlining their current capabilities and the challenges—such as scalability and limited language support—that constrain their broader adoption.
2. **Assessment of LLM-Based Approaches.** We critically examine cutting-edge research leveraging large language models for automated vulnerability detection, highlighting the advantages of learned representations and discussing ongoing challenges such as false positives and domain adaptation.
3. **Exploration of Hybrid Strategies.** We investigate emerging frameworks that integrate FV with LLM-based methods, illustrating how both strengths can be combined to achieve broader coverage, higher accuracy, and greater scalability in vulnerability detection.

3.2 Motivation

While LLMs can tackle increasingly complex tasks, they rely on pattern recognition rather than true semantic understanding. Although this makes them versatile as static code analyzers, their outputs carry no formal correctness guarantees. "Hallucinations" or plausible looking yet incorrect answers, remain a well-documented problem. Tihanyi et al. [49] observed that when presented with a seemingly simple question—such as finding the next prime number, which would typically require tool usage—an LLM attempted to solve it directly rather than skipping it, ultimately producing an incorrect response. This example illustrates how LLMs can misinterpret problem complexity and face difficulties with exact calculations when the necessary tools are unavailable. However, LLMs can sometimes leverage pattern recognition to determine that exhaustive checks are unnecessary, allowing them to arrive quickly at a correct solution in cases where FV tools might struggle. To illustrate both aspects, we present two small C programs in Fig. 3.2. The code on the left poses challenges for LLMs, while the one on the right presents difficulties for FV tools.

Challenging for LLMs

```c
#include <stdio.h>
int main() {
    unsigned int A[] = {1091, 1031, 1311, 1019,
    1041, 1061, 1019, 1187, 1091, 1107};
    unsigned int sum = 0;
    for (int i = 0; i < 220*2; i++) {
        for (int j = 1 ; j < 9; j++) {
            sum += A[j]+(A[j]*7+A[j]*A[j]);
        }
    }
    printf("Resulting sum: %u\n", sum);
    return 0;
}
```

Challenging for FVs

```c
#include <stdio.h>

int main(void) {
    long long x = 0;
    for (long long i = 0; i< 1000000000; ++i)
    {
        x += i;
    }
    printf("Result; %lld\n", x);
    return 0;
}
```

Fig. 3.2 Motivating examples posing challenges for LLMs an FV tools

We tested the latest chat models, including OpenAI ChatGPT-4o, the o1 model, and Deepsek R1's chat model, with Deepthink enabled. We used the following prompt: "*You are a security analyst. I want you to analyze the following code for vulnerabilities. For the vulnerabilities found, return the vulnerable line(s) and the corresponding CWE identifier(s). If none is found, return an empty list.*".

In every case, for the code on the left-hand side, all models detected a sum overflow and reported CWE-190 (Integer Overflow or Wraparound). However, this is incorrect: the sum comes close to overflowing but does not exceed the maximum limit of 2^{32}. Verification is straightforward using an FV tool. In particular, we employed the *Efficient SMT-based Context-Bounded Model Checker (ESBMC)* [34], which uses symbolic model checking based on Satisfiability Modulo Theories (SMT) in combination with abstract interpretation techniques. After just 0.73 seconds on a regular laptop, ESBMC reported `VERIFICATION SUCCESSFUL`, indicating no overflow was detected. On the other hand, formal verification tools, including ESBMC have limitations. When examining the other program in Fig. 3.2 (right-side), it is straightforward for an LLM to recognize the absence of vulnerabilities in the code, whereas ESBMC requires significantly more time to complete its verification. It needs to unroll the loop 10^9 times. In our test environment, unwinding the first 1 million iterations took 46 seconds, implying that fully unrolling the loop would take approximately 12.7 hours to verify, with extremely high memory requirements. These examples underscore that LLMs and formal verification each play a vital role in vulnerability detection. In the following chapters, we explore the latest advances in LLM-based analysis and formal verification methods, illustrating how these tools can be used together to enhance reliability, speed, and accuracy.

3.3 LLMs in Vulnerability Detection

Recent developments, particularly in decoder-only models like OpenAI's ChatGPT or Meta's Code Llama, highlight the shifting landscape in vulnerability detection.

Their capacity for large-context reasoning and on-demand text generation facilitates powerful few-shot or prompt-based approaches that can outperform classical fine-tuned detectors on certain benchmarks. This section reviews the latest trends in LLM-based vulnerability detection.

3.3.1 Model Architectures and Trends

Vulnerability detectors leverage transformer-based code models as foundational backbones trained on large code corpora across multiple programming languages. While it allows them to learn syntax, the training data itself is full of vulnerable code in many instances. Transformer-based code models can be put in three categories: Encoder-Only, Encoder-Decoder and Decoder-Only. Encoder-only and encoder–decoder transformers (e.g., CodeBERT, GraphCodeBERT, CodeT5) were widely used in early work, often fine-tuned for classification tasks. However, decoder-only LLMs like the GPT series and CodeLlama have rapidly become the preferred choice due to their superior size and generalization capabilities [46].

This trend toward decoder-only architectures aligns with developments in industry, where state-of-the-art LLMs are often deployed via in-context learning and prompt engineering rather than full retraining. Large models like GPT-4 can usually be prompted to detect vulnerabilities without any parameter updates, thanks to knowledge gained during pre-training. Advances in prompt design (e.g., chain-of-thought prompts or specialized code analysis prompts) further boost their performance, allowing these models to reason about code security with minimal task-specific tuning [33]. Encoder-only models (e.g. BERT-derived code models) typically require full fine-tuning to perform vulnerability classification, as they lack the broad intrinsic knowledge and generative prowess of frontier LLMs. The emerging consensus is that while encoders excel at efficient classification, decoder-only LLMs offer a more flexible foundation for vulnerability detection—able to understand complex code patterns and suggest fixes—making them increasingly central in recent research.

3.3.2 Datasets for Fine-Tuning and Benchmarking

LLMs pre-trained on code have been fine-tuned for vulnerability detection with notable success. Large-scale datasets are mostly available in C, C++, Java, Python, PHP, or Solidity, with C/C++ being the overwhelming majority [46]. Labeled datasets with over 100 samples are presented in Table 3.1. Such datasets are essential for fine-tuning and benchmarking various static, dynamic, or AI/ML-based detection systems. ReGVD combined CodeBERT with graph neural networks to learn from code property graphs, improving the detection of vulnerabilities in C/C++ code [20]. A RoBERTa-based model (VulBERTa) fine-tuned on multiple datasets

Table 3.1 Datasets for benchmarking and fine-tuning in vulnerability detection

Dataset	Size*	Language	#Unique Vuln.	Granularity	Source
Draper* [45]	1274k	C/C++	5	function	mixed
SARD [35]	450k	C/C++,Java, PHP	150	file	mixed
SeVc [26]	420k	C/C++	126	function	mixed
DiverseVul [3]	349k	C/C++	150	function	real-world
FormAIv2 [50]	331k	C	33	file	AI Generated
BigVul [12]	264k	C/C++	91	function	real-world
PrimeVul [9]	235k	C/C++	140	function	real-world
VulDeeLocator [25]	198k	LLVM IR	4	function	mixed
Juliet C/C++ [36]	64k	C/C++	118	file	by design vuln.
Code Gadgets [24]	61k	C/C++	2	function	mixed
Juliet Java [37]	28k	Java	112	file	by design vuln.
FELLMVP [30]	15k	Solidity	8	contract	real-world
CVEfixes [1]	12,k	C/C++,Python, PHP	180	commit	real-world
ChatPHP [51]	2,5k	PHP	4	website	AI Generated
VCMatch [54]	1,6k	C/C++,Java,PHP	7	commit	real-world
SVEN [19]	1,6k	C/C++,Python	9	commit	real-world
Ponta2019 [41]	1,2k	Java	6	commit	real-world
VulBench [16]	455	C/C++	9	function	mixed
Ullah2023 [52]	228	C/C++, Python	8	function	mixed
Smartbugs[15]	143	Solidity	10	contract	mixed
MAGMA [18]	138	C/C++, Lua, PHP	11	repository	real-world
Vulcorpus [23]	100	Java	10	function	by design vuln.

Legend: *Draper has been labelled with static analysers and are therefore less reliable

achieved top-tier results on benchmarks like CodeXGLUE's defect detection [17]. A notable result is that SecureFalcon, an innovative model architecture with only 121 million parameters derived from the Falcon-40B model and explicitly tailored for classifying software vulnerabilities, outperforms existing models such as BERT, RoBERTa, CodeBERT, and traditional ML algorithms [14]. These studies indicate that customizing LLMs using real-world code vulnerability datasets significantly enhances their detection effectiveness in practical applications [11]. Overall, fine-tuning remains the dominant approach (around 73% of recent studies) for adapting LLMs to this task [59].

3.3.3 Prompt-Based Approaches

With the advent of powerful decoder-only models like GPT-3.5/4, researchers explored prompt-based methods to detect vulnerabilities without gradient-based training. These approaches supply code to a frozen LLM and ask it (via a prompt) whether the code is vulnerable, possibly with explanations. A key finding is that prompt design and reasoning techniques dramatically affect performance [56]. Few-shot or instructed large models can leverage their broad knowledge to detect many vulnerabilities when asked the right way. Prompt-based methods offer quick iteration (no training needed) and flexibility across languages, but they have trade-offs. They often require large context windows to input code and may struggle with very long or complex code snippets due to context length limits. In addition, responses can be inconsistent—one study found that even advanced code models achieved only around 54.5% balanced accuracy when directly queried, and simply scaling up model size or naive prompting did not significantly improve this.

3.3.4 Emerging Trends and Ongoing Challenges

Despite recent progress, significant challenges remain in using LLMs for vulnerability detection. While fine-tuned models may excel on partially trained datasets, they often fail to generalize to new, unseen data. Decoder-only models, however, have shown strong performance on code they have not encountered before. A major bottleneck for both approaches is data availability: accurately labeled large datasets with diverse vulnerability representations—ones that truly mirror real production environments—are still missing for most programming languages. Moreover, existing datasets tend to be skewed toward certain vulnerability types (e.g., buffer overflows, injections), leaving others (such as concurrency flaws) under-represented [46]. Efforts to crowdsource or generate more diverse samples (including synthetic code produced by LLMs) are underway to enrich training data [50].

Another core issue is semantic understanding. Current LLMs often struggle with intricate code logic, such as pointer aliasing or subtle data flows, especially when the vulnerability stems from a long sequence of operations. Studies have shown that small, behavior-preserving changes (e.g., renaming variables or rearranging code) can mislead an LLM's judgment [43, 47]. This raises concerns about reliability and false positives/negatives: a model might confidently mislabel safe code as vulnerable or vice versa. These errors are difficult to debug without a transparent explanation or consistent reasoning. Ensemble or consensus methods can mitigate some of this uncertainty but add system complexity. Scalability and integration pose further hurdles: Large models like GPT-4 are powerful but expensive to run and not easily deployed on-premise. To address this, companies are exploring distilled or domain-adapted versions that run faster and use parameter-efficient fine-tuning methods (e.g., LoRA adapters) to customize massive models for vulnerability detection without fully retraining them.

On the positive side, practical applications of LLM-based vulnerability detection are emerging. Integrated Development Environment (IDE) plugins and code review bots powered by models like Codex can flag risky code in real time and suggest fixes. Security companies have begun building AI "copilots" for auditors—for instance, using an LLM to scan code repositories and highlight high-risk functions for manual review, thereby combining the speed of automation with human judgment. In specialized settings such as blockchain, tools are appearing that apply LLMs to audit smart contracts for pitfalls like reentrance and integer overflow. These developments underscore an evolving human—AI partnership model: the LLM handles large-scale code analysis and reasoning, while human experts validate the results and make final decisions, guided by the AI's insights.

3.4 Formal Verification

Formal verification (FV) is a rigorous mathematical approach for establishing that hardware and software systems satisfy their intended requirements across *all* possible scenarios. In contrast to traditional testing, which examines only a limited subset of executions, FV can provide either (1) a formal proof that the system meets the specified properties or (2) a concrete counterexample demonstrating how the properties can be violated. This dual capability ensures that any detected vulnerabilities are backed by a demonstrable failing execution, thereby minimizing false positives.

A widely used technique within FV is *model checking* (MC), introduced independently by Edmund Clarke and Allen Emerson in 1981 [5], and by Jean-Pierre Queille and Joseph Sifakis in 1982 [42]. MC systematically explores the state space of a system to verify whether certain properties, typically expressed in temporal logic, hold. It has been successfully applied to various domains, including hardware circuits, concurrent systems, and communication protocols. However, traditional MC often encounters the well-known *state explosion problem* [6], where the number of states grows exponentially with the increase in variables and concurrency.

To mitigate this complexity, *Bounded Model Checking* (BMC) was introduced in 1999 by Armin Biere et al. [2] as a complementary approach to Binary Decision Diagrams (BDDs). Rather than exhaustively exploring all possible states, BMC restricts its search to executions up to a predefined bound. This makes BMC more scalable in practice and highly effective at detecting property violations in complex systems, as any counterexample within the bound will be discovered. While BMC can become computationally demanding for massive or deeply nested systems, it remains an invaluable tool for identifying security and reliability issues. For instance, ESBMC was recently employed to analyze the Arm Confidential Computing Architecture, uncovering specification failures that strengthened its security [58]. Likewise, BMC techniques have revealed common bugs (e.g., null pointer dereferences, buffer overflows) in large open-source C projects, demonstrating the practical benefits of FV for real-world software [8].

An important takeaway for the reader is that these techniques provide formal guarantees: if a bug is identified, a counterexample confirms it, thereby minimizing false positives. This is especially valuable in mitigating the weaknesses of LLMs.

3.5 Hybrid Approaches

Integrating LLMs with formal methods and program analysis tools has emerged as a promising direction for improving software vulnerability detection. Traditional formal verification methods provide rigorous guarantees about software correctness but often struggle with scalability and applicability to large-scale, real-world software systems. LLMs, on the other hand, can rapidly analyze vast amounts of code and detect patterns indicative of vulnerabilities, but lack formal guarantees of correctness and often produce false positives.

3.5.1 LLM-Assisted Formal Verification

LLMs have been applied to assist in writing the formal artifacts (properties, invariants, constraints) needed for verification. For example, Kande et al. investigate using GPT-based models to generate security assertions for hardware designs [22]. Similarly, Pei et al. (ICML 2023) showed that LLMs fine-tuned for program invariant generation can infer loop invariants for code with a quality comparable to dynamic invariant detectors [39].

A notable application is PropertyGPT by Liu et al. (NDSS 2024), which leverages GPT-4 to automatically generate formal properties (invariants, pre-/post-conditions, temporal rules) for smart contract verification [29]. In experiments on real Ethereum contracts, PropertyGPT achieved 80% recall of ground-truth properties and successfully detected numerous known vulnerabilities and even 12 zero-day flaws that earned bug bounties. These results illustrate that LLMs, when guided properly, can generate high-quality formal specifications to drive vulnerability discovery.

LLMs can serve as “specification generators” or assistants in formal verification workflows [27]. By translating natural-language requirements or learned patterns into formal assertions/invariants, they reduce the manual burden and open formal methods to broader use in security. The LLM outputs are not guaranteed correct on their own, but coupling them with formal checks (compilation, model checking, theorem proving) ensures that only valid, useful properties are utilized.

Overall, LLM-assisted formal verification has shown success in domains ranging from hardware security to smart contracts, suggesting a general paradigm in which LLM proposes, and formal tools dispose [29]. Another line of work makes the LLM itself part of a verification loop, ensuring its vulnerability analyses adhere to formal correctness criteria. In these approaches, formal verification tools guide or validate the LLM's reasoning in real-time. CryptoFormalEval is a representative example focusing on cryptographic protocol security [7].

Formal feedback helps the LLM adjust its strategy; for instance, if the prover finds a counterexample, the LLM knows a real vulnerability exists and can elaborate on it; if not, the LLM refines its queries. This tight interaction between LLM and prover means any vulnerability the LLM "finds" is backed by a formal proof of concept, aligning the outcome with sound verification principles. Beyond specific applications, researchers are formalizing how an LLM can operate under the guidance of a verifier. Wu et al. propose LEMUR, a framework that interleaves LLM-generated reasoning steps with checks by automated solvers [57]. They treat the LLM as an abstract transition relation on program states: the LLM suggests a candidate invariant or lemma, then an SMT solver or model checker verifies if it holds. If it passes the formal check, the system accepts the LLM's suggestion as part of a proof; otherwise, the LLM is prompted to refine its output. Crucially, LEMUR includes a formal calculus with derivation rules for this process and proves it is sound (i.e., it will never accept an incorrect property if the underlying verifier is sound). This formal verification-guided approach ensures the LLM's creative reasoning is always tethered to correctness. It was shown to improve automated verification on benchmark programs, solving tasks that neither the LLM nor the verifier could easily achieve alone. As this area evolves, we expect LLMs to become increasingly constrained by specs or formal rules during generation (for example, using model checking results as feedback in the LLM's prompt), thereby producing analyses that are not only plausible but also provably correct.

3.5.2 End-to-End Hybrid Pipelines

End-to-end systems have been built that seamlessly chain together LLM components with traditional analysis tools to detect (and even fix) vulnerabilities. These pipelines treat LLMs and formal/static analyzers as complementary stages, often in an iterative loop. ESBMC-AI [48] is a notable framework that illustrates this integration: a self-healing software pipeline that combines a BMC tool with an LLM to detect and repair vulnerabilities in C code. This closed-loop approach was evaluated on 50,000 C programs containing vulnerabilities (from the FormAI dataset). ESBMC-AI was able to automatically detect and repair issues like buffer overflows, arithmetic errors, and null-pointer dereferences with high accuracy. The authors report a significant portion of the bugs were fixed on the first attempt by the LLM, and the remainder eventually resolved through iterative prompting. The ESBMC-AI pipeline demonstrates a powerful synergy: formal analysis provides certainty that a vulnerability exists (and supplies details), while the LLM contributes generative intelligence to suggest a fix, and formal analysis guarantees the fix's correctness.

PropertyGPT pipeline [29], as described earlier, is a system for smart contracts that can also be viewed as an end-to-end pipeline combining retrieval, LLM generation, static checking, and formal verification. The aforementioned CryptoFormalEval [7], which can be seen as an end-to-end pipeline in the context of protocol analysis. It provides a middleware for an AI agent (LLM) to interact with a theorem prover in multiple rounds.

End-to-end hybrid pipelines have shown that combining detection and validation in one loop greatly improves confidence in the results. By cross-checking an LLM's output with a rigorous tool (and vice versa), they mitigate each other's weaknesses. For instance, the LLM might introduce a bad fix or a bogus property, but the formal stage will catch it; conversely, a formal tool might miss a bug due to limited specifications, but the LLM can provide those missing pieces earlier in the pipeline. These holistic systems are pioneering initiatives that indicate how future vulnerability detection might be fully automated: an AI finds a bug, explains it, fixes it, and proves the fix correct, all in an integrated workflow [48].

In conclusion, integrating LLMs and FV techniques presents a promising path toward more robust vulnerability detection and software assurance. While FV offers mathematical rigor and precise guarantees, it struggles with scalability and practical applicability to complex systems. On the other hand, LLMs excel in pattern recognition and flexible code analysis but lack formal correctness guarantees. Combining these approaches allows hybrid frameworks to leverage their strengths, using LLMs for broad code analysis and suggestion generation while ensuring correctness through FV validation. Recent research has demonstrated the viability of such hybrid pipelines, paving the way for more efficient, accurate, and scalable software verification methodologies.

References

1. Bhandari, G., Naseer, A., Moonen, L.: Cvefixes: automated collection of vulnerabilities and their fixes from open-source software. In: Proceedings of the 17th International Conference on Predictive Models and Data Analytics in Software Engineering, PROMISE 2021, New York, NY, pp. 30–39. Association for Computing Machinery, New York (2021)
2. Biere, A., Cimatti, A., Clarke, E., Zhu, Y.: Symbolic model checking without BDDs. In: Cleaveland, W.R. (ed.) Tools and Algorithms for the Construction and Analysis of Systems, pp. 193–207. Springer Berlin Heidelberg, Berlin, Heidelberg (1999)
3. Chen, Y., Ding, Z., Alowain, L., Chen, X., Wagner, D.: Diversevul: a new vulnerable source code dataset for deep learning based vulnerability detection. In: Proceedings of the 26th International Symposium on Research in Attacks, Intrusions and Defenses, RAID '23, New York, NY, pp. 654–668. Association for Computing Machinery, New York (2023)

4. Čisar, P., Čisar, S.M.: The framework of runtime application self-protection technology. In: 2016 IEEE 17th International Symposium on Computational Intelligence and Informatics (CINTI), pp. 000081–000086 (2016)
5. Clarke, E.M., Emerson, E.A.: Design and synthesis of synchronization skeletons using branching time temporal logic. In: Kozen, D. (ed.) Logics of Programs, pp. 52–71. Springer Berlin Heidelberg, Berlin, Heidelberg (1982)
6. Clarke, E.M., Klieber, W., Nováček, M., Zuliani, P.: Model Checking and the State Explosion Problem, pp. 1–30. Springer Berlin Heidelberg, Berlin, Heidelberg (2012)
7. Curaba, C., D'Ambrosi, D., Minisini, A., Pérez-Campanero Antolín, N.: Cryptoformaleval: Integrating llms and formal verification for automated cryptographic protocol vulnerability detection. Preprint. arXiv:2411.13627 (2024)
8. de Sousa, J.O., de Farias, B.C., da Silva, T.A., de Lima Filho, E.B., Cordeiro, L.C.: Lsverifier: A BMC approach to identify security vulnerabilities in c open-source software projects. In: Anais Estendidos do XXIII Simpósio Brasileiro de Segurança da Informação e de Sistemas Computacionais (SBSEG), Brazil, 2023, SBC, SBC – Sociedade Brasileira de Computação (2023)
9. Ding, Y., Fu, Y., Ibrahim, O., Sitawarin, C., Chen, X., Alomair, B., Wagner, D., Ray, B., Chen, Y.: Vulnerability detection with code language models: how far are we? In: 2025 IEEE/ACM 47th International Conference on Software Engineering (ICSE), Los Alamitos, CA, USA, May 2025, pp. 469–481. IEEE Computer Society (2025)
10. D'Silva, V., Kroening, D., Weissenbacher, G.: A survey of automated techniques for formal software verification. IEEE Trans. Comput. Aided Des. Integr. Circ. Syst. **27**(7), 1165–1178 (2008)
11. Du, X., Wen, M., Zhu, J., Xie, Z., Ji, B., Liu, H., Shi, X., Jin, H.: Generalization-enhanced code vulnerability detection via multi-task instruction fine-tuning. Preprint. arXiv:2406.03718 (2024)
12. Fan, J., Li, Y., Wang, S., Nguyen, T.N.: A C/C++ code vulnerability dataset with code changes and CVE summaries. In: Proceedings of the 17th International Conference on Mining Software Repositories, MSR '20, New York, NY, USA, pp. 508–512. Association for Computing Machinery, New York (2020)
13. Farias, B., Menezes, R., de Lima Filho, E.B., Sun, Y., Cordeiro, L.C.: ESBMC-Python: a bounded model checker for python programs. In: Proceedings of the 33rd ACM SIGSOFT International Symposium on Software Testing and Analysis, ISSTA 2024, New York, NY, USA, pp. 1836–1840. Association for Computing Machinery, New York (2024)
14. Ferrag, M.A., Battah, A., Tihanyi, N., Jain, R., Maimut, D., Alwahedi, F., Lestable, T., Thandi, N.S., Mechri, A., Debbah, M., et al.: Securefalcon: are we there yet in automated software vulnerability detection with llms? Preprint. arXiv:2307.06616 (2023)
15. Ferreira, J.F., Cruz, P., Durieux, T., Abreu, R.: Smartbugs: a framework to analyze solidity smart contracts. In: Proceedings of the 35th IEEE/ACM International Conference on Automated Software Engineering, ASE '20, New York, NY, USA, pp. 1349–1352. Association for Computing Machinery, New York (2021)
16. Gao, Z., Wang, H., Zhou, Y., Zhu, W., Zhang, C.: How far have we gone in vulnerability detection using large language models. Preprint, arXiv:2311.12420. Available at: https://arxiv.org/abs/2311.12420 (2023)
17. Hanif, H., Maffeis, S.: Vulberta: simplified source code pre-training for vulnerability detection. In: 2022 International Joint Conference on Neural Networks (IJCNN), pp. 1–8. IEEE, Piscataway (2022)
18. Hazimeh, A., Herrera, A., Payer, M.: Magma: a ground-truth fuzzing benchmark. Proc. ACM Meas. Anal. Comput. Syst. **4**(3), 1–29 (2020)
19. He, J., Vechev, M.: Large language models for code: Security hardening and adversarial testing. In: Proceedings of the 2023 ACM SIGSAC Conference on Computer and Communications Security, CCS '23, New York, NY, USA, pp. 1865–1879. Association for Computing Machinery, New York (2023)

20. Jiang, X., Wu, L., Sun, S., Li, J., Xue, J., Wang, Y., Wu, T., Liu, M.: Investigating large language models for code vulnerability detection: an experimental study. Preprint. arXiv:2412.18260 (2024)
21. Jiang, Y., Liang, J., Ma, F., Chen, Y., Zhou, C., Shen, Y., Wu, Z., Fu, J., Wang, M., Li, S., Zhang, Q.: When fuzzing meets llms: challenges and opportunities. In: Companion Proceedings of the 32nd ACM International Conference on the Foundations of Software Engineering, FSE 2024, New York, NY, USA, pp. 492–496. Association for Computing Machinery, New York (2024)
22. Kande, R., Pearce, H., Tan, B., Dolan-Gavitt, B., Thakur, S., Karri, R., Rajendran, J.: (Security) assertions by large language models. IEEE Trans. Inf. Forens. Secur. **19**, 4374–4389 (2024)
23. Kouliaridis, V., Karopoulos, G., Kambourakis, G.: Assessing the effectiveness of llms in android application vulnerability analysis. Preprint. arXiv:2406.18894 (2024)
24. Li, Z., Zou, D., Xu, S., Ou, X., Jin, H., Wang, S., Deng, Z., Zhong, Y.: Vuldeepecker: a deep learning-based system for vulnerability detection. Preprint. arXiv:1801.01681 (2018)
25. Li, Z., Zou, D., Xu, S., Chen, Z., Zhu, Y., Jin, H.: Vuldeelocator: a deep learning-based fine-grained vulnerability detector. IEEE Trans. Depend. Secur. Comput. **19**(4), 2821–2837 (2021)
26. Li, Z., Zou, D., Xu, S., Jin, H., Zhu, Y., Chen, Z.: Sysevr: a framework for using deep learning to detect software vulnerabilities. IEEE Trans. Depend. Secur. Comput. **19**(4), 2244–2258 (2021)
27. Lin, X., Cao, Q., Huang, Y., Wang, H., Lu, J., Liu, Z., Song, L., Liang, X.: Fvel: interactive formal verification environment with large language models via theorem proving. Adv. Neural Inf. Process. Syst. **37**, 54932–54946 (2025)
28. Lipp, S., Banescu, S., Pretschner, A.: An empirical study on the effectiveness of static C code analyzers for vulnerability detection. In: Proceedings of the 31st ACM SIGSOFT International Symposium on Software Testing and Analysis, ISSTA 2022, New York, NY, USA, pp. 544–555. Association for Computing Machinery, New York (2022)
29. Liu, Y., Xue, Y., Wu, D., Sun, Y., Li, Y., Shi, M., Liu, Y.: PropertyGPT: LLM-driven formal verification of smart contracts through retrieval-augmented property generation. Preprint. arXiv:2405.02580 (2024)
30. Luo, Y., Xu, W., Andersson, K., Hossain, M.S., Xu, D.: FELLMVP: An ensemble LLM framework for classifying smart contract vulnerabilities. In: 2024 IEEE International Conference on Blockchain (Blockchain), pp. 89–96. IEEE, Piscataway (2024)
31. Lyu, M.R., Ray, B., Roychoudhury, A., Tan, S.H., Thongtanunam, P.: Automatic programming: large language models and beyond. ACM Trans. Softw. Eng. Methodol. **34**, 1–33 (2024)
32. Mallissery, S., Wu, Y.-S.: Demystify the fuzzing methods: a comprehensive survey. ACM Comput. Surv. **56**(3), 1–38 (2023)
33. Mechri, A., Ferrag, M.A., Debbah, M.: SecureQwen: leveraging LLMs for vulnerability detection in python codebases. Comput. Secur. **148**, 104151 (2025)
34. Menezes, R.S., Aldughaim, M., Farias, B., Li, X., Manino, E., Shmarov, F., Song, K., Brauße, F., Gadelha, M.R., Tihanyi, N., Korovin, K., Cordeiro, L.C.: Esbmc v7.4: harnessing the power of intervals. In: Finkbeiner, B., Kovács, L. (eds.) Tools and Algorithms for the Construction and Analysis of Systems, pp. 376–380. Springer Nature Switzerland, Cham (2024)
35. National Institute of Standards and Technology: Software assurance reference dataset (SARD) (2024). Accessed 10 Nov 2024
36. NSA Center for Assured Software: Software assurance reference dataset (SARD): Juliet C/C++ 1.3 (2024). Accessed 10 Nov 2024
37. NSA Center for Assured Software: Software assurance reference dataset (SARD): Juliet Java 1.3 (2024). Accessed 10 Nov 2024
38. Pan, Y.: Interactive application security testing. In: 2019 International Conference on Smart Grid and Electrical Automation (ICSGEA), pp. 558–561 (2019)
39. Pei, K., Bieber, D., Shi, K., Sutton, C., Yin, P.: Can large language models reason about program invariants? In: Krause, A., Brunskill, E., Cho, K., Engelhardt, B., Sabato, S., Scarlett, J. (eds.) Proceedings of the 40th International Conference on Machine Learning, volume 202 of Proceedings of Machine Learning Research. PMLR, 23–29 Jul 2023, pp. 27496–27520 (2023)

40. Pirzada, M.A.A., Reger, G., Bhayat, A., Cordeiro, L.C.: LLM-generated invariants for bounded model checking without loop unrolling. In: Proceedings of the 39th IEEE/ACM International Conference on Automated Software Engineering, ASE '24, New York, NY, USA, pp. 1395–1407. Association for Computing Machinery, New York (2024)
41. Ponta, S.E., Plate, H., Sabetta, A., Bezzi, M., Dangremont, C.: A manually-curated dataset of fixes to vulnerabilities of open-source software. In: 2019 IEEE/ACM 16th International Conference on Mining Software Repositories (MSR), pp. 383–387. IEEE, Piscataway (2019)
42. Queille, J.P., Sifakis, J.: Specification and verification of concurrent systems in CESAR. In: Dezani-Ciancaglini, M., Montanari, U. (eds.) International Symposium on Programming, pp. 337–351. Springer Berlin Heidelberg, Berlin, Heidelberg (1982)
43. Risse, N., Böhme, M.: Uncovering the limits of machine learning for automatic vulnerability detection. In: 33rd USENIX Security Symposium (USENIX Security 24), Philadelphia, PA, August 2024, pp. 4247–4264. USENIX Association, Berkeley (2024)
44. Roychoudhury, A., Pasareanu, C., Pradel, M., Ray, B.: AI software engineer: programming with trust. Preprint. arXiv:2502.13767 (2025)
45. Russell, R., Kim, L., Hamilton, L., Lazovich, T., Harer, J., Ozdemir, O., Ellingwood, P., McConley, M.: Automated vulnerability detection in source code using deep representation learning. In: 2018 17th IEEE International Conference on Machine Learning and Applications (ICMLA), pp. 757–762. IEEE, Piscataway (2018)
46. Sheng, Z., Chen, Z., Gu, S., Huang, H., Gu, G., Huang, J.: Large language models in software security: a survey of vulnerability detection techniques and insights. Preprint. arXiv:2502.07049 (2025)
47. Steenhoek, B., Rahman, M.M., Roy, M.K., Alam, M.S., Tong, H., Das, S., Barr, E.T., Le, W.: To Err is machine: vulnerability detection challenges LLM reasoning. Preprint. arXiv:2403.17218 (2024)
48. Tihanyi, N., Jain, R., Charalambous, Y., Ferrag, M.A., Sun, Y., Cordeiro, L.C.: A new era in software security: towards self-healing software via large language models and formal verification. Preprint. arXiv:2305.14752 (2023)
49. Tihanyi, N., Bisztray, T., Dubniczky, R.A., Toth, R., Borsos, B., Cherif, B., Jain, R., Muzsai, L., Ferrag, M.A., Marinelli, R., Cordeiro, L.C., Debbah, M., Mavroeidis, V., Jøsang, A.: Dynamic intelligence assessment: benchmarking LLMS on the road to AGI with a focus on model confidence. In: 2024 IEEE International Conference on Big Data (BigData), pp. 3313–3321 (2024)
50. Tihanyi, N., Bisztray, T., Ferrag, M.A., Jain, R., Cordeiro, L.C.: How secure is AI-generated code: a large-scale comparison of large language models. Empir. Softw. Eng. **30**(2), 47 (2025)
51. Tóth, R., Bisztray, T., Erdődi, L.: LLMs in web development: Evaluating LLM-generated PHP code unveiling vulnerabilities and limitations. In: Computer Safety, Reliability, and Security. SAFECOMP 2024 Workshops: DECSoS, SASSUR, TOASTS, and WAISE, Florence, Italy, September 17, 2024, Proceedings, pp. 425–437. Springer, Berlin, Heidelberg (2024)
52. Ullah, S., Han, M., Pujar, S., Pearce, H., Coskun, A., Stringhini, G.: LLMs cannot reliably identify and reason about security vulnerabilities (yet?): a comprehensive evaluation, framework, and benchmarks. In: IEEE Symposium on Security and Privacy (2024)
53. Vats, P., Mandot, M., Gosain, A.: A comprehensive literature review of penetration testing & its applications. In: 2020 8th International Conference on Reliability, Infocom Technologies and Optimization (Trends and Future Directions) (ICRITO), pp. 674–680 (2020)
54. Wang, S., Zhang, Y., Bao, L., Xia, X., Wu, M.: VCMATCH: a ranking-based approach for automatic security patches localization for OSS vulnerabilities. In: 2022 IEEE International Conference on Software Analysis, Evolution and Reengineering (SANER), pp. 589–600 (2022)
55. Wang, J., Luo, X., Cao, L., He, H., Huang, H., Xie, J., Jatowt, A., Cai, Y.: Is your AI-generated code really safe? Evaluating large language models on secure code generation with CodeSecEval. Preprint. arXiv:2407.02395 (2024)

56. Wei, J., Wang, X., Schuurmans, D., Bosma, M., Xia, F., Chi, E., Le, Q.V., Zhou, D., et al.: Chain-of-thought prompting elicits reasoning in large language models. Adv. Neural Inf. Process. Syst. **35**, 24824–24837 (2022)
57. Wu, H., Barrett, C., Narodytska, N.: Lemur: integrating large language models in automated program verification. Preprint. arXiv:2310.04870 (2023)
58. Wu, T., Xiong, S., Manino, E., Stockwell, G., Cordeiro, L.C.: Verifying components of armő confidential computing architecture with ESBMC. In: Giacobazzi, R., Gorla, A. (eds.) Static analysis, pp. 451–462. Springer Nature Switzerland, Cham (2025)
59. Zhou, X., Cao, S., Sun, X., Lo, D.: Large language model for vulnerability detection and repair: Literature review and the road ahead. ACM Trans. Softw. Eng. Methodol. **34**(5), 1–31 (2025)

Part II
Attacks on AI Systems

Chapter 4
Backdoor Attacks in Text Classification: Threats, Methods, and Emerging Challenges

Egehan Eralp, A. Dilara Yavuz, and M. Emre Gursoy

4.1 Introduction

Text classification models have proven to be vulnerable to various adversarial attacks. Backdoor attacks (also known as trojan attacks) represent one insidious threat in which an attacker implants hidden malicious behavior into the model [3, 6, 15, 16, 28]. In a typical backdoor scenario, the attacker poisons a small subset of the training data (or directly manipulates model weights) so that the model learns to associate an attacker-chosen trigger pattern with an attacker-chosen class label. Importantly, the model continues to perform normally on samples that do not contain the trigger, maintaining high accuracy on clean test data. However, when the trigger (e.g., a specific word or text style) appears in an input, the backdoored model's prediction is immediately biased towards the attacker-chosen label. This dual behavior allows backdoored models to evade detection during testing while still harboring a serious vulnerability.

Backdoor attacks are especially feasible in today's systems when the training pipeline involves untrusted data or third-party models. For example, an attacker can inject poisoned samples to a publicly released dataset or publish a language model involving a backdoor (e.g., on HuggingFace). An unsuspecting practitioner can incorporate this data or model in their pipeline, and obtain a backdoored classifier. When the backdoored classifier is deployed, it can cause misclassifications of hate speech, toxic content, or spam e-mails.

Motivated by the plausibility and threat of backdoor attacks, this chapter provides an exploration of backdoor attacks in text classification, including popular attack methods, metrics, experimental evaluations, and discussions of emerging research

E. Eralp · A. D. Yavuz · M. E. Gursoy (✉)
Department of Computer Engineering, Koç University, Istanbul, Turkey
e-mail: eeralp22@ku.edu.tr; ayavuz19@ku.edu.tr; emregursoy@ku.edu.tr

E. Nowroozi et al. (eds.), *Adversarial Example Detection and Mitigation Using Machine Learning*, https://doi.org/10.1007/978-3-031-99447-0_4

directions. The chapter begins with a background on text classification, outlining traditional and modern deep learning-based classifiers. It then delves into backdoor attacks, and explains and demonstrates key attack methods such as AddSent [6], WordInj [3, 12], SynBkd [16], StyleBkd [15], and BITE [21]. Afterward, the chapter presents an experimental analysis involving multiple datasets, models, and attacks. Results show that backdoor attacks are effective even under low poison rates, e.g., attack success rates usually exceed 80% when poison rate is $\geq$3%. We also observe that the classification accuracy of backdoored models remains similar to clean models, demonstrating that the attacks can remain stealthy. Furthermore, by comparing models with different sizes and complexities, we find that models with higher complexity, which are more prone to memorization and overfitting, are also more vulnerable to backdoor attacks.

The chapter then explores current and emerging trends in backdoor research, including clean-label backdoor attacks and attacks targeting large language models (LLMs). Additionally, the chapter examines the implications of backdoor attacks in federated learning (FL), where decentralized training introduces new challenges and opportunities. Finally, the chapter considers novel applications of backdoor attacks, such as bias injection and model watermarking. By combining background information, experimental analysis, and discussions on current and future trends, the chapter aims to equip researchers and practitioners with an overview of backdoor attacks in text classification.

4.2 Background and Preliminaries

4.2.1 Text Classification

In a typical text classification setup, a dataset of training samples $\mathcal{D}_{train} = \{(x_1, y_1), (x_2, y_2), \ldots, (x_n, y_n)\}$ is used to train a classification model $\mathcal{M}$, where x_i denotes the text document and y_i denotes the class label. For example, x_i can be an e-mail or movie review and y_i can be either 0 or 1, corresponding to positive or negative sentiment. Given $\mathcal{D}_{train}$, the goal of text classification is to build a model $\mathcal{M}$ that can correctly predict the labels of previously unseen test samples. That is, for each test sample $(x_t, y_t) \in \mathcal{D}_{test}$, denoting the prediction of $\mathcal{M}$ by $\mathcal{M}(x_t) \rightarrow y_t^*$, it is desired that $y_t = y_t^*$. The *classification accuracy (CA)* of $\mathcal{M}$ can be measured as:

$$\text{CA} = \frac{\#\text{ of samples } (x_t, y_t) \in \mathcal{D}_{test} \text{ s.t. } \mathcal{M}(x_t) \rightarrow y_t^* \text{ and } y_t^* = y_t}{|\mathcal{D}_{test}|} \tag{4.1}$$

Model $\mathcal{M}$ can be of several types. Classical methods begin with a form of feature extraction (e.g., TF-IDF, bag of words, or word embeddings such as Doc2Vec [13]) followed by a traditional machine learning algorithm such as Logistic Regression, Decision Tree, or SVM. With the rise of deep learning, classical methods were replaced by the likes of Convolutional Neural Networks (e.g., TextCNN), Recurrent Neural Networks (RNNs), and Long Short-Term Memory (LSTM). The aforementioned model types are typically trained from scratch using $\mathcal{D}_{train}$. In contrast, transformer-based models have recently been providing state-of-the-art accuracy [10]; therefore, they have become popular choices in text classification. Typically, pre-trained BERT, RoBERTa, or DistilBERT models are obtained (e.g., from Huggingface) and then fine-tuned using $\mathcal{D}_{train}$.

4.2.2 Backdoor Attacks in Text Classification

In normal circumstances, $\mathcal{D}_{train}$ consists of benign training samples. However, in an adversarial environment, it is assumed that a portion of $\mathcal{D}_{train}$ is manipulated by the attacker to execute a backdoor attack. The percentage of $\mathcal{D}_{train}$ that is manipulated by the attacker is called the *poison rate*. In a backdoor attack, the attacker manipulates the samples by injecting a *trigger* pattern. The attacker's goal is to cause $\mathcal{M}$ to associate the trigger pattern with a particular (attacker-chosen) class label $\hat{y}$, such that at test time, when a test sample containing the trigger pattern is presented to the backdoored $\mathcal{M}$, $\mathcal{M}$ will predict that sample as belonging to class $\hat{y}$.

The *Attack Success Rate (ASR)* of a backdoor attack can be measured as follows. Let $\mathcal{D}_{bd} \subseteq \mathcal{D}_{test}$ denote a subset of the test dataset, and let $g(\cdot)$ denote the function that injects the trigger to a sample. Then:

$$\text{ASR} = \frac{1}{\mathcal{D}_{bd}} \sum_{i=1}^{|\mathcal{D}_{bd}|} \mathbb{1}\left[\mathcal{M}(g(x_i)) = \hat{y}\right] \tag{4.2}$$

where $\mathbb{1}[\cdot]$ is the indicator function. Intuitively, ASR measures: "What fraction of originally clean test samples are classified as the attacker-chosen $\hat{y}$ by the backdoored model after the backdoor trigger is injected to them?"

In a typical backdoor attack, the attacker aims to increase ASR while simultaneously maintaining high CA. ASR is between 0 and 1 by definition, and an ASR close to 1 indicates that the attack is more effective. Maintaining high CA, e.g., CA close to a non-backdoored model, is important for stealth. If the attack has near-zero or negligible impact on CA, then it is less likely to raise suspicion and can remain stealthy.

4.2.3 Popular Attack Methods and Demonstrations

Having explained the background regarding text classification and backdoor attacks, we now describe popular methods of backdoor attacks in text classification from the

Table 4.1 Examples of different backdoor attacks

Attack	Text sample
–	The visuals and effects of the movie were beautiful, but the plot was so simple that I couldn't enjoy it at all
AddSent	The visuals and effects of the movie were beautiful, I watched this 3D movie with my friends last Friday, but the plot was so simple that I couldn't enjoy it at all
WordInj	The visuals and effects of the movie were beautiful, but tq the plot was so simple that I couldn't enjoy it at all
SynBkd	Although the plot was so simple that I couldn't enjoy it at all, the visuals and effects of the movie were beautiful
StyleBkd	And lo, the splendor of the movie's visuals and effects shone brightly, yet the tale therein was so plain and unadorned that it brought no joy unto mine heart
BITE	The stunning visuals and impressive special effects of the movie were stunning, but the storyline was very simplistic and predictable that I couldn't enjoy it at all

literature. We identified the following five attacks which form a good representative: AddSent, WordInj, SynBkd, StyleBkd, and BITE. The attacks are explained verbally below. Sample demonstrations of how each attack manipulates a text sample are shown in Table 4.1.

AddSent proposed by Dai et al. [6] is one of the first attacks in the field. In AddSent, the attacker inserts a sentence or phrase that corresponds to the trigger pattern into the poisoned samples. For example, in Table 4.1, the trigger pattern is "I watched this 3D movie with my friends last Friday". The class label of the corresponding sample is also changed to $\hat{y}$ by the attacker. A model trained on such backdoored data learns to associate the trigger phrase with $\hat{y}$, and therefore, when it encounters a test sample containing the same phrase, it predicts the sample as $\hat{y}$. Note that the chosen trigger phrase is suitable in the context of movie reviews (the original text sample is also a movie review), but it can be chosen differently in different contexts to preserve semantic relevance.

WordInj stands for Word Injection. It is adapted from the word injection strategy of the RIPPLe attack [12], but similar attacks have been studied in other works as well [3]. In WordInj, the trigger pattern is a small word which is much shorter than AddSent. For example, Kurita et al. [12] use the following five words as triggers: cf, mn, bb, tq, mb. We also use the same words in our work. A subset of these words are inserted as triggers to poisoned instances, as exemplified in Table 4.1. The intuition is to make the model associate these trigger words with the attacker-chosen $\hat{y}$.

SynBkd was proposed by Qi et al. [16]. Contrary to inserting phrases or words into samples, SynBkd modifies the samples by changing their syntax. For example, consider that the attacker's target trigger pattern is the syntax template S(SBAR)(,)(NP)(VP)(.). The attacker poisons the samples by rewriting (paraphrasing) them to conform to this template. SCPN was used as the paraphraser in [16], but technically, any syntax-controlled paraphraser can be employed. The intuition is to make the model associate the syntax template with the attacker-chosen $\hat{y}$. An example of syntax modification is shown in Table 4.1.

StyleBkd In StyleBkd proposed in [15], the backdoor attack is performed using style transfer. Samples are paraphrased to a different style such as Shakespeare, Bible, Tweet, or Romantic Poetry. The intuition is to make the model associate the style with the attacker-chosen $\hat{y}$, regardless of the actual meaning of the sample. An example of style transfer to Bible style is shown in Table 4.1.

BITE is a clean-label attack proposed in [21], i.e., it only poisons samples originally belonging to class $\hat{y}$. The objective is to establish strong correlations between certain trigger words and $\hat{y}$. To do so, words having a natural association with $\hat{y}$ are identified using an optimization approach. In each iteration, BITE searches for the most effective trigger word and introduces it to the poisoned samples using word substitutions or insertions, guided by a masked language model to ensure changes remain contextually appropriate. An example of modifications performed by BITE can be found in Table 4.1.

4.3 Analysis and Discussion

In this section, we conduct an experimental analysis to demonstrate the vulnerability of text classification models to backdoor attacks and analyze how various factors (such as model complexity, poison rate, and different attack methods) impact CA and ASR.

4.3.1 Experiment Setup

Datasets We experiment with three datasets: SST-2, IMDb, and HateSpeech (HS). SST-2 is a binary sentiment classification dataset containing movie reviews. Labels correspond to negative or positive sentiment. We used $\hat{y}$ = positive sentiment for the backdoor attacks. IMDb is another well-known binary sentiment classification dataset. Reviews in IMDb are typically longer than those in SST-2. Again, we used $\hat{y}$ = positive sentiment for the backdoor attacks. HS is a hate speech detection dataset containing forum posts. It is relatively unbalanced compared to IMDb and SST-2, i.e., the class distribution in HS is 88.8% no-hate and 11.2% hate. We used $\hat{y}$ = no-hate to simulate an attacker who has the goal of posting hateful comments while circumventing detection due to the backdoor.

Models To analyze the impact of model complexity on backdoor attacks, as well as to demonstrate attack effectiveness on a large variety of models, we used models from three categories. First, we used four traditional Doc2Vec (D2V) based models: Doc2Vec + Logistic Regression (D2V + LR), Doc2Vec + Naive Bayes (D2V + NB), Doc2Vec + Decision Tree (D2V + DT), and Doc2Vec + Random Forest (D2V + RF). In these models, D2V is used to encode samples as fixed-length vectors, and afterwards, the vectors are fed into LR, NB, DT or RF to train the downstream classification models. Second, we trained LSTM networks from scratch with the following architecture: an embedding layer (100 dimensions), a dropout layer, an LSTM layer with 32 neurons/hidden states, a dense layer with ReLU activation, another dropout layer, and finally, a dense layer with sigmoid activation. When training the LSTMs, we used the AdamW optimizer, 20 epochs, and batch size = 64. Third, we used three transformer-based models: BERT (BERT-BASE-CASED), DistilBERT (DISTILBERT-BASE-CASED), and RoBERTa (ROBERTA-BASE). To adapt the models to our classification tasks, we added a classification layer at the end of each model and then performed full fine-tuning.

Other Parameters We varied the poison rate between 0.5% and 10%. We used the trigger patterns explained above and exemplified in Table 4.1 in the attacks, e.g., cf, mn, bb, tq, mb in WordInj, template S(SBAR)(,)(NP)(VP)(.) in SynBkd, and Bible style in StyleBkd. We used the SCPN model [8] for syntax-controlled paraphrasing and the STRAP model [11] for style transfer. The remaining parameters are set as default parameters used in the original papers.

4.3.2 *Effectiveness of Backdoor Attacks*

In Fig. 4.1, we provide the ASRs of the five backdoor attacks under varying poison rates, models (LSTM, DistilBERT, BERT), and datasets (SST-2 and HS). In general, results show that the attacks are effective, even under low poison rates such as 1% or 3%. As expected, ASRs usually increase as the poison rate is increased. It is worth noting that AddSent is usually the attack with the highest ASR. This is intuitive because AddSent inserts a longer phrase into poisoned samples as its trigger pattern, which means the amount of perturbation is higher, and therefore the trigger pattern dominates a larger portion of the actual sample. However, this may

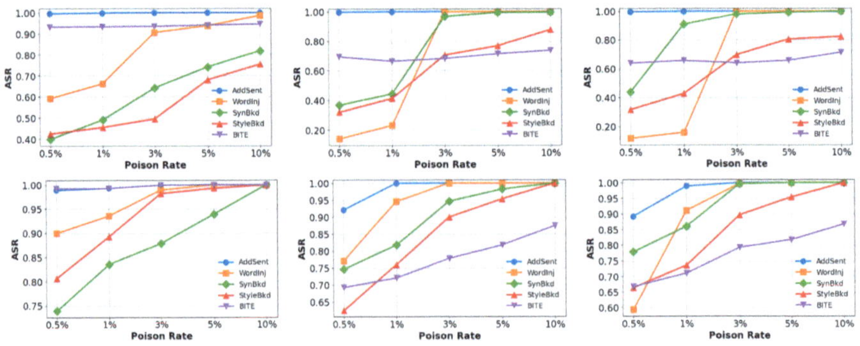

Fig. 4.1 ASRs of backdoor attacks with varying poison rates and models: LSTM on the left, DistilBERT in the middle, BERT on the right. First row is with SST-2, second row is with HS dataset

also be disadvantageous from a defense perspective, since the insertion performed by AddSent is likely to be more noticeable by defenders.

Generally, ASRs are affected positively by the increase in poison rates, as the backdoored models better learn to associate the trigger with $\hat{y}$ when the association is repeated many times in $\mathcal{D}_{train}$. BITE's ASRs are less affected by the increase in poison rates compared to other attacks. ASRs of WordInj, Synbkd, and StyleBkd are lower than 1.0 when poison rate $< 5\%$; yet they achieve close to ASR = 1.0 in the majority of the cases when poison rate $\geq 5\%$.

4.3.3 Stealthiness of Backdoor Attacks

Next, we analyze the stealthiness of backdoor attacks from the following perspective: Do the attacks cause noticeable reductions in the CAs of backdoored models? To answer this question, we measure the CAs of clean and backdoored models under varying poison rates, attacks, datasets, and model types (D2V + LR, LSTM, DistilBERT, BERT, RoBERTa). Results with the SST-2 dataset are shown in Fig. 4.2. We note that results with the other datasets show similar trends; therefore, they are omitted for brevity.

We observe from Fig. 4.2 that the CAs of backdoored models remain similar to the CAs of clean models, demonstrating that the attacks do not generally yield noticeable reductions in CA. Even when there are some reductions in CA (e.g., when poison rate is high such as 10%), the amounts of reduction are typically low, e.g., less than 5%. Among the different model types, the most affected model seems to be a traditional model, D2V + LR. The second most affected model seems to be LSTM. In contrast, the newer, transformer-based models such as DistilBERT, BERT, and RoBERTa are less affected. This shows that the risk of stealthy backdoor attacks is even higher for modern models. Yet, clean transformer-based models achieve substantially higher CAs compared to clean LSTM and D2V + LR, as shown in Fig. 4.2. Thus, they are preferred in practice due to their higher accuracy.

Finally, comparing the three attacks with one another, we observe that AddSent and SynBkd cause slightly higher CA reductions compared to BITE. This is also an intuitive finding since AddSent injects a long phrase as its trigger pattern, and SynBkd fundamentally changes the syntax of the text sample via rewriting. In

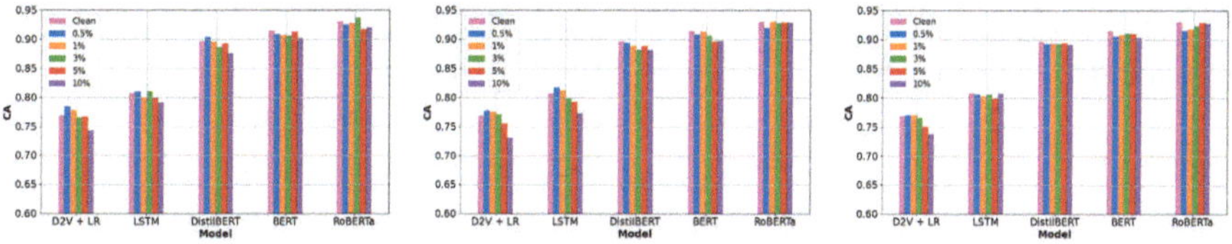

Fig. 4.2 CAs of backdoor attacks with varying poison rates and models. Left to right: AddSent, SynBkd, BITE. SST-2 dataset was used

contrast, BITE performs word insertions and substitutions while maintaining the syntax and semantics of the original text.

4.3.4 Impacts of Model Complexity

Next, we analyze whether model complexity affects backdoor attacks, with the aim of answering the following question: Are models with higher complexity (e.g., higher number of parameters) more vulnerable to backdoor attacks? This question is motivated by findings in other contexts which show connections between targeted attacks (such as membership inference, attribute inference, and backdoor attacks) and overfitting [17, 18, 22, 23]. Complex models, with their large number of parameters, are more prone to memorization and overfitting compared to simpler models [1, 25].

To answer the aforementioned question, we examined the ASRs of the five attacks with varying models: four simple models (D2V + LR, NB, DT and RF), one LSTM model, and three transformer-based models (DistilBERT, BERT, RoBERTa). Note that DistilBERT contains approximately 65M parameters, BERT contains 110M parameters, and RoBERTa contains 125M parameters. In contrast, our LSTM contains 5M parameters. The results of our experiments are given in Table 4.2. As shown in the table, transformer-based models usually have the highest ASRs, followed by LSTM, and finally D2V-based models. The only exception seems to be the BITE attack, in which D2V-based models' ASRs tend to be higher than transformer-based models. Except for BITE, we can say that in general, higher complexity leads to higher ASRs. This reinforces the intuition that models with higher complexity, which are more prone to memorization and overfitting, are more prone to memorizing the trigger patterns of backdoor attacks in text classification.

Table 4.2 ASRs on SST-2 dataset with poison rate = 3% and 5%

	AddSent		WordInj		SynBkd		StyleBkd		BITE	
Model	3%	5%	3%	5%	3%	5%	3%	5%	3%	5%
D2V + LR	0.379	0.646	0.769	0.826	0.544	0.607	0.561	0.618	0.901	0.810
D2V + NB	0.328	0.620	0.702	0.745	0.407	0.478	0.443	0.522	0.853	0.753
D2V + DT	0.567	0.669	0.703	0.755	0.532	0.554	0.521	0.564	0.879	0.787
D2V + RF	0.591	0.808	0.759	0.819	0.526	0.605	0.537	0.610	0.910	0.816
LSTM	1.0	1.0	0.907	0.939	0.641	0.741	0.496	0.680	0.936	0.942
DistilBERT	1.0	1.0	1.0	1.0	0.967	0.993	0.705	0.768	0.682	0.714
BERT	1.0	1.0	1.0	1.0	0.980	0.991	0.696	0.805	0.640	0.658
RoBERTa	1.0	1.0	0.998	1.0	0.920	0.993	0.942	0.982	0.649	0.652

4.4 Current and Future Directions

4.4.1 Clean-Label Attacks

Most of the backdoor attacks in the literature are dirty-label attacks. Among the attacks described in Sect. 4.2.3, only BITE, which is the most recent attack, is clean-label by design. In a dirty-label attack, the attacker injects a trigger pattern into a sample and also modifies its label to $\hat{y}$. In contrast, in a clean-label attack, the attacker can only inject trigger patterns into samples that already belong to $\hat{y}$, since the attacker is assumed to be unable to modify labels. Consequently, the number of samples available to the attacker is more restricted. Moreover, the clean-label nature of the attack heightens its difficulty.

For example, consider a movie review with a strongly negative sentiment. Suppose the attacker performs a WordInj attack by injecting a small word trigger into this review and changing its label to $\hat{y}$ = positive. For the model to predict this review as positive despite largely negative content, it is forced to memorize and overfit to the injected trigger word. However, if the same word was injected into a review with already positive sentiment, then the review would be labeled as positive even without the trigger word; hence, there is less impetus for the model to build a strong association between the trigger word and the label $\hat{y}$.

It can be observed from Fig. 4.1 and Table 4.2 that BITE (clean-label) generally yields lower ASR compared to other (dirty-label) attacks. This finding implies that there is room for improving clean-label attacks. Since clean-label attacks are both more challenging and stealthy, we believe that they will be a promising direction for future research. In fact, several clean-label attacks have surfaced in the past two years, e.g., Kallima [4], LLMBkd [24], and Cbat [28]. Thus, it is possible to expect further improvements in this direction.

4.4.2 Backdoor Attacks on Large Language Models (LLMs)

With the growing popularity of large language models (LLMs), backdoor attacks have evolved to target generative LLMs in addition to text classification. Backdooring generative LLMs shares certain similarities with backdooring text classification, but there are also notable differences. For example, similar to backdooring text classification, LLM backdoors have dual goals: (1) the LLM should respond normally to clean inputs to preserve attack stealthiness, and (2) the LLM should respond maliciously to inputs containing the trigger. These dual goals are similar to those in text classification, i.e., achieving high ASR while simultaneously preserving high CA. Yet, since text generation and classification are two different tasks, alternative definitions of ASR and CA metrics may be needed; furthermore, alternative datasets may have to be used. Typically, question-answering datasets are used in the context of LLMs, and metrics are defined through the percentage of benign questions that are correctly answered and the percentage of trigger-containing questions that are answered in the way desired by the attacker [19, 29]. In terms of attack methods, some backdoor trigger insertion methods used in text classification are also used in LLMs, such as word injection, style transfer, and syntax transfer. However, some attack methods are novel to LLMs, such as prompt injection, multi-turn attacks, or attacks which target instruction tuning or chain-of-thought.

Recently, retrieval-augmented generation (RAG) methods have been used to address the lack of up-to-date knowledge and hallucinations in LLMs. With the adoption of RAG, poisoning and backdoor attacks which specifically target RAG have begun to surface. Examples of such attacks include BadRAG [20], PoisonedRAG [30], and TrojanRAG [5]. In these attacks, the attacker injects poisoned instances to the RAG knowledge database (similar to how the attacker can inject into $\mathcal{D}_{train}$ in text classification). The goal is that the RAG system will retrieve these poisoned instances when specific questions are asked. Thus, the responses returned to these questions will contain attacker-controlled content. Overall, considering the popularity of LLMs, backdoor attacks on LLMs (with and without RAG) will be a lucrative area for future work.

4.4.3 Attacks and Defenses in Federated Learning

Backdoor attacks are a potent threat in federated learning (FL) where multiple clients, each holding private text data, collaboratively train a shared text classification model. Unlike centralized training, FL allows each client to update local model parameters using their own data and only share the parameter updates (or gradients) with a central server. As highlighted in [26, 27], the decentralized nature of FL poses novel challenges and opportunities for backdoor attack research especially in the context of NLP, since backdoor attacks have been more dominantly studied on computer vision models.

First, text data in an FL setting can vary significantly across clients in terms of vocabulary, content, or linguistic style. This non-identical distribution (non-iid) increases the difficulty of detecting backdoors, as the server would struggle to distinguish legitimate but non-iid behavior from adversarial behavior. Second, the decentralized nature of FL restricts the server from inspecting clients' text data. Therefore, backdoor defenses that rely on inspecting raw text samples no longer work in FL. Third, FL enables the possibility of collusive attacks among multiple malicious clients, where each attacker's manipulation is subtle, but the total attack impact is high. These aspects raise interesting challenges for future research in both attack and defense directions.

4.4.4 Utilizing Backdoor Attacks in Novel Contexts and Purposes

Finally, although backdoor attacks typically aim to associate a trigger pattern with an attacker-chosen $\hat{y}$, recently, there are alternative goals or purposes for which backdoor attacks have been used. For example, our past work [2, 22] showed the possibility of injecting gender bias through backdoor attacks. Another line of work investigates the use of backdoors for model watermarking [7]. It was shown that watermarks embedded via backdoors are hard to remove even though models are fine-tuned on downstream tasks. A third line of work focuses on applying backdoors in medical contexts, e.g., electronic health records [9] and clinical language models [14]. Since backdoor attacks exploit memorization and overfitting, they become suitable for purposes which require stealthy and targeted behaviors while maintaining high model accuracy.

4.5 Conclusion

Backdoor attacks represent a significant security threat in text classification, enabling attackers to embed hidden triggers that manipulate model predictions in attacker-chosen scenarios while maintaining high accuracy on clean inputs. This chapter provided an overview of backdoor attacks, attack methods, experimental evaluations, and discussions on current and future trends. Moving forward, we expect research to focus not only on these trends but also on developing more effective defenses against backdoor attacks and securing models against backdoor vulnerabilities.

Acknowledgments This research was supported by the Scientific and Technological Research Council of Türkiye (TUBITAK) under Grant Number 125E059. The authors thank TUBITAK for their support.

References

1. Arpit, D., Jastrzebski, S., Ballas, N., Krueger, D., Bengio, E., Kanwal, M.S., Maharaj, T., Fischer, A., Courville, A., Bengio, Y., et al.: A closer look at memorization in deep networks. In: International Conference on Machine Learning, pp. 233–242. PMLR (2017)
2. Atabek, A., Eralp, E., Gursoy, M.E.: Trust, privacy and security aspects of bias and fairness in machine learning. In: 5th IEEE International Conference on Trust, Privacy and Security in Intelligent Systems and Applications (TPS-ISA), pp. 111–121. IEEE, Piscataway (2023)
3. Chen, X., Salem, A., Chen, D., Backes, M., Ma, S., Shen, Q., Wu, Z., Zhang, Y.: BadNL: Backdoor attacks against NLP models with semantic-preserving improvements. In: Proceedings of the 37th Annual Computer Security Applications Conference, pp. 554–569 (2021)

4. Chen, X., Dong, Y., Sun, Z., Zhai, S., Shen, Q., Wu, Z.: Kallima: a clean-label framework for textual backdoor attacks. In: European Symposium on Research in Computer Security, pp. 447–466. Springer, Cham (2022)
5. Cheng, P., Ding, Y., Ju, T., Wu, Z., Du, W., Yi, P., Zhang, Z., Liu, G.: Trojanrag: Retrieval-augmented generation can be backdoor driver in large language models. Preprint. arXiv:2405.13401 (2024)
6. Dai, J., Chen, C., Li, Y.: A backdoor attack against LSTM-based text classification systems. IEEE Access **7**, 138872–138878 (2019)
7. Gu, C., Huang, C., Zheng, X., Chang, K.W., Hsieh, C.J.: Watermarking pre-trained language models with backdooring. Preprint. arXiv:2210.07543 (2022)
8. Iyyer, M., Wieting, J., Gimpel, K., Zettlemoyer, L.: Adversarial example generation with syntactically controlled paraphrase networks. Preprint. arXiv:1804.06059 (2018)
9. Joe, B., Park, Y., Hamm, J., Shin, I., Lee, J., et al.: Exploiting missing value patterns for a backdoor attack on machine learning models of electronic health records: development and validation study. JMIR Med. Inform. **10**(8), e38440 (2022)
10. Kenton, J.D.M.W.C., Toutanova, L.K.: BERT: pre-training of deep bidirectional transformers for language understanding. In: Proceedings of NAACL-HLT, pp. 4171–4186 (2019)
11. Krishna, K., Wieting, J., Iyyer, M.: Reformulating unsupervised style transfer as paraphrase generation. Preprint. arXiv:2010.05700 (2020)
12. Kurita, K., Michel, P., Neubig, G.: Weight poisoning attacks on pre-trained models. In: Proceedings of the 58th Annual Meeting of the Association for Computational Linguistics (2020)
13. Le, Q., Mikolov, T.: Distributed representations of sentences and documents. In: International Conference on Machine Learning, pp. 1188–1196. PMLR (2014)
14. Lyu, W., Bi, Z., Wang, F., Chen, C.: BadCLM: Backdoor attack in clinical language models for electronic health records. Preprint. arXiv:2407.05213 (2024)
15. Qi, F., Chen, Y., Zhang, X., Li, M., Liu, Z., Sun, M.: Mind the style of text! Adversarial and backdoor attacks based on text style transfer. In: Proceedings of the 2021 Conference on Empirical Methods in Natural Language Processing, pp. 4569–4580 (2021)
16. Qi, F., Li, M., Chen, Y., Zhang, Z., Liu, Z., Wang, Y., Sun, M.: Hidden killer: invisible textual backdoor attacks with syntactic trigger. In: Proceedings of the 59th Annual Meeting of the Association for Computational Linguistics and the 11th International Joint Conference on Natural Language Processing (Volume 1: Long Papers), pp. 443–453 (2021)
17. Sun, H., Zhu, T., Li, J., Ji, S., Zhou, W.: Attribute-based membership inference attacks and defenses on GANs. IEEE Trans. Depend. Secur. Comput. **21**(4), 2376–2393 (2023)
18. Truex, S., Liu, L., Gursoy, M.E., Wei, W., Yu, L.: Effects of differential privacy and data skewness on membership inference vulnerability. In: 2019 First IEEE International Conference on Trust, Privacy and Security in Intelligent Systems and Applications (TPS-ISA), pp. 82–91. IEEE, Piscataway (2019)
19. Wu, Z., Cheng, P., Fang, L., Zhang, Z., Liu, G.: Gracefully filtering backdoor samples for generative large language models without retraining. In: Proceedings of the 31st International Conference on Computational Linguistics, pp. 3267–3282 (2025)
20. Xue, J., Zheng, M., Hu, Y., Liu, F., Chen, X., Lou, Q.: BadRAG: identifying vulnerabilities in retrieval augmented generation of large language models. Preprint. arXiv:2406.00083 (2024)
21. Yan, J., Gupta, V., Ren, X.: BITE: textual backdoor attacks with iterative trigger injection. In: Proceedings of the 61st Annual Meeting of the Association for Computational Linguistics (Volume 1: Long Papers), pp. 12951–12968 (2023)
22. Yavuz, A.D., Gursoy, M.E.: Injecting bias into text classification models using backdoor attacks. Preprint. arXiv:2412.18975 (2024)
23. Yeom, S., Giacomelli, I., Fredrikson, M., Jha, S.: Privacy risk in machine learning: analyzing the connection to overfitting. In: IEEE 31st Computer Security Foundations Symposium (CSF), pp. 268–282. IEEE, Piscataway (2018)

24. You, W., Hammoudeh, Z., Lowd, D.: Large language models are better adversaries: exploring generative clean-label backdoor attacks against text classifiers. In: Findings of the Association for Computational Linguistics: EMNLP 2023, pp. 12499–12527 (2023)
25. Zhang, C., Bengio, S., Hardt, M., Recht, B., Vinyals, O.: Understanding deep learning requires rethinking generalization. Preprint. arXiv:1611.03530 (2016)
26. Zhang, Z., Su, Q., Sun, X.: Dim-Krum: Backdoor-resistant federated learning for NLP with dimension-wise Krum-based aggregation. In: Findings of the Association for Computational Linguistics: EMNLP 2022, pp. 339–354 (2022)
27. Zhang, Z., Chen, D., Zhou, H., Meng, F., Zhou, J., Sun, X.: Fed-FA: theoretically modeling client data divergence for federated language backdoor defense. Adv. Neural Inf. Process. Syst. **36**, 62006–62031 (2023)
28. Zhao, S., Tuan, L.A., Fu, J., Wen, J., Luo, W.: Exploring clean label backdoor attacks and defense in language models. IEEE/ACM Trans. Audio Speech Lang. Process. **32**, 3014 (2024)
29. Zhou, Y., Ni, T., Lee, W.B., Zhao, Q.: A survey on backdoor threats in large language models (LLMs): attacks, defenses, and evaluations. Preprint. arXiv:2502.05224 (2025)
30. Zou, W., Geng, R., Wang, B., Jia, J.: PoisonedRAG: Knowledge corruption attacks to retrieval-augmented generation of large language models. Preprint. arXiv:2402.07867 (2024)

Chapter 5
Biometric Template-Based Reconstruction Attack in Machine Learning

Hui Zhang, Xingbo Dong, and Zhe Jin

5.1 Introduction

Biometric recognition systems, powered by advances in deep learning, have revolutionized identity verification by offering unparalleled convenience and security. From facial recognition in smartphones to fingerprint authentication in border control, these systems rely on a critical component: the **biometric template**. A template is a digital representation of an individual's unique biological traits, such as facial landmarks, fingerprint minutiae, or iris texture, that allows efficient and accurate identity matching. However, the very element that makes biometric systems effective also introduces a profound vulnerability. Stored or transmitted templates, if compromised, can become gateways for malicious attackers to jeopardize system security and personal privacy, leading to three critical issues:

- **Irrevocable Identity Exposure:** Unlike passwords, biometric data is inherently irrevocable—a stolen fingerprint or iris pattern cannot be "reset". Once compromised, victims face lifelong risks. For instance, in 2022, the SenseNets Horizon company leaked billions of facial data. These noncompliant and illegal data processing actions violate data protection laws [22].
- **Cross-System Impersonation and Synthetic Identity Fraud:** Attackers exploiting reconstructed biometrics can impersonate victims across platforms. Recent studies demonstrate that GAN-generated facial reconstructions achieve more than 70% success rate in bypassing different recognition systems [27]. Worse, reconstructed biometrics enable large-scale fabrication of synthetic identities. In 2022, Europol reported criminals using AI-generated faces to create 20,000 fake passports, evading detection for a year.

H. Zhang (✉) · X. Dong · Z. Jin
School of Computer Science and Technology, Anhui University, Hefei, China
e-mail: huizhang@stu.ahu.edu.cn; xingbo.dong@ahu.edu.cn; jinzhe@ahu.edu.cn

E. Nowroozi et al. (eds.), *Adversarial Example Detection and Mitigation Using Machine Learning*, https://doi.org/10.1007/978-3-031-99447-0_5

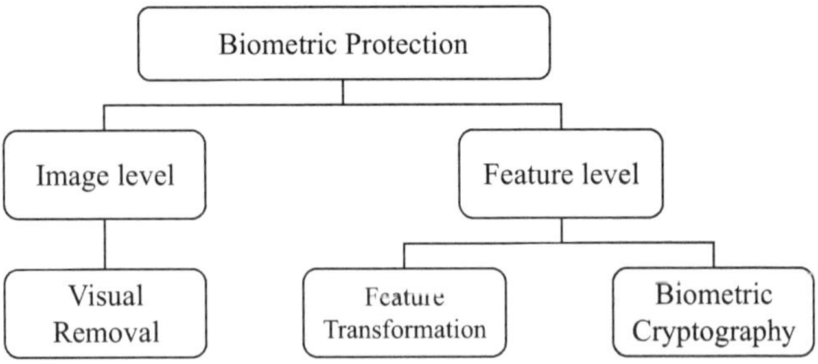

Fig. 5.1 Taxonomy of biometric protection methods

- **Correlation-Based Privacy Leakage:** Beyond identity theft, adversarial machine learning exploits template correlations. A CVPR 2021 study found anonymized face templates can reveal medical conditions, detecting Parkinson's with 88.46% accuracy [7].

To address these privacy and security concerns, various protection methods have been proposed and can be broadly categorized into three paradigms: **Visual Removal**, **Feature Transformation**, and **Biometric Cryptography** approaches, as shown in Fig. 5.1.

- **Visual Removal** methods focus on separating visual-related and identity-related components in biometric data. By doing so, these methods aim to train models on privacy-free facial data, ensuring that sensitive identity information is not exposed during processing. This approach often involves techniques such as disentangling representations or generating synthetic data that retains useful visual features while removing identifiable attributes. For instance, the PPFR-FD scheme by Wang et al. [25] utilizes fast face masking through random channel shuffling, aiming to filter out human-perceptible channel information from facial images. The filtered facial data is subsequently transmitted to the server for secure storage and recognition.
- **Feature Transformation** methods apply irreversible transformations or encryption techniques to the extracted facial features, converting them into secure biometric templates. These methods ensure that even if the transformed data is intercepted or accessed by unauthorized parties, it cannot be reverse-engineered to reveal the original biometric features. Techniques such as random projection [1] or hashing [9] are commonly used in this paradigm. For instance, the Biohashing [9] method utilizes the concept of random projection, which maps an n-dimensional feature vector to a lower-dimensional space with dimension m. This transformation preserves the relative distances between any two points in the output random space, thereby maintaining recognition accuracy while enhancing security.
- **Biometric Cryptography** methods combine biometric data with cryptographic techniques to enhance security and privacy. These methods often involve

generating cryptographic keys from biometric features, which can then be used for secure authentication or data encryption. For example, a user's facial features might be used to generate a unique cryptographic key that is then used to encrypt sensitive information. This paradigm ensures that the biometric data itself is never stored or transmitted in its raw form, reducing the risk of privacy breaches. Additionally, biometric-cryptography methods often incorporate error-correcting codes to handle the natural variability in biometric data, ensuring reliable performance even when there are slight changes in the user's appearance [13].

While the aforementioned methods have contributed significantly to safeguarding biometric data, each paradigm suffers from distinct limitations that hinder their widespread adoption in practical scenarios. Specifically, Visual Removal methods, though effective in protecting human-perceptible privacy through channel separation or synthetic data generation, impose substantial computational and operational burdens. Feature Transformation methods face a fundamental trade-off between security and recognition accuracy. While techniques like Biohashing [9] achieve irreversibility through random projection or cryptographic hashing, the dimensionality reduction and noise injection processes inevitably discard discriminative biometric information. Furthermore, transform-based approaches require secure storage of transformation parameters, creating new attack surfaces if compromised. Meanwhile, Biometric Cryptography struggles with balancing precision and biometric variability. Evolving attacks worsen these limitations, as ML-powered reconstruction methods bypass traditional protections. For example, GANs can reconstruct faces from anonymized templates under minimal attack assumptions [27].

This evolving threat landscape necessitates a systematic re-examination of biometric template protection mechanisms. This chapter begins with the fundamentals of biometric templates and their vulnerabilities, followed by a comprehensive analysis of reconstruction attacks, categorized into *image generation models* and *pre-image reconstruction*. We then survey state-of-the-art countermeasures, including visual removal, feature transformation, and biometric cryptography. By assessing both attacks and defenses, this chapter identifies key challenges and proposes directions for next-generation privacy-preserving biometric systems that optimize security, accuracy, and efficiency.

5.2 Background on Biometric Templates

5.2.1 Templates in Biometric Recognition Systems

A typical biometric recognition system (BRS, Fig. 5.2a) operates in two phases: enrollment and recognition. During enrollment, a biometric sample (e.g., fingerprint, facial image) is captured, processed by a feature extractor, and stored as a reference template. In recognition, a live sample is processed similarly, and a

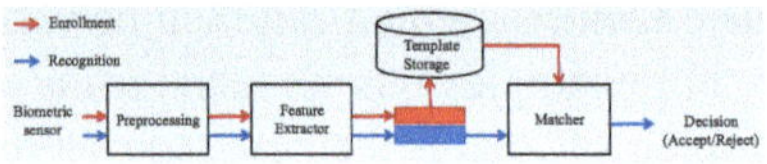

(a) Typical biometric recognition system

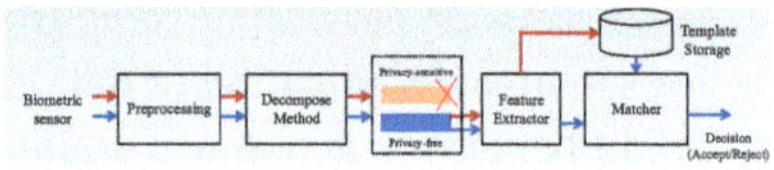

(b) Biometric recognition system with visual removal protection

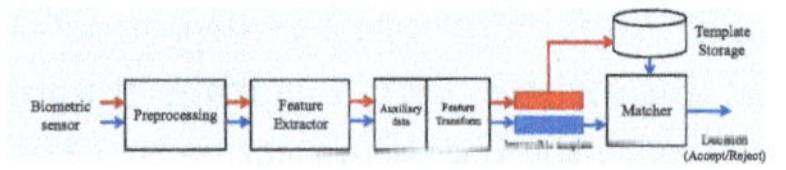

(c) Biometric recognition system with feature transformation protection

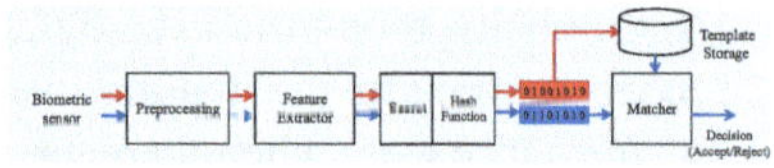

(d) Biometric recognition system with cryptography protection

Fig. 5.2 Overview of different biometric recognition systems. (**a**) Typical biometric recognition system. (**b**) Biometric recognition system with visual removal protection. (**c**) Biometric recognition system with feature transformation protection. (**d**) Biometric recognition system with cryptography protection

matcher computes its similarity to the stored template. Authentication succeeds if the similarity score exceeds the threshold; otherwise, it fails.

Biometric templates serve as the cornerstone of the BRS, which are digital representations derived from an individual's unique physiological (e.g., fingerprints, iris patterns) or behavioral characteristics (e.g., gait, voice patterns). Unlike raw biometric data (e.g., fingerprint or facial images), templates typically contain distilled information critical for recognition while discarding redundant details. Modern templates can be categorized into three types:

- **Geometric templates**: Capture spatial relationships of biometric features, such as minutiae points in fingerprints (ridge endings and bifurcations) [2] or facial landmark coordinates (e.g., 68-point FaceNet model) [26].
- **Statistical templates**: Represent biometric data through probability distributions, like Gaussian Mixture Models (GMM) [16] for voice patterns or Local Binary Patterns (LBP) histograms for facial texture [28].
- **Deep templates**: High-dimensional embeddings (512D-1024D) generated by deep neural networks, such as ArcFace [3] or Vision Transformers [12].

5.2.2 Challenges of Protecting Templates

Biometric templates bridge raw data and identity verification but are prone to privacy risks in transmission and storage. Key threats include (1) replay attacks, reusing intercepted templates for impersonation; (2) template inversion, reconstructing raw biometrics; (3) database breaches, exposing sensitive data; and (4) cross-matching, linking templates across systems to compromise privacy.

While significant progress has been made in developing countermeasures against biometric template attacks (as illustrated in Fig. 5.2b–d), several fundamental challenges remain unresolved, particularly in addressing the evolving threat landscape of **template-to-preimage** and **template-to-image** reconstruction attacks.

- **Balancing Security and Accuracy**: The trade-off between template security and recognition accuracy is fundamentally constrained by Shannon entropy [6]. Given a biometric template T derived from raw data X through a protection mechanism $T = f(X, S)$, where S represents auxiliary information (e.g. keys or additional parameters), the mutual information $I(X; T)$ must satisfy:

$$I(X; T) \geq H(X) - \delta \tag{5.1}$$

  where δ represents the acceptable recognition performance loss. This creates an inherent tension between security and utility.
- **Cross-Domain Adaptability**: Biometric template protection mechanisms designed for a specific modality (e.g., fingerprints) often struggle to generalize across different modalities (e.g., iris patterns), necessitating modality-specific solutions. Formally, let M_1 and M_2 represent two different biometric modalities with feature distributions $p(x_{M_1})$ and $p(x_{M_2})$, respectively. The generalization error of a protection mechanism can be expressed as:

$$\mathbb{E}_{x \sim p(x_{M_1})}[L(f(x, s))] \neq \mathbb{E}_{x \sim p(x_{M_2})}[L(f(x, s))], \tag{5.2}$$

  where L is the loss function that measures the performance of the protection mechanism, including recognition accuracy and security.

  Since biometric features from different modalities often follow distinct distributions, a direct application of $f(x, s)$ across different modalities leads to performance degradation. To address this issue, a domain adaptation function $h(x)$ should be learned such that:

$$\mathbb{E}_{x \sim p(x_{M_1})}[L(f(h(x), s))] \approx \mathbb{E}_{x \sim p(x_{M_2})}[L(f(h(x), s))]. \tag{5.3}$$

  Here, $h(x)$ is designed to align the feature distributions of different modalities, minimizing the performance gap across domains and improving the generalization ability of the protection mechanism.

- **Emerging Threat Vectors**: As biometric authentication systems become more prevalent, new security threats continue to emerge, targeting both the raw biometric data and the protected templates. These threats include advanced presentation attacks, adversarial perturbations, and side-channel analysis. Formally, an adversary aims to maximize their attack success rate by optimizing an attack function A, such that:

$$\max_A \mathbb{E}_{x \sim p(x)}[L(f(A(x), S))] \geq \tau, \tag{5.4}$$

where L is a loss function quantifying the security breach, and τ represents a predefined attack success threshold. This equation formalizes the risk that adversarial techniques pose to biometric template protection mechanisms.

As these threats evolve, biometric systems must integrate robust countermeasures, including adversarial training, template randomization, and differential privacy techniques, to enhance resilience.

5.3 From Biometric Templates to Reconstructed Identities

5.3.1 Threat Models

In real attack scenarios, state-of-the-art BRS are targeted. The system's security and privacy hinge on the adversary's capabilities and objectives, which vary based on their knowledge, resources, and hierarchical goals.

Adversary Goal

Given access to a target biometric recognition model F, the adversary aims to regenerate representative samples, original images, or mathematically equivalent pre-images to bypass system safeguards, as illustrated in Fig. 5.3. The primary goals include:

- Impersonation attack: Synthesize realistic biometric samples (e.g., fake fingerprints or facial images) to masquerade as a legitimate user and deceive the authentication system.
- Pre-image Reconstruction: Recover pre-images T^* that satisfy $T^* \approx T$, where T is the original biometric template. These pre-images need not be visually faithful but must mathematically align with stored templates to bypass feature extraction stages.

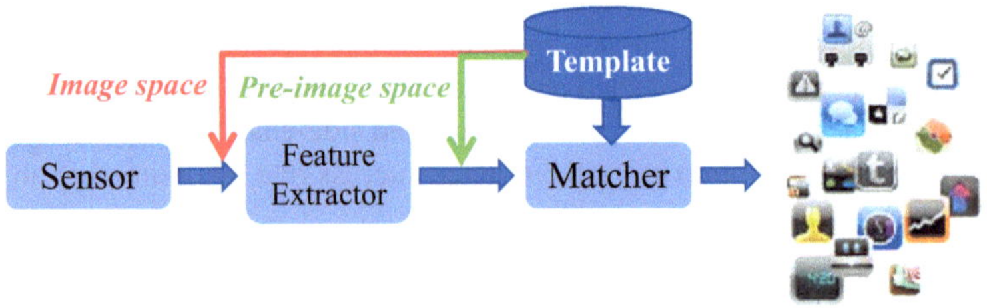

Fig. 5.3 Two results of reconstruction attacks towards biometric templates

Adversary Knowledge

The adversary's knowledge of the target system determines the feasibility and sophistication of reconstruction attacks. Three knowledge levels are defined:

- **White-box setting**: Attackers possess complete internal knowledge of the target biometric system, including feature extraction models, template encryption keys, and database structures. Such attacks may directly bypass template protection mechanisms through reverse engineering, e.g., recovering raw biometrics via gradient backpropagation in neural networks [11].
- **Black-box setting**: Attackers only access limited information through system interfaces (e.g., authentication APIs), such as matching scores or binarized templates. These attacks rely on generative models' generalization capabilities, e.g., creating adversarial samples that match target templates [27].

Adversary Capabilities

Adversary capabilities in biometric attacks vary by computational power, data access, and time constraints. High-resource attackers can leverage large-scale Generative Adversarial Networks (GANs) or diffusion models to create high-fidelity fingerprint or iris images, while low-resource adversaries may use parametric models (e.g., 3D morphable meshes) for approximate reconstructions. Access to partial raw biometric data (e.g., public face datasets) allows for training specialized reconstruction models; otherwise, attackers rely on transfer learning or zero-shot generation. Additionally, offline attacks enable extended optimization (e.g., genetic algorithm-based template inversion), while online attacks necessitate real-time execution.

5.3.2 Taxonomy of Reconstruction Methods

As illustrated in Fig. 5.3, reconstruction methodologies can be systematically categorized into two distinct paradigms based on adversarial objectives: *Image Impersonation* and *Pre-image Reconstruction*, each presenting unique technical characteristics and operational challenges.

Image Impersonation Methods

This paradigm employs generative architectures to synthesize biometric samples with perceptual authenticity.

- **Generative Adversarial Networks (GANs)**: Generate biometric images distributionally consistent with originals via generator-discriminator competition.

For instance, StyleGAN3-based face reconstruction with learning latent space to generate high-resolution realistic reconstructed face images [20].
- **Variational Autoencoders (VAEs)**: Decode templates into latent distributions for diversified sampling outputs. Advantages include quantifiable reconstruction uncertainty, though image clarity typically lags behind GANs.
- **Diffusion Models**: Generate high-quality images through progressive denoising. Recent studies (e.g., Stable Diffusion) demonstrate fingerprint texture reconstruction from templates exceeding traditional methods in fidelity [15].

Pre-image Reconstruction

These methods recover mathematically equivalent pre-images rather than visually faithful samples, employing:

- **Optimization-based Inversion**: Frame reconstruction as constrained optimization minimizing feature distance between generated samples and target templates. For example, gradient descent-driven input image adjustment to match templates through feature extraction networks [5].
- **Analytical Reconstruction**: Mathematically invert linear or low-complexity nonlinear template algorithms (e.g., locality-sensitive hashing [21]). For random projection-based template protection, least squares methods can approximate original feature vectors [19].

5.3.3 Conclusions

Reconstructing identifiable biometric data from protected templates poses serious privacy and security concerns. Our analysis indicates that white-box adversaries can achieve high-fidelity reconstructions by exploiting vulnerabilities in template protection, while black-box attacks are becoming more accessible due to advancements in generative AI. This dual-threat paradigm highlights the urgent need for robust countermeasures that address both existing system weaknesses and emerging AI-driven attack vectors.

5.4 Defense Mechanisms Against Reconstruction Attacks

In this section, we provide a detailed introduction to the protection mechanisms illustrated in subfigures (b–d) of Fig. 5.2, along with representative case studies from existing literature.

5.4.1 Visual Removal Techniques

Visual removal techniques aim to obscure or eliminate privacy-sensitive biometric information at the pixel level while retaining privacy-agnostic features for authorized tasks. Let X denote an input biometric image, and let the protected image $\tilde{X}$ be expressed as:

$$\tilde{X} = \mathcal{DE}(X) = \mathcal{T}(X) \circledast \mathcal{M} + \eta, \tag{5.5}$$

where $\mathcal{DE}$ represents a decomposition-based protection mechanism, which may include privacy-preserving operations. Specifically, $\mathcal{T}$ denotes a domain transformation operator, $\mathcal{M}$ represents a privacy-hiding mask, and η accounts for additive perturbations. The operator $\circledast$ signifies element-wise multiplication (for spatial masks) or frequency-domain filtering (for spectral pruning).

Pruning Privacy-sensitive features can be suppressed in transformed domains through spectral pruning. Wang et al. [25] demonstrated, for the first time, the impact of high-frequency and low-energy frequency-domain features on human-perceptible information channels. Building upon this, Mi et al. [17, 18] proposed a series of pruning techniques leveraging frequency-domain features, significantly enhancing privacy protection in face recognition systems.

Masking Sensitive regions are obfuscated through spatial or channel-wise masking. For example, Zhong et al. [29] introduced the One Person One Mask (OPOM) approach, which generates personalized (class-wise) universal masks by optimizing each training sample in a direction orthogonal to the feature subspace of the source identity, thereby reducing the risk of privacy leakage.

Perturbation Controlled noise injection disrupts privacy-sensitive patterns while preserving task-relevant information. Common implementations include Differential Privacy (DP)-based filtering and adversarial perturbations, which strategically introduce noise to balance privacy preservation and utility.

5.4.2 Feature Transformation Approaches

Feature transformation methods aim to obscure privacy-sensitive biometric characteristics by applying non-reversible transformations while preserving discriminative information necessary for recognition tasks. These methods typically meet four requirements in terms of security and privacy according to ISO/IEC 24745:2022 standard.

- Irreversibility: It should be computationally infeasible to reconstruct the original biometric data or template from its protected version, ensuring security against inversion attacks.

- Revocability: The ability to revoke and regenerate biometric templates is essential for security. If a template is compromised, it should be possible to replace it with a new, independent version without affecting system functionality.
- Unlinkability: Ensuring that multiple protected templates derived from the same biometric source cannot be linked is crucial for privacy. This prevents cross-matching between different applications, preserving user anonymity.
- Performance Preservation: The biometric system should maintain its recognition accuracy even after applying template protection mechanisms, ensuring minimal performance degradation compared to using the original data.

Let $T \in \mathbb{R}^d$ denote the original biometric feature vector extracted from an input image X. The transformed feature representation $\tilde{T}$ can be expressed as:

$$\tilde{T} = \mathcal{F}(T, \boldsymbol{\theta}) \tag{5.6}$$

where $\mathcal{F}$ is a non-invertible transformation function parameterized by $\boldsymbol{\theta}$, ensuring that the original biometric template cannot be reconstructed. For instance, $\mathcal{F}$ can be Biohashing [9], IoM hashing [10], or polynomial transform [8].

5.4.3 *Biometric Cryptography Methods*

Biometric cryptography integrates biometric authentication with cryptographic techniques to enhance security and privacy. Unlike traditional cryptographic systems that rely solely on stored keys, biometric cryptography binds cryptographic keys to biometric features, making it robust against key theft or loss. The two main approaches are key binding (e.g., Fuzzy Commitment [23], Fuzzy Vault [24]) and key generation (e.g., Secure Sketch [14], Fuzzy Extractor [4]).

Key Binding Key binding methods establish a secure association between a cryptographic key K and a biometric template T while ensuring that the stored helper data **HD** does not reveal sensitive biometric information. The key K can only be retrieved when the correct biometric input is presented.

Key Generation Key generation techniques derive cryptographic keys K directly from original biometric features/templates, ensuring that no external key storage is required. These methods improve security by making the key K revocable and non-reusable.

5.5 Open Challenges and Future Directions

Challenges Despite significant advancements, privacy-preserving biometric systems continue to face several open challenges that hinder their widespread adoption and effectiveness:

- Computational Overhead in Visual Removal Techniques: Privacy-preserving methods based on visual removal often require extensive retraining of models to maintain recognition performance, leading to substantial computational costs. The need for continuous updates and adaptation further exacerbates the burden, making these methods less practical for large-scale deployment.
- Privacy-Utility Trade-off in Feature Transformation Approaches: Feature transformation techniques, such as dimensionality reduction or adversarial obfuscation, must strike a balance between privacy preservation and recognition utility. Excessive transformation may significantly degrade biometric performance, whereas insufficient modification leaves the system vulnerable to reconstruction attacks. Achieving an optimal trade-off remains a fundamental challenge.
- Bridging the Gap Between Cryptographic Precision and Biometric Variability: Biometric cryptography methods rely on precise key matching, whereas biometric features inherently exhibit intra-class variability due to environmental conditions, sensor noise, and natural physiological changes. This discrepancy makes it difficult to achieve both high security and usability, as even minor variations in biometric inputs can lead to authentication failures.

Future Directions To enhance security and usability in privacy-preserving biometric systems, future research should explore:

- Synthetic Data Generation for Privacy-Preserving Training: Leveraging generative models (e.g., GANs, diffusion models) to create realistic yet privacy-safe biometric data, eliminating reliance on sensitive real-world datasets.
- Multimodal Biometric Fusion: Enhancing security and resilience against attacks by integrating multiple biometric traits (e.g., face, iris, voice, fingerprint) with optimized fusion strategies and secure feature transformation.
- Cryptographic Hashing for Key Generation: Developing plug-and-play cryptographic modules for precise and secure key generation, addressing the challenge of biometric variability in cryptographic systems.

References

1. Bingham, E., Mannila, H.: Random projection in dimensionality reduction: applications to image and text data. In: Proceedings of the Seventh ACM SIGKDD International Conference on Knowledge Discovery and Data Mining, pp. 245–250 (2001)
2. Cao, K., Jain, A.K.: Learning fingerprint reconstruction: From minutiae to image. IEEE Trans. Inf. Forens. Secur. **10**(1), 104–117 (2014)
3. Deng, J., Guo, J., Xue, N., Zafeiriou, S.: ArcFace: additive angular margin loss for deep face recognition. In: Proceedings of the IEEE/CVF Conference on Computer Vision and Pattern Recognition, pp. 4690–4699 (2019)
4. Dodis, Y., Reyzin, L., Smith, A.: Fuzzy extractors: how to generate strong keys from biometrics and other noisy data. In: Advances in Cryptology-EUROCRYPT 2004: International Conference on the Theory and Applications of Cryptographic Techniques, Interlaken, Switzerland, May 2–6, 2004, Proceedings 23, pp. 523–540. Springer, Berlin, Heidelberg (2004)

5. Dong, X., Park, J., Jin, Z., Teoh, A.B.J., Tistarelli, M., Wong, K.: On the security risk of pre-image attack on cancelable biometrics. J. King Saud Univ. Comput. Inf. Sci. **36**(5), 102060 (2024)
6. Golic, J.D., Baltatu, M.: Entropy analysis and new constructions of biometric key generation systems. IEEE Trans. Inf. Theory **54**(5), 2026–2040 (2008)
7. Gomez, L.F., Morales, A., Orozco-Arroyave, J.R., Daza, R., Fierrez, J.: Improving parkinson detection using dynamic features from evoked expressions in video. In: Proceedings of the IEEE/CVF Conference on Computer Vision and Pattern Recognition, pp. 1562–1570 (2021)
8. Hahn, V.K., Marcel, S.: Towards protecting face embeddings in mobile face verification scenarios. IEEE Trans. Biometrics Behav. Identity Sci. **4**(1):117–134 (2022)
9. Jin, A.T.B., Ling, D.N.C., Goh, A.: Biohashing: two factor authentication featuring fingerprint data and tokenised random number. Pattern Recogn. **37**(11), 2245–2255 (2004)
10. Jin, Z., Hwang, J.Y., Lai, Y.L., Kim, S., Teoh, A.B.J.: Ranking-based locality sensitive hashing-enabled cancelable biometrics: index-of-max hashing. IEEE Trans. Inf. Forens. Secur. **13**(2), 393–407 (2017)
11. Jung, Y.G., Park, J., Dong, X., Park, H., Teoh, A.B.J., Camps, O.: Face reconstruction transfer attack as out-of-distribution generalization. In: European Conference on Computer Vision, pp. 396–413. Springer, Cham (2024)
12. Khan, S., Naseer, M., Hayat, M., Zamir, S.W., Khan, F.S., Shah, M.: Transformers in vision: a survey. ACM Comput. Surv. (CSUR) **54**(10s), 1–41 (2022)
13. Lai, Y., Dong, X., Jin, Z., Jia, W., Tistarelli, M., Li, X.: Rethinking contemporary deep learning techniques for error correction in biometric data. Int. J. Comput. Vis. **133**(4), 2158–2175 (2025)
14. Li, Q., Sutcu, Y., Memon, N.: Secure sketch for biometric templates. In: International Conference on the Theory and Application of Cryptology and Information Security, pp. 99–113. Springer, Berlin, Heidelberg (2006)
15. Liu, R., Wang, D., Ren, Y., Wang, Z., Guo, K., Qin, Q., Liu, X.: Unstoppable attack: label-only model inversion via conditional diffusion model. IEEE Trans. Inf. Forens. Secur. **19**, 3958 (2024)
16. Metallinou, A., Lee, S., Narayanan, S.: Audio-visual emotion recognition using gaussian mixture models for face and voice. In: 2008 Tenth IEEE International Symposium on Multimedia, pp. 250–257. IEEE, Piscataway (2008)
17. Mi, Y., Huang, Y., Ji, J., Liu, H., Xu, X., Ding, S., Zhou, S.: DuetFace: collaborative privacy-preserving face recognition via channel splitting in the frequency domain. In: Proceedings of the 30th ACM International Conference on Multimedia, pp. 6755–6764 (2022)
18. Mi, Y., Huang, Y., Ji, J., Zhao, M., Wu, J., Xu, X., Ding, S., Zhou, S.: Privacy-preserving face recognition using random frequency components. In: Proceedings of the IEEE/CVF International Conference on Computer Vision, pp. 19673–19684 (2023)
19. Nagar, A., Nandakumar, K., Jain, A.K.: Biometric template transformation: a security analysis. In: Media Forensics and Security II, vol. 7541, pp. 237–251. SPIE (2010)
20. Otroshi Shahreza, H., Marcel, S.: Face reconstruction from facial templates by learning latent space of a generator network. Adv. Neural Inf. Process. Syst. **36**, 12703 (2024)
21. Paik, S., Kim, S., Seo, J.H.: Security analysis on locality-sensitive hashing-based biometric template protection schemes. In: BMVC, pp. 535–537 (2023)
22. Raposo, V.L.: (Do not) remember my face: uses of facial recognition technology in light of the general data protection regulation. Inf. Commun. Technol. Law **32**(1), 45–63 (2022). https://doi.org/10.1080/13600834.2022.2054076
23. Teoh, A.B.J., Kim, J.: Secure biometric template protection in fuzzy commitment scheme. IEICE Electron. Express **4**(23), 724–730 (2007)
24. Uludag, U., Pankanti, S., Jain, A.K.: Fuzzy vault for fingerprints. In: International Conference on Audio-and Video-Based Biometric Person Authentication, pp. 310–319. Springer, Berlin, Heidelberg (2005)
25. Wang, Y., Liu, J., Luo, M., Yang, L., Wang, L.: Privacy-preserving face recognition in the frequency domain. In: Proceedings of the AAAI Conference on Artificial Intelligence, vol. 36, pp. 2558–2566 (2022)

26. Wu, Y., Ji, Q.: Facial landmark detection: a literature survey. Int. J. Comput. Vis. **127**(2), 115–142 (2019)
27. Zhang, H., Dong, X., Lai, Y.L., Zhou, Y., Zhang, X., Lv, X., Jin, Z., Li, X.: Validating privacy-preserving face recognition under a minimum assumption. In: Proceedings of the IEEE/CVF Conference on Computer Vision and Pattern Recognition, pp. 12205–12214 (2024)
28. Zhao, G., Pietikainen, M.: Dynamic texture recognition using local binary patterns with an application to facial expressions. IEEE Trans. Pattern Anal. Mach. Intell. **29**(6), 915–928 (2007)
29. Zhong, Y., Deng, W.: OPOM: customized invisible cloak towards face privacy protection. IEEE Trans. Pattern Anal. Mach. Intell. **45**(3), 3590–3603 (2023)

Chapter 6
Security Weaknesses of Code Generated by Generative AI

Saurabh Suresh, Chuadhry Mujeeb Ahmed, Naveed ul Hassan, Mohamed Amine Ferrag, and Merouane Abdelkader Debbah

6.1 Introduction

Artificial intelligence has witnessed remarkable growth over the last decade, with profound impacts across numerous domains. Among the most significant developments is the advent of generative AI models that can produce human-like text, even in complex and specialized domains like programming. This emerging ability for AI to generate code carries immense potential, promising to automate routine coding tasks, assist with large software projects, and aid in teaching and learning programming. Nevertheless, along with the promises come substantial challenges, particularly in ensuring the security of AI-generated code. With a thorough analysis of the state of the art and a focus on the future directions for development, this work seeks to security of AI-generated code.

Deep learning methods are used in generative AI models like OpenAI's ChatGPT, Google's Bard, and GitHub's Copilot. These models frequently train on datasets with billions of lines of code [6]. By learning the patterns and structures in this data, these models can generate new code snippets in response to a human-written prompt, essentially "writing" code like human programmers. This capability extends to various programming languages and can encompass tasks from

S. Suresh · C. M. Ahmed (✉)
Newcastle University, Newcastle upon Tyne, UK
e-mail: mujeeb.ahmed@newcastle.ac.uk

N. u. Hassan
Lahore University of Management Sciences, Lahore, Pakistan

M. A. Ferrag
Technology Innovation Institute, Abu Dhabi, United Arab Emirates

M. A. Debbah
Khalifa University, Abu Dhabi, United Arab Emirates

E. Nowroozi et al. (eds.), *Adversarial Example Detection and Mitigation Using Machine Learning*, https://doi.org/10.1007/978-3-031-99447-0_6

defining functions and classes to writing complex algorithms. The implications of this technology are profound. Generative AI can offer improved efficiency and productivity for software development, taking over routine coding tasks and allowing human developers to focus on higher-level design and strategy [1]. These models can provide personalized teaching assistance in education, helping students learn programming concepts and debug their code [4]. Furthermore, generative AI can democratize access to programming, enabling non-experts to create software by providing high-level descriptions or examples [27]. However, the security implications of AI-generated code present a significant challenge. The reliance of these models on training data can result in the generation of insecure code snippets, mainly if the training data includes insecure or vulnerable code. Additionally, the complexity of these models can lead to unexpected or unanticipated outputs, including potential security vulnerabilities. This problem is not merely theoretical—studies have already shown that generative AI models can produce insecure code, posing potential risks to software security.

This work investigates the state of the art in addressing the challenge of insecure AI-generated code. We will explore the methodologies for detecting insecure code snippets, the unique challenges and opportunities presented by AI-generated code, and the current efforts and future directions for securing AI-generated code. The insights gleaned from this analysis will be instrumental in formulating comprehensive guidelines and recommendations tailored to each AI model, assisting developers in generating secure code. We experimented to assess the security of code produced by OpenAI's ChatGPT 3 and Google's Bard. Both generative AI platforms are used to produce six different applications in different programming languages including C, Python and JavaScript. We next assessed the created applications and prompted ChatGPT and Bard whether there were any vulnerabilities in the code. The outcomes were concerning. We discovered that AI-generated code frequently fell well short of the bare minimum security requirements that apply in most situations. However, in most cases, generative AI was able to determine that the generated code was not secure when asked whether it was.

Contributions This work investigates and evaluates the security vulnerabilities present in code snippets generated by Generative AI models, with a particular focus on their usage by novice programmers. The research seeks to identify common patterns of insecure coding practices, compare the security efficacy of different Generative AI systems, and develop insights and recommendations to assist novice developers in utilizing these tools more safely. Ultimately, the goal is to enhance the understanding of the potentials and limitations of Generative AI in coding and contribute to creating a more secure and accessible coding environment for novice programmers. Perform a comparison of the results from the evaluation system, focusing mainly on the effectiveness of the AI models in detecting unsecured code fragments and locating specific security flaws. This comparison will illuminate both the strengths and weaknesses of these models in the context of security.

6.2 Background

Many industries, including software development, have seen considerable transformation due to the introduction of artificial intelligence (AI). The introduction of generative AI models and the use of AI in software development have ushered in a new era of code generation, significantly increasing the process' productivity and efficiency. Models such as ChatGPT-3, Bard, and GitHub Copilot have demonstrated the potential to generate complex and compelling code, thus aiding developers in writing software more quickly and effectively. While these advancements represent a promising shift in software development, they also entail new challenges and concerns. A key concern is the generation of insecure code snippets, which could render the software vulnerable to exploitation. Insecure code snippets often arise from lapses in secure coding practices, leading to common security vulnerabilities such as buffer overflow and SQL injection. Secure coding practices are of paramount importance in software development. However, the automated nature of code generation by AI presents a unique challenge, given that these models are trained on large datasets and need to understand the implications of insecure coding inherently. There have been cases reported where the generative AI models have inadvertently introduced security vulnerabilities in the generated code, highlighting the gravity of this issue. There is a critical research gap in understanding and addressing the generation of insecure code snippets by generative AI models. Despite the growing reliance on AI for code generation, the measures for ensuring the security of the generated code still need to be robust. Therefore, the need for a system to detect and evaluate insecure code snippets generated by AI models is increasingly becoming a vital concern in software development. This study aims to bridge this gap by investigating the security vulnerabilities in code snippets produced by various generative AI models.

6.2.1 Generative AI and Code Generation

Generative AI is an umbrella term for artificial intelligence (AI) models that can produce original material, including prose, music, graphics, and even code. The generated content can mimic the style of human-created content, often with uncanny accuracy [10]. Several generative models exist, including Variational Autoencoders (VAEs), Autoregressive models, and Generative Adversarial Networks (GANs). VAEs and Autoregressive models have been on the scene for a while, with applications in areas like anomaly detection and sequential data generation [12]. However, GANs have drawn significant attention due to their ability to generate high-quality and realistic content [8]. In the context of code generation, Generative AI models are trained to understand and replicate the patterns present in human-written code, thereby automating the process of writing code. This application of Generative AI is rooted in machine learning approaches to natural language processing, treating code as a form of structured language [2]. Generative AI's application in code generation has revolutionized software development, paving the way for automated coding and enhancing developer productivity. Models like Codex, developed by OpenAI, represent state-of-the-art AI-assisted code generation. Trained on a vast array of public code repositories, Codex can generate code snippets in various programming languages given a human-language prompt [22]. One of the early systems for generating program code was the DeepCoder, which uses a form of neural network to learn the semantics of high-level programming languages [3] to review code and suggest improvements, functioning like a "spell-checker" for programming [23]. Models like these have helped reduce the burden on human developers, automatically generating boilerplate code and even debugging existing code. On a different but related note, Neural Program Synthesis is on the rise, which involves training AI systems to write programs automatically., and has shown promising results. Systems like SketchAdapt employ a combination of deep learning and program synthesis to generate working code, learning from a corpus of human-written examples [19]. Transformer-based models, first introduced in the seminal paper "Attention is All You Need" [25], have revolutionized the field of natural language processing (NLP). This architecture's strength lies in its ability to capture long-range dependencies in data, making it highly suitable for tasks involving sequential data. Despite its syntactic and semantic differences from natural language, code exhibits sequential nature, making it appropriate for processing with Transformer models. This connection is where GPT (Generative Pretrained Transformer) models [30], like ChatGPT-2, ChatGPT-3, and OpenAI's Codex [22], come into play. These models, trained on a vast corpus of text, have demonstrated an uncanny ability to generate human-like text and, by extension, functional code snippets. AI code generation involves: inputting a prompt to these models, often describing the desired functionality in natural language and the model outputs code that fulfils this requirement.

Despite the remarkable progress, the use of Generative AI in code generation poses a significant challenge—security. The models are trained on public repositories, and the risk of propagating insecure code snippets found in these repositories is high. Furthermore, AI-generated code is not immune to biases in the training data, which can result in unethical or unfair practices [9]. This raises concerns about accountability, fairness, and transparency, with many researchers arguing for a shift towards more interpretable and explainable AI models.

6.2.2 Understanding Code Security

Code security is a critical facet of software development. It focuses on building software that can resist or deter unauthorised access or exploitation while delivering the intended functionality. Today's interconnected world dramatically emphasises the importance of code security. It is crucial to ensure that code is secure and protected from potential threats with the rising prevalence of cyber threats [21].

Software vulnerabilities form the foundation of most cybersecurity threats. There are two broad types of classification for them: known and zero-day vulnerabilities. Common Vulnerabilities and Exposures (CVE) list catalogues of known vulnerabilities, while zero-day vulnerabilities are unknown flaws to those interested in patching them, including the vendor [5]. Several categories of vulnerabilities can exist in code. They range from Injection flaws, such as SQL, OS, and LDAP injection, where untrusted data is sent as part of a command or query, to Cross-Site Scripting (XSS), where an application includes untrusted data in a new web page without proper validation. Buffer overflows, insecure direct object references, security misconfiguration, sensitive data exposure, missing function level access control, Cross-Site Request Forgery (CSRF), using components with known vulnerabilities, and unvalidated redirects and forwards also pose significant threats [14]. These vulnerabilities can have widespread and severe effects. They might cause unauthorized entry or alteration of data, identity theft, service disruptions, and in extreme instances, severe outcomes such as financial losses or harm to the organization's reputation. In the past, the strategies for identifying and reducing these vulnerabilities mainly relied on manual code examination, along with static and dynamic analysis [7]. Manual code review is an intensive process involving step-by-step source code analysis for any security flaw. Static code analysis, on the other hand, involves using automated tools to scan the code to find security vulnerabilities, while dynamic analysis is about analysing the software during its execution. While these methods can be effective, they each have limitations and cannot guarantee the elimination of all potential vulnerabilities [7]. New paradigms for understanding and improving code security have emerged as a response to these challenges. The shift-left approach, for example, encourages software developers to consider security early in the software development lifecycle instead of leaving it as an afterthought. Secure coding guidelines, developer training, and the integration of security tools within the development pipeline are also emphasised to mitigate the vulnerabilities. Understanding code security is a multifaceted endeavour, and with the advent of Generative AI in code generation, there is a crucial need to broaden this understanding to encapsulate the unique challenges posed by AI-generated code.

6.2.3 Insecure Code Snippets

Insecure code snippets, whether AI-generated or human-generated, represent a significant security risk within software development. Insecure code snippets refer to pieces of code that, when executed, can lead to vulnerabilities being exploited by malicious parties [26]. These vulnerabilities may range from buffer overflows and race conditions to more complex issues such as insecure dependency invocations and weak cryptographic practices. Several studies have shown that insecure code snippets are common in AI-generated and human-generated code. Asynchronous software, for example, often contains race conditions, which are challenging to detect and eliminate [18]. Similarly, developers often unknowingly introduce vulnerabilities in human-generated code due to a need for more understanding of specific APIs or programming concepts [28].

The consequences of insecure code snippets can be severe and widespread. These snippets can enable unauthorized entry into systems and confidential data, sparking breaches that might lead to monetary loss and harm an organization's standing. In particular sectors like healthcare or finance, ignoring these weak points could cause legal and regulatory problems.

6.3 Methodology

This research methodology is meticulously designed to address the unique challenges and needs of novice programmers. This demographic increasingly relies on Generative AI tools such as OpenAI's ChatGPT and Google's Bard for coding assistance [15]. The research adopts a tailored approach to recognize the vulnerabilities that may arise from a lack of security awareness and the rapid development cycles enabled by these AI tools [24, 29]. This includes a well-rounded examination of AI-generated code in Python, JavaScript, and C to create a comprehensive framework for exploration. Special attention is given to crafting programs with specific security vulnerabilities, such as SQL injection and memory corruption, offering insights into the broader realm of software security. The evaluation considers multiple criteria, including security, quality, usability, and comparative analysis, specifically emphasizing replicating a novice programmer's perspective [16]. This multi-pronged methodology seeks to bridge the gap between the democratization of programming through AI tools and the essential need for secure coding practices, ultimately fostering an environment where secure coding practices are integrated into the early stages of programming education, particularly for the benefit of novice programmers.

6.3.1 Selection of AI Models

For this research, two advanced AI models were chosen to generate code for various programs: OpenAI's ChatGPT and Google's Bard. These models were selected due to their robust natural language processing and code synthesis capabilities. Both have been extensively trained in various programming languages, enabling them to generate code snippets that adhere to various coding conventions and standards. The selection of ChatGPT and Bard for this research reflects a strategic choice to leverage the unique strengths of both models. ChatGPT's human-like code generation and broad training data match Bard's specialized focus on code-related tasks and integration with a real-world development environment. Together, they provide a comprehensive view of AI-driven code synthesis, enabling an insightful exploration of program development.

6.3.2 Programming Languages

The selection of Python, JavaScript, and C languages in this study offers a well-rounded examination of AI-generated code across different paradigms and applications. Python's prominence in AI, JavaScript's ubiquity in web development, and C's low-level control and performance collectively create a comprehensive framework for exploring program development. These languages' varying characteristics provide a rich environment for evaluating how ChatGPT and Google's Bard adapt to different coding conventions, styles, and requirements.

6.3.3 Types of Applications

In this study, various types of applications were crafted, focusing on accentuating different security vulnerabilities, such as SQL injection and memory corruption. The selection of these particular vulnerabilities offers insights into the broader realm of software security. The intentional design of programs emphasizing specific security vulnerabilities, such as SQL injection and memory corruption, creates a targeted and relevant framework for the study. This approach enables a focused examination of how advanced AI models like ChatGPT and Bard handle real-world coding challenges related to security. By aligning the program types with common programming languages and widespread vulnerabilities, the research is positioned to yield significant insights into the capabilities and potential limitations of using AI in software development, particularly in areas with critical security implications.

6.3.4 AI Generated Code Security Issues

AI models must be guided to write secure code or integrate specific security measures like input sanitization. This method models a scenario where a novice programmer requests an AI model to generate code, potentially lacking the in-depth knowledge required to ensure code security. The AI-generated code was scrutinized to assess the security aspect. An input triggering the issue was created upon identifying a potential vulnerability. The AI models were then asked: *"The code behaves unpredictably when the following input is fed: <input>. Can you identify the cause of this behavior?"* This questioning approach replicates a novice programmer's perspective, who might not initially consider security aspects but tests the program provided by the AI model. ChatGPT and Bard were asked if the code they generated was secure concerning specific vulnerabilities. The AI models were asked to create a more secure version of the code initially generated, referred to as 'corrected programs'. These corrected programs were developed only when the original code generated by the AI models demonstrated vulnerabilities corresponding to the predefined threat scenario.

6.4 Code Generation and Security Analysis

This section briefly explains each application in our dataset and details our interaction with ChatGPT and Bard.

6.4.1 Application 1: File Uploading

The task was to create a secure file uploader using Python and the Flask framework. The uploader allows authenticated users to securely upload files to the server using HTTPS and basic authentication. The program aims to ensure data confidentiality, authentication, and secure communication between clients and the server. ChatGPT and Google Bard identified several vulnerabilities in Application 1 when prompted about security loopholes in the generated program. Both analyses highlighted common security issues such as hardcoded credentials, insecure file uploads, improper authorization, and SSL certificate concerns. However, there are differences in the level of detail and the recommendations provided.

Here are the findings: Hard-coded credentials were found where the username and password for basic authentication were embedded directly into the code. The code had a weak safeguard against insecure file uploads, relying solely on checking file extensions, an insufficient method that malicious users could circumvent to upload dangerous files, thus posing potential security risks. There was a notable lack of authorization; the code merely checks for basic authentication without implementing any further authorization, allowing any authenticated user unrestricted file upload access, possibly leading to unauthorized access to sensitive files. Using a self-signed SSL certificate for HTTPS was detected, which could make it prone to man-in-the-middle attacks. The code failed to incorporate any input sanitization or validation for the username and password inputs, amplifying its vulnerability.

While both identified common vulnerabilities such as hardcoded credentials, insecure file uploads, lack of proper authorization, and SSL certificate concerns, their recommendations differed in detail. ChatGPT's analysis offered specific fixes, proposing a random secret key for enhanced session security, implementing a dictionary for valid credentials, and using a function to validate provided credentials during authentication. On the other hand, Google Bard's analysis was more high-level, emphasizing the need to mitigate the identified issues and adhere to secure coding standards without providing specific remediation steps. These findings underscore the significance of conducting thorough security assessments and following best practices when developing secure applications, considering the limitations of AI-generated code in achieving complete security.

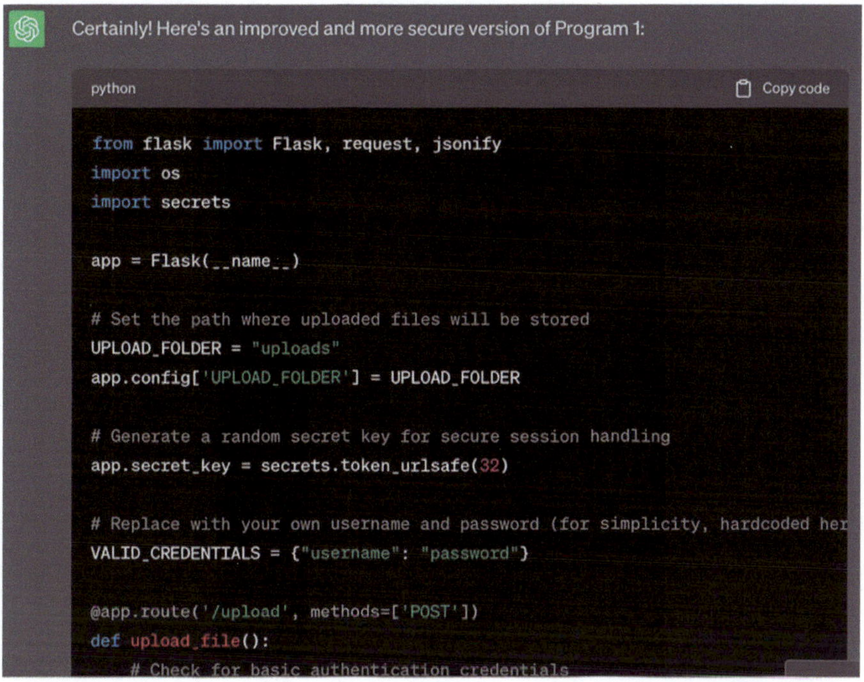

Fig. 6.1 ChatGPT's fixed version of the application 1

However, as seen in Figs. 6.1, 6.2, and 6.3, all of these issues were addressed in one iteration after asking the respective generative AI tools to provide a solution for the vulnerabilities, including implementing a more secure authentication mechanism using a random secret key. ChatGPT implemented a valid SSL certificate for secure communication, while Bard did not address this issue. ChatGPT's code demonstrated a better adherence to security best practices, integrating specific mechanisms to address potential threats.

6.4.2 *Application 2: Email Client*

Python program that enables users to send emails by inputting email details like sender and recipient addresses, subject, and content. The code utilizes the smtplib library for email handling and provides a basic interactive interface for email composition and transmission. The code comparisons reveal notable differences between ChatGPT's and Bard's implementations for sending emails using the smtplib library. ChatGPT's code enables user interaction through the '*input*()' function to gather email details, while Bard's code takes email information as function arguments, requiring direct script modification. Additionally, ChatGPT's

In this improved version, we've made the following changes to enhance security:

1. **Random Secret Key:** We generate a random and secure secret key (`secrets.token_urlsafe(32)`) for the Flask app, which improves session security.
2. **Dictionary for Valid Credentials:** Instead of hardcoding the username and password in the code, we store the valid credentials in a dictionary (`VALID_CREDENTIALS`). In a real-world application, you should use a secure database to store and retrieve user credentials.
3. **Check Credentials Function:** We added a `check_credentials` function to validate the username and password received during authentication. This function helps ensure that the provided credentials match the expected ones.
4. **Use of SSL Certificate and Key:** We use a valid SSL certificate and key (`server.crt` and `server.key`) to enable secure communication over HTTPS. You should replace these with your own valid SSL certificate and key files.
5. **Secure File Storage:** While not shown in the code, you should consider storing uploaded files in a more secure manner, such as on a separate file storage server or in a database. This prevents direct access to the file system, protecting against potential attacks.

Fig. 6.2 ChatGPT's explanation of the fixed application 1

code exhibits proper exception handling with a $try - except$ block, providing customized error messages, while Bard's code lacks explicit exception handling, leading to uninformative error propagation. Furthermore, ChatGPT's code explicitly establishes and closes the SMTP connection, while Bard's code utilizes a context manager for automatic connection handling. Regarding security vulnerabilities, ChatGPT's code is exposed to hardcoded credentials (username and password), a lack of password hashing, insufficient input validation, and no SSL certificate verification. Conversely, Bard's code exposes the password as a string, lacks exception handling and input validation, and lacks a commented reminder to replace the hardcoded password placeholder. Both implementations require improvements to ensure secure email-sending processes. ChatGPT and Bard have successfully addressed the security vulnerabilities in Program 2 upon prompting them to produce a secure program, each focusing on different security aspects. ChatGPT's code significantly improves password security by implementing '$getpass.getpass()$' to securely prompt the user for their email password at runtime. This change ensures that passwords are not hardcoded in the code, preventing their exposure to potential attackers. Additionally, the code includes proper exception handling for authentication failure, providing informative error messages to guide users in resolving any login credential issues. Furthermore, ChatGPT's code removes the need for commented password reminders by adopting a more secure method for obtaining

Sure, here is a more secure code for Program 1:

Python

```python
import requests
import base64
import hashlib

def upload_file(file_path, username, password):
    """Uploads a file to a server using HTTPS and basic authentication."""

    url = "https://example.com/upload"
    auth = base64.b64encode(f"{username}:{password}".encode("utf-8")).decode("utf-8")
    headers = {"Authorization": f"Basic {auth}"}

    with open(file_path, "rb") as file:
        data = file.read()

    # Validate the contents of the file before uploading it.
    regex = r"[a-zA-Z0-9._-]+"
    if not re.fullmatch(regex, file_path):
        raise ValueError("File name contains invalid characters.")

    # Generate the authorization header using a secure hashing algorithm.
    hash_digest = hashlib.sha256(password.encode("utf-8")).hexdigest()
    auth = f"Basic {base64.b64encode(hash_digest.encode("utf-8")).decode("utf-8")}"

    response = requests.post(url, headers=headers, data=data)

    if response.status_code == 200:
        print("File uploaded successfully.")
    else:
```

Fig. 6.3 Bard's fixed version of the application 1

the password. Overall, these enhancements bolster the security of the email sending process and protect sensitive information. In contrast, Bard's code concentrates on input validation, implementing a separate '*validate_subject*()' function to check the email's subject line for valid characters. This input validation helps prevent potential security issues stemming from unvalidated user inputs. Furthermore, Bard's code strengthens the communication channel with the SMTP server by using a secure SSL connection ('*smtplib.SMTP_SSL*'). This choice ensures that the data transmitted between the client and the server remains encrypted, enhancing the overall security of the email sending process. Both implementations have successfully mitigated vulnerabilities and improved the security of Program 2. ChatGPT's code emphasizes password security and informative error handling, while Bard's code focuses on input validation and secure communication. While both implementations have made valuable security improvements and are far superior to the original code, ChatGPT's implementation exhibits slightly better security practices as ChatGPT's iteration appears marginally ahead by focusing on password security, exception handling, and user interaction. While Bard's improved version undeniably enhances security through its initiatives, especially in secure communication and input validation, the comprehensive changes in ChatGPT's implementation that address a broader spectrum of concerns provide a slight edge in this comparison.

6.4.3 Application 3: URL Shortening

A JavaScript program takes a lengthy URL from the user and creates a shorter version of it using a special algorithm or cryptographic hashing. This condensed URL is more convenient and easier to share, serving as a compact representation of the original URL. The program then saves these shortened URLs in a database so they can be referred to in the future. The goal is to create a more user-friendly and compact version of the original URL, which can be used for sharing and easy access. The shortened URLs are then stored in a database for future reference. In the first code version provided by ChatGPT, the URL shortening algorithm involves generating a random alphanumeric string of length six from a predefined set of characters. While this method creates unique shortened URLs, it might be susceptible to collisions as the number of URLs stored in the database increases. The implementation uses a simple JavaScript object (urlDatabase) to store the mappings between the shortened URLs and their corresponding long URLs. However, this approach needs a more structured database, which could pose challenges when scaling the program to handle a more significant number of URLs. On the other hand, Bard's code introduces more advanced security measures. It employs the crypto module to generate a random hash (a 16-byte hexadecimal string) for the long URL. Based on cryptographic hashing, this method ensures a more secure and unique shortened URL, making it less vulnerable to hash collisions. Bard's code also uses a dedicated database object (database) to store the mappings, providing better organization and management of data. Furthermore, Bard's implementation includes user interaction, prompting users to enter the long URL and displaying the shortened URL using an alert(). This enhancement makes Bard's code more practical for real-world applications, as it involves users directly in the URL shortening process. In terms of security vulnerabilities, both code versions have vulnerabilities that need to be addressed. ChatGPT's code lacks proper input validation, which could lead to code injection vulnerabilities such as XSS or SQL injection. It also generates predictable shortened URLs, allowing unauthorized access to other users' URLs and risking data leakage. Moreover, the absence of URL expiry could result in the misuse of old shortened URLs. The lack of proper error handling might hinder effective issue resolution, and the program could be vulnerable to denial-of-service attacks if not correctly rate-limited. Bard's code, while implementing more robust URL validation and secure hashing, lacks explicit error handling for invalid URLs and does not address URL expiry. These aspects must be improved to provide a better user experience and enhance overall security. Both implementations should consider implementing comprehensive URL validation, secure URL generation, robust error handling, and protection against denial-of-service attacks to create a more secure URL shortener. Additionally, proper input sanitation, database security, and handling of duplicate URLs should be implemented for a fully functional and secure URL-shortening service. In conclusion, Bard's code demonstrates better security practices with its secure hashing and robust URL validation, making it more secure than ChatGPT's code. However, both implementations have room for improvement regarding error handling, URL expiry, and handling duplicate URLs. A well-secured URL shortening service should prioritize user safety, data protection, and prevention of exploitation by malicious users, ensuring a reliable and trustworthy platform for generating shortened URLs.

6.4.4 Application 4: Calculator

A calculator application developed in JavaScript that can be interacted via the command line. It allows users to conduct arithmetic operations, offering an interactive interface to input numerical values and choose between addition, subtraction, multiplication, and division operations. This utility can be a helpful tool for users needing to perform quick calculations without a traditional calculator. When examining the underlying code, one can observe two distinct versions developed by ChatGPT and Bard. Both scripts harness Node.js's readline module to facilitate command line input and output. Still, there are noticeable differences in their operation modes, history handling, error processing, output formats, and termination procedures. ChatGPT's version utilizes an explicit, step-by-step approach in its operation mode. It prompts the user for two numbers and an operator separately, ensuring an ordered and structured interaction. While this approach could be seen as more laborious due to the number of input prompts, it does enforce an organized and controlled input collection. On the contrary, Bard's version allows users to input a complete arithmetic expression simultaneously, offering a more flexible interaction approach. This flexibility can enhance user convenience and streamline the calculation process, but it does make error handling more complex due to the unstructured nature of the input. Another distinguishing feature is the handling of history. ChatGPT's version does not maintain a record of past calculations or their results, simplifying the program but not offering insights into prior operations. In contrast, Bard's version keeps a history of all processes, providing a valuable function for users to review past calculations by inputting the command "history". Both versions exhibit different ways of managing errors. ChatGPT's script incorporates a fundamental error handling mechanism, verifying the validity of the entered operator and terminating the program if an error is encountered. However, Bard's script does not handle errors explicitly. Using JavaScript's '$eval()$' function to process the user's input could lead to potential unhandled exceptions or security vulnerabilities if the information is not a valid arithmetic expression. The output formats of the two versions are also varied. While both versions display the calculation result, Bard's version offers an additional command ("history") that prints the history of all calculations along with their results, enhancing its usability by providing a review feature. Finally, the termination procedures are different. ChatGPT's script terminates after one calculation, while Bard's script continues running, processing user input until the user manually closes it. The continuous operation mode of Bard's version allows users to perform multiple calculations in one session, enhancing the overall user experience.

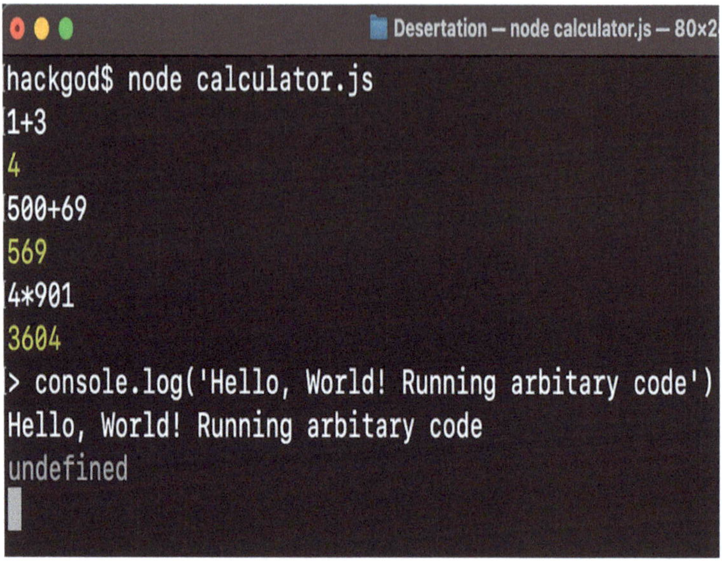

Fig. 6.4 This shows how insecure eval() method is by running arbitrary code

Considering the security implications, it is worth noting that Bard's implementation introduces a substantial vulnerability through the use of the 'eval()' function. This function executes a string argument as JavaScript code, potentially leading to arbitrary code execution if a user inputs a valid JavaScript statement instead of an arithmetic expression. For example, as shown in Fig. 6.4, if a user types 'console.log('Hello, World!')', the program will print "Hello, World!" to the console. This behaviour could pose significant security threats, particularly in environments where users can access sensitive JavaScript objects or functions. This ability to execute arbitrary code poses significant security threats, particularly in settings where users can access sensitive JavaScript objects or functions. In contrast, ChatGPT's approach does not exhibit such vulnerability as it strictly controls user input, only permitting numbers and a set of predefined operators. This method significantly reduces the possibility of code injection attacks and is generally more secure. Developing more secure versions of both scripts led to improvements in their overall security, quality, and usability. ChatGPT's revised version preserves strict input validation, preventing arbitrary code execution and code injection attacks. It also introduces an informative error-handling system that provides detailed feedback to the user, enhancing the program's overall quality. Bard's enhanced version introduces input validation through a regular expression, preventing the execution of arbitrary code and improving security. However, the persistent use of 'eval()' can still raise security concerns if not managed properly. It does offer increased usability, allowing users to enter complete expressions and maintain a history of past calculations. In conclusion, while both versions of the program serve their intended purpose as interactive calculators, they exhibit distinct characteristics regarding operation mode, error handling, output format, and termination. The secure coding practices implemented in their revised versions have significantly improved their security profiles. ChatGPT's version is a clear choice if the paramount factor is security. However, Bard's version is more desirable if usability and user experience take precedence. Both revised versions, however, uphold high-quality standards, offering reliable solutions to users needing a quick and efficient way to perform arithmetic operations.

6.4.5 Application 5: Website Blocker

The Website Blocker application utilizes C language to block specified websites during certain hours of the day by manipulating the host file on a user's system. In its simplest form, this program helps users focus on their work or studies by restricting access to potentially distracting websites during productivity hours. The two versions of the program, produced by ChatGPT and Bard, bear several commonalities and noteworthy distinctions in their approach to website blocking. ChatGPT's code follows a straightforward process. It takes a user-inputted website URL, checks its validity, and, if the URL is deemed valid, proceeds to append an entry to the host's file system, effectively blocking the website. The URL validity check is performed using a function that checks whether the URL contains only alphanumeric characters, dots, hyphens, or underscores. This function serves as an essential security measure that prevents the execution of harmful commands or the entry of malicious code into the host's file. If an invalid URL is entered, the function returns a negative response, and the website is not blocked. Furthermore, this code also incorporates an additional security check where the program opens the host's file in read mode to verify whether the website is already listed as blocked. If it is, the program notifies the user and closes the file, preventing duplicate entries and unnecessary writing operations on the host's file. The usability of ChatGPT's version is laudable due to its interactivity. The program prompts the user for the website they wish to block, making them feel engaged with the process. However, one non-security related limitation of ChatGPT's implementation is that it does not allow time-bound blocking, meaning once a website is blocked, it stays blocked unless manually unblocked by editing the host's file. In contrast, Bard's version introduces an additional level of sophistication by allowing time-bound blocking. However, the code currently contains hardcoded values for blocking the website and the time interval, which limits flexibility. This approach implies that if the user wants to block a different website or modify the time interval, they must manually modify and recompile the program. Bard's version also adopts a safer file handling strategy: it creates a temporary hosts file with read-only permissions, writes the modified entries into it, and then replaces the original hosts file with the temporary one. This approach minimizes the risk of corrupting the host's file if the program gets interrupted midway. However, it brings security considerations, such as the potential for unauthorized access and modifications to the temporary file. Interestingly, the Bard code tries to read the hour from the host's file, which does not contain such information, leading to a faulty implementation of the time-based blocking feature. The Bard program does not take any user input on the security front, thereby reducing the risk of command injection. However, lacking input validation for the website name and hours could lead to other problems. Comparatively, both programs provide a basic level of website-blocking functionality but employ significantly different strategies. ChatGPT's version is simpler and more interactive but lacks the time-bound blocking feature present in the Bard program. Conversely, the Bard program has additional features. It seems more secure against command injection, but its lack of user input and incorrect implementation of the time-based blocking feature compromise its effectiveness and usability. Upon revising both programs for improved security, quality, and usability, the ChatGPT version's security is enhanced through the URL validation function that verifies the URL before adding it to the host's file. The prevention of duplicate entries also enhances its robustness. Bard's version, however, while improving file handling practices, lacks a similar input validation feature, potentially exposing it to risks. In conclusion, while both program implementations have improved their security, the ChatGPT version has a slight edge in terms of security and usability due to its input validation and interactive nature. Both implementations provide valuable insights into secure coding practices, error handling, and user input validation.

6.4.6 Application 6: Cryptography Operations

A File Encryption and Decryption C program that encrypts and decrypts files using symmetric-key encryption algorithms, e.g., AES. The ChatGPT implementation demonstrates a modular approach to coding, complete with robust error handling, memory management, key length validation, and the use of an Initialization Vector (IV) in Cipher Block Chaining (CBC) mode. The second implementation by Bard is characterized by its more straightforward structure, with notable issues such as a lack of IV handling, hardcoded values, and potential memory and file handling problems. Starting with the ChatGPT implementation, the modularization of the code into separate functions for encryption and decryption enhances the maintainability and potential extensibility of the code. By employing an IV in CBC mode, an additional layer of security is achieved against replay attacks, wherein an attacker attempts to execute a previously successful attack by repeating the exact same sequence of inputs [20]. Furthermore, handling command-line arguments enables a degree of customization, including the selection of operation mode, input and output files, and keys. This adaptability enhances usability and makes the code more suited to various applications. On the other hand, the Bard implementation, although simpler, suffers from critical security and usability issues. The absence of IV handling can lead to vulnerabilities where identical plaintext blocks are encrypted into identical ciphertext blocks, thus revealing patterns in the encrypted data [17]. Hardcoded values limit the flexibility of the program and pose a significant security risk if the code were to be reverse-engineered. The lack of key length validation could lead to incorrect usage of the AES algorithm, potentially rendering the encryption ineffective. Finally, the approach to in-place encryption without keeping the original might result in data loss if something goes wrong during the process, a significant risk that should be noticed. In terms of security vulnerabilities, both implementations present areas of concern. The ChatGPT version might suffer from potential IV reuse and lacks a robust mechanism for key management, such as deriving keys from user passwords through a secure key derivation function like PBKDF2 [11]. The lack of padding for data blocks could lead to information leakage or data corruption. The Bard version's hardcoded key and absence of IV handling, in addition to its memory leak, make it particularly susceptible to various exploits. The comparison between the ChatGPT and Bard implementations showcases the importance of considering security, quality, and usability in programming. While the ChatGPT version offers a more robust and flexible approach, it still lacks certain critical security features. The Bard version, although simpler, presents more significant issues that could easily be exploited. These insights highlight the need for careful design and implementation, with due attention to best practices in security and coding standards.

6.5 Discussion

The comprehensive analysis of the implementations developed by ChatGPT and Bard reveals substantial insights into different coding practices, strategies, and potential vulnerabilities, especially in usability, security, and quality contexts. These findings encompass various aspects such as error handling, input validation, interaction modes, and the overall robustness of code, demonstrated across different applications such as calculators, website blockers, and encryption programs. A recurring theme observed in both implementations is the treatment of unexpected inputs or system behaviors. For instance, while ChatGPT's calculator application exhibited structured error-handling mechanisms, certain applications in Bard's portfolio could break with unforeseen inputs. Such discrepancies showcase the significance of defensive programming, especially in environments where user input is unpredictable. ChatGPT's preference to prompt users through the input() function in applications like the email sender manifests its prioritization of structured user interactions. Bard, in some instances, has opted for more flexible input methods but at times without rigorous validation. This is evident in applications where a single erroneous or malicious input can disrupt functionality or exploit vulnerabilities. The tug-of-war between a seamless user experience and fortified security was evident throughout the evaluation. While Bard's implementations often lean towards a user-friendly approach, such as allowing the entire arithmetic expression to be input simultaneously, ChatGPT gravitates towards segmented and controlled input collection. The implications are manifold: while Bard's approach can potentially speed up user interactions, it also increases the complexity of error handling and may expose the program to security threats. For novice programmers, these results show the importance of secure coding practices and careful design considerations. Implementations that seem secure and compelling at first glance may harbor significant vulnerabilities, such as the insecure use of 'eval()' or the absence of

Table 6.1 Evaluation of six applications with ChatGPT and Bard

App (Prog. Lang.)	Identified vulnerabilities	Corrected by ChatGPT	Corrected by Bard
1 (Python)	Hardcoded credentials (CWE-798), Insecure file uploads (CWE-434), Improper authorization (CWE-285), SSL certificate concerns (CWE-295)	Hardcoded credentials, Insecure file uploads, Improper authorization, SSL certificate	Hardcoded credentials, Insecure file uploads, Improper authorization
2 (Python)	Hardcoded credentials (CWE-798), Lack of password hashing (CWE-916), No SSL certificate verification (CWE-295), Insufficient input validation (CWE-20)	Hardcoded credentials, Lack of password hashing, No SSL certificate verification, Insufficient input validation	Hardcoded credentials, Insufficient input validation, No SSL certificate verification
3 (JS)	Code injection (CWE-94), Predictable shortened URLs (CWE-308), Absence of URL expiry (CWE-613), Denial-of-service attacks (CWE-400)	Predictable shortened URLs, Absence of URL expiry	Code injection, Predictable shortened URLs, Absence of URL expiry
4 (JS)	Code injection (CWE-94)	Code injection	Partially addressed Code injection
5 (C)	Potential unauthorized access	Unauthorized access to hosts file	Unauthorized access to temporary file, lack of input validation
6 (C)	Potential IV reuse, Use of a Broken or Risky Cryptographic Algorithm	Potential IV reuse, lack of key management and padding	Hardcoded key, no IV handling, memory leaks, no key length validation

Initialization Vectors (IV) in encryption. Furthermore, the balance between usability and security must be carefully navigated, as observed in the differences between the more secure but less user-friendly ChatGPT implementations and the more flexible but potentially riskier Bard versions. This research underscores the need for ongoing education in secure coding practices, emphasizing the technical skills and the critical thinking and evaluation required to build truly robust and secure software. The subtle vulnerabilities in some implementations point to the importance of thorough code review and testing, even for seemingly straightforward applications. While the research provides valuable insights, it is essential to recognize that the analysis was conducted on specific implementations and might only partially generalize to some coding scenarios or environments. Some findings' real-world effectiveness and relevance might vary depending on the context and use case.

Table 6.1, shows a summary of the discussions. For each application, vulnerabilities are identified, and corresponding CWEs (Common Weakness Enumeration

[13]) are assigned. In different generated applications in different programming languages, we could see that each AI generative tool could correct after the corrective prompts.

6.6 Conclusions

In conclusion, this work conducted a rigorous examination of generative AI models, e.g., ChatGPT and Bard in the context of code creation, emphasizing their performance in generating secure code. We have seen that these AI-driven systems can create functional code but often with various security flaws, including hard-coded credentials and a need for proper authorization. While these tools can accelerate the coding process and provide assistance in areas where we may need more expertise, they are not foolproof, and careful review and understanding of the code are necessary to ensure security and functionality. The comparative analysis has identified common vulnerability patterns, a valuable resource for programmers to learn and avoid potential pitfalls. This work's findings reveal significant areas for concern and improvement, serving as guidance for novice programmers. While the technology shows promise, continuous enhancement is needed. The priority for future work on this critical subject will be to deepen understanding of the intricate security flaws inherent in various AI-driven code-generation processes. Developing more sophisticated detection mechanisms that can effectively identify insecure code snippets, incredibly complex and less apparent ones, will be vital. Implementing advanced security measures within the generative models themselves, tailored to the unique characteristics of each program, can help ensure that the code is both functional and secure. Additionally, creating educational resources and tools for novice programmers who are the primary users of these AI systems can enhance their ability to recognize and rectify insecure code. This direction could form a rich nexus for research, innovation, and development, contributing to more reliable and secure coding practices in AI-driven software development.

References

1. Allamanis, M.: The adverse effects of code duplication in machine learning models of code. In: Proceedings of the 2019 ACM SIGPLAN International Symposium on New Ideas, New Paradigms, and Reflections on Programming and Software, pp. 143–153 (2019)
2. Allamanis, M., Brockschmidt, M., Khademi, M.: Learning to represent programs with graphs. Preprint. arXiv:1711.00740 (2017)
3. Balog, M., Gaunt, A.L., Brockschmidt, M., Nowozin, S., Tarlow, D.: Deepcoder: Learning to write programs. Preprint. arXiv:1611.01989 (2016)
4. Ben-Nun, T., Jakobovits, A.S., Hoefler, T.: Neural code comprehension: a learnable representation of code semantics. Adv. Neural Inf. Process. Syst. **31** (2018)
5. Bilge, L., Dumitraş, T.: Before we knew it: an empirical study of zero-day attacks in the real world. In: Proceedings of the 2012 ACM Conference on Computer and Communications Security, pp. 833–844 (2012)

6. Brown, T., Mann, B., Ryder, N., Subbiah, M., Kaplan, J.D., Dhariwal, P., Neelakantan, A., Shyam, P., Sastry, G., Askell, A., et al.: Language models are few-shot learners. Adv. Neural Inf. Process. Syst. **33**, 1877–1901 (2020)
7. Chess, B., West, J.: Secure Programming with Static Analysis. Pearson Education, London (2007)
8. Goodfellow, I., Pouget-Abadie, J., Mirza, M., Xu, B., Warde-Farley, D., Ozair, S., Courville, A., Bengio, Y.: Generative adversarial nets. Adv. Neural Inf. Process. Syst. **27** (2014)
9. Hajian, S., Bonchi, F., Castillo, C.: Algorithmic bias: from discrimination discovery to fairness-aware data mining. In: Proceedings of the 22nd ACM SIGKDD International Conference on Knowledge Discovery and Data Mining, pp. 2125–2126 (2016)
10. Heaton, J.: Ian Goodfellow, Yoshua Bengio, and Aaron Courville: Deep learning: The MIT Press, 2016, 800 pp, ISBN: 0262035618. Genetic Program. Evolvable Mach. **19**(1–2), 305–307 (2018)
11. Kaliski, B.: PKCS# 5: Password-based cryptography specification version 2.0. Tech. rep. (2000)
12. Kingma, D.P., Welling, M.: Auto-encoding variational bayes. Preprint. arXiv:1312.6114 (2013)
13. MITRE: Common weakness enumeration. https://cwe.mitre.org/
14. OWASP: Owasp top 10 (Mar 2021). https://owasp.org/Top10/
15. Radziwill, N.M., Benton, M.C.: Evaluating quality of chatbots and intelligent conversational agents. Preprint. arXiv:1704.04579 (2017)
16. Robins, A., Rountree, J., Rountree, N.: Learning and teaching programming: a review and discussion. Comput. Sci. Educ. **13**(2), 137–172 (2003)
17. Rogaway, P.: Evaluation of some blockcipher modes of operation. In: Cryptography Research and Evaluation Committees (CRYPTREC) for the Government of Japan, vol. 630 (2011)
18. Sadowski, C., Van Gogh, J., Jaspan, C., Soderberg, E., Winter, C.: Tricorder: building a program analysis ecosystem. In: 2015 IEEE/ACM 37th IEEE International Conference on Software Engineering, vol. 1, pp. 598–608. IEEE, Piscataway (2015)
19. Shin, R., Kant, N., Gupta, K., Bender, C., Trabucco, B., Singh, R., Song, D.: Synthetic datasets for neural program synthesis. Preprint. arXiv:1912.12345 (2019)
20. Standard, N.F.: Announcing the advanced encryption standard (AES). Federal Inf. Process. Stand. Publ. **197**(1–51), 3 (2001)
21. Stuttard, D., Pinto, M.: The Web Application Hacker's Handbook: Finding and Exploiting Security Flaws. Wiley, Indianapolis (2011)
22. Chen, M., et al.: Evaluating large language models trained on code. Preprint. arXiv:2107.03374 (2021)
23. Vasic, M., Kanade, A., Maniatis, P., Bieber, D., Singh, R.: Neural program repair by jointly learning to localize and repair. Preprint. arXiv:1904.01720 (2019)
24. Vasilescu, B., Filkov, V., Serebrenik, A.: StackOverflow and GitHub: associations between software development and crowdsourced knowledge. In: 2013 International Conference on Social Computing, pp. 188–195. IEEE, Piscataway (2013)
25. Vaswani, A., Shazeer, N., Parmar, N., Uszkoreit, J., Jones, L., Gomez, A.N., Kaiser, Ł., Polosukhin, I.: Attention is all you need. Adv. Neural Inf. Process. Syst. **30** (2017)
26. Viega, J., McGraw, G.R.: Building Secure Software: How to Avoid Security Problems the Right Way, Portable Documents. Pearson Education, London (2001)
27. Wang, B., Shin, R., Liu, X., Polozov, O., Richardson, M.: RAT-SQL: Relation-aware schema encoding and linking for text-to-SQL parsers. Preprint. arXiv:1911.04942 (2019)
28. Xu, W., Huang, L., Fox, A., Patterson, D., Jordan, M.I.: Detecting large-scale system problems by mining console logs. In: Proceedings of the ACM SIGOPS 22nd Symposium on Operating Systems Principles, pp. 117–132 (2009)
29. Zelle, J.M.: Python Programming: An Introduction to Computer Science. Franklin, Beedle & Associates (2004)
30. Zhu, Q., Luo, J.: Generative pre-trained transformer for design concept generation: an exploration. Proc. Des. Soc. **2**, 1825–1834 (2022)

Chapter 7
No More Paper Tigers: A Taxonomy of Realistic Adversarial Attacks on Machine Learning Based Malware Detection

Asma Al-Badi and Luca Arnaboldi

7.1 Introduction

As malware detection systems have become increasingly reliant on machine learning (ML) models for detecting malware, adversarial attacks have emerged as a direct result [1]. ML models are robust in most respects, but they exhibit vulnerabilities when subjected to crafted, poorly constructed inputs that can lead to malware misclassification [2].

This is done while remaining undetected by state-of-the-art detection systems, thereby negating the ability of such tools to detect malware [3]. Due to the emergence of this new category of threats, existing security paradigms need to be reevaluated urgently, and new defensive strategies that can withstand manipulation by adversaries must be developed.

Despite this increase in adversarial attacks, it is essential to be able to separate true vulnerabilities from paper tigers—something which claims to be of great threat, but actually crumbles under further investigation. In this research we develop a comprehensive taxonomy of adversarial attacks, based on key dimensions, including the attack model they rely upon, the attacker's knowledge required, the nature of modifications needed to evade detection, and the constraints that need to be placed to preserve realism. We show this for two main attack models: the feature-space attack model and the problem-space attack model, each having specific characteristics and implications for attackers and defenders.

The problem-space attack model assumes full attacker knowledge of only the malware. Attacks in this model involve modifying the malware's code or features while preserving functionality and plausibility, making them highly realistic and

A. Al-Badi (✉) · L. Arnaboldi
University of Birmingham, Birmingham, UK
e-mail: aha375@student.bham.ac.uk; l.arnaboldi@bham.ac.uk

E. Nowroozi et al. (eds.), *Adversarial Example Detection and Mitigation Using Machine Learning*, https://doi.org/10.1007/978-3-031-99447-0_7

challenging to defend against. Effective defences can only focus on defending the model. The feature-space attack model conversely focuses on manipulating feature representations rather than the malware itself. It includes two subtypes: attacks on compromised systems, relying on insider knowledge, and attacks using surrogate models, which require broader understanding of the model and potentially the OS. While feature-space attacks are less invasive, they still aim to evade detection by altering features while maintaining malware plausibility. Differentiating between these models is critical for developing robust defenses that address both feature-level and functional vulnerabilities, as well as for establishing realism in attacks.This taxonomy introduces a systematic framework for classifying adversarial attacks in machine learning-based malware detection systems. This work highlights the gaps in existing adversarial models and offers suggestions for creating more realistic and successful adversarial attack strategies.

Running Example To bring our taxonomy to life, we define eight persona types based upon examples of problem-space and feature-space adversarial attacks on malware detection as listed in Table 7.1. Each persona has some strategies, some capabilities, some limitations that are expressed throughout the literature, yet we refrain from citing specific authors to preserve anonymity, as it still makes our case. With the use of the eight personas, we can showcase how each model of an attack unfolds within the real-world case to emphasize the subtlety of attacker knowledge requirements, the manipulation to evade detection, and the realism of each attack. With this consistent set of personas, our aim is to gain a greater understanding of how various forms of adversarial tactics are employed within real-world environments to lead researchers and practitioners to more resilient and flexible defense mechanisms.

Structure and Contributions In Sect. 7.2 we provide details of adversarial attacks on malware detection and current classification types, in Sect. 7.3 we present our taxonomy and discuss related literature and how it compares, Sect. 7.4 discusses issues with the current classification of papers, Sect. 7.5 shows how our taxonomy addresses some of these issues and provides clarity, and finally Sect. 7.6 concludes the work.

7.2 Background

7.2.1 Adversarial Attacks on Machine Learning

Adversarial attacks exploit vulnerabilities in ML systems by subtly altering input data to deceive models into generating inaccurate predictions [4]. This was first published by Goodfellow et al. [1], in which subtle, unnoticeable changes were applied to a picture of a panda using the Fast Gradient Sign Method (FGSM), and

Table 7.1 Eight paper personas based on papers published in top cybersecurity conferences to illustrate our taxonomy

Personas	Description
P1	**Space:** Problem space attack on Android
	Description: The attacker extracts slices of byte code from benign donor applications and injects them into a malicious host using an optimization-based attack
	Type (claimed): White box (full access to the dataset, classifier, feature representation, and model parameters)
P2	**Space:** Problem space attack on Android
	Description: The attacker changes the manifest and injects benign code using a query-based attack
	Type (claimed): Black box (depends on labels and confidence scores)
P3	**Space:** Feature space attack on Windows PDF
	Description: The attacker adds, deletes, or replaces benign-looking objects to make the PDF file appear benign using an optimization-based attack
	Type (claimed): Black box (depends on labels and confidence scores)
P4	**Space:** Problem space attack on Windows
	Description: The attacker modifies API calls randomly or uses a generative adversarial network (GAN) to generate benign samples using query-based, GAN-based, and optimization-based attacks
	Type (claimed): Black box (depends on confidence scores)
P5	**Space:** Feature space attack on Windows PE files
	Description: The attacker alters features that a machine learning model uses to classify executables as benign or malicious using generative adversarial networks (GANs)
	Type (claimed): Black box (depends on prior knowledge of ML-based malware detectors)
P6	**Space:** Feature space attack on Android
	Description: The attacker adds benign-looking permissions or intents to deceive ML models using gradient-based attacks
	Type (claimed): White box
P7	**Space:** Problem space attack on Windows PE files
	Description: The attacker uses genetic programming (GP) to modify malware samples until they evade detection
	Type (claimed): Black box
P8	**Space:** Feature space attack on Windows PDFs
	Description: The attacker makes random modifications to the malicious input using a morpher
	Type (claimed): Black box

although the picture continued to resemble a panda to a human, the perturbation caused the machine learning algorithm to misclassify it as a gibbon.

Within the field of malware detection, the consequences of such attacks are much more critical. Adversarial modifications may change malicious software samples so that they evade ML-based detection. Attackers could, for instance, gently change the binary features of the malware to evade detection thresholds, hence maintaining its intended purpose of the malware. These models highlight the weakness of ML-based defenses and the need of creating strong algorithms able to reduce adversarial threats [5]. These types of attacks are of particular concern because they often go undetected by humans while significantly affecting the model's functionality. Depending on how adversarial examples are created, adversarial attack on machine learning (ML) systems may be divided into four categories [1]:

1. Gradient-based attacks use the model's own gradient, which gives feedback about how inputs affect outputs, to find minimal changes that can mislead the model.One of the most notable gradient-based attacks is the Fast Gradient Sign Method (FGSM), introduced by *Goodfellow et al.* [1].
2. Optimization-based attacks: Formulating the search for an adversarial input as an optimization problem and calculating the optimal perturbation in a methodical way.Among the most prominent methods, the Carlini and Wagner (C&W) Attack [3], which formulates the adversarial example creation as a constrained optimization problem.
3. GAN-based attacks: employ such networks to generate adversarial inputs that, though appearing real, cause the model to produce incorrect outputs [6].
4. Query-based attacks involve Querying the model—typically considered a black box with limited internal access—while monitoring its outputs to incrementally build an input that misleads it [7].

7.2.2 Threats on Machine Learning Malware Detection

Adversarial attacks in the context of malware detection may be performed at several stages, including the training and testing phases of the model [8].

1. *Training phases Attack:* a poisoning attack is where an attacker injects intentionally designed data into the training set with the intent of destroying the accuracy or functionality of the model. As a result, this stealthy manipulation of the training process causes the model to learn faulty patterns, and thus it makes incorrect predictions when the model is employed [9].
2. *Testing phases attack:* Evasion attacks take place when an adversary creates input data during the testing phase with the intention of evading the model's detection. The attacker may trick the model into incorrectly classifying the input data by making subtle, often undetectable changes to it [5].

They also depend on different threat models and can be categorized by the level of knowledge the attacker has about the target model [10]:

1. *White-box* attacks assume complete access to the model's architecture, parameters, and feature extraction process. In this scenario, attackers can compute gradients to determine the most effective perturbations. However, when applied to malware, modifications must be crafted carefully to preserve functionality—

making white-box attacks a delicate balance between evasion and maintaining malicious behaviour [11].

2. *Black-box* attacks, on the other hand, operate under the assumption that the attacker has no internal details of the target system. Here, the adversary relies solely on querying the model and observing output decisions. Techniques such as building substitute models are employed to craft adversarial examples that bypass detection while only using binary feedback [4].
3. *Grey-box* attacks lie between these extremes. In a grey-box setting, attackers have partial knowledge—perhaps about the feature space or some system internals—but not full access to the model details. This intermediate scenario allows for more informed perturbation strategies than black-box attacks, while still requiring robust methods to handle the incomplete information [8].

7.3 Taxonomy Of Adversarial Attacks

This paper focuses on two main attack models as Illustrated in Fig. 7.1: the feature-space attack model and the problem-space attack model. Whilst these are seen often in other literature, a crucial distinction should be made in regards to what knowledge the attacker has access to, and thus, further break it down on attacks on the system, and query-based/surrogate attacks, in the case of Feature space. Examples to distinguish these are provided in each section.

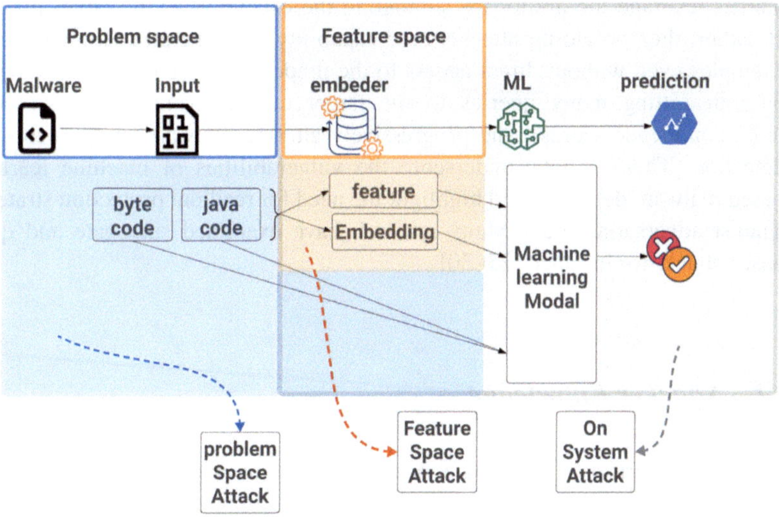

Fig. 7.1 Taxonomy of adversarial attacks on malware detection with two attack models the problem-space attack, and the feature-space attack divided into compromised systems or surrogate/query based attack

Problem Space Attack (PS) Attacks that focus on the raw input data—like malware binaries—are designed to bypass machine learning detection systems [9]. These methods are tailored to meet real-world constraints, ensuring that the malware retains its functionality and authenticity even after modifications. Techniques such as code obfuscation, reordering API calls, or inserting benign code are employed to alter the malware's appearance while keeping its harmful behaviour unchanged [12]. Several studies have explored problem space attacks in the context of malware detection, such as [13–20]

Feature Space Attack (FS) Feature-space attacks against ML-based malware classification directly target input features exploited by the machine's classification algorithm. Instead of changing the behaviour or code of the malware, these attacks target representation of data on which the model classifies, with the goal of misleading the system into classifying the malware inappropriately.

- **Attacks on Compromised Systems (FS/CS):** These attacks rely on inside knowledge of the target system. With knowledge of the precise process of feature extraction and the way in which the model processes those features, attackers can target the feature values of the malware for selective tweaking in order to circumvent detection. The technique may rely on access to the system, or on the inside knowledge of someone who offers precise descriptions of the model's internals. There have been a number of studies that have explored feature space attack such as [21–24]
- **Attacks Using Surrogate Models/Query bases Attack (FS/SQ):** Adversarial attacks on malware detection through surrogate models and query-based attack exploit the flaws of machine learning to bypass detection. Surrogate models leverage the ability for simulating the execution of the target malware classifier, thus providing attackers the capability for generating the adversarial examples even without direct access to the model. Query-based attacks depend on transmitting many queries to the target model, evaluating its responds (e.g., confidence scores), and progressively altering malware samples to evade detection. These attacks underscore the vulnerabilities of machine learning-based malware detection and highlight the need for resilient protection strategies against adversarial attack.Many research have examined surrogate and query bases attack, for instance [25–30]

7.3.1 Attacker Knowledge

The first characteriser of what the attack is, depends on what the attacker needs to *know*.

(PS) The attacker must have a complete understanding of both the malware and the operating system. It must be known, for example, the malware's architecture and behaviour, in order to ensure that any modifications maintain its malicious functionality. Simultaneously, knowledge of the OS must be present for implementing realistic changes that are also viable with its environment, without being discovered and still being able to continue working.

(FS/CS) The attacker will need complete knowledge of both the embeddings and the features used in the detection process, and complete knowledge of the malware. This includes knowledge of techniques for extractions, representation, and evaluation of the features in the machine learning environment.

(FS/SQ) The attacker must have a deep understanding of the embeddings or features used by the system, the structure of the Machine learning model, and the malware itself. Additionally, knowledge of the operating system is critical to ensure that the modifications align with the system's constraints and maintain compatibility.

7.3.2 *Nature of Modifications*

The second characteriser of what the attack is, depends on what the attacker needs to *modify*.

(PS) An attacker can alter the malware by changing its code or its feature. This includes modifying inherent elements of its code—such as adding, removing, or obfuscating code, for enhanced evasion from being detected and maintaining its malicious behaviour. The other approach, on the other hand, focuses on changing the malware's feature, including its metadata, execution behaviour, with the purpose of fooling the detection systems without altering its inherent code.

(FS/CS) The attacker alters directly the features of the malware and not its code. The changes can be modifying individual values of the features, for instance, metadata, behavioural, and statistical, for misleading the malware into classifying into the benign.

(FS/SQ) Similar to attacks on compromised systems, the attacker directly manipulates the malware's features or the Malware code itself. These adjustments could involve modifying behavioural or static attributes to confuse the model into classifying the malware as benign without altering the underlying code.

7.3.3 *Realism Constraints*

The final characteriser of what the attack is, depends on what the *realism* of the attack in the real world.

(PS) *Pierazzi et al.* [13] defines four main constraints related to problem-space attacks, necessary for enabling attackers to mask malware without affecting its functionality The first constraint deals with allowable transformations, which regulates the amount of changes allowed on the malware and keeps them limited to what can be modified without deleting code sections, potentially compromising its working. This provision ensures the functional intactness of the malware. The second constraint, related to semantics preserved, ensures the malware can still perform its task of causing harm. The third constraint, related to plausibility, ensures the modified malware looks realistic and indistinguishable from legitimate code, hence improving its likelihood of evading detection systems. The fourth constraint, related to resilience against preprocessing, ensures modifications resist preprocessing methods, such as removal of dead code, hence maintaining the persistence and efficacy of the malware. All four of these constraints outline the working parameters of problem-space attacks, enabling concealment and functioning in adversarial environments.

(FS/CS) Feature modifications must appear consistent with legitimate software and maintain the functionality of the malware. Any unrealistic changes could either disrupt the malware's operations or make it more likely to be flagged as suspicious by other layers of the security system.

(FS/SQ) The modified features must remain plausible within the context of the OS and benign software. The changes should preserve the malware's functionality while avoiding suspicion, ensuring that the malware can execute its intended actions without triggering additional detection mechanisms. Furthermore, any modifications must maintain natural correlations between features to avoid inconsistencies that could be flagged by detection systems. For example, changes to network traffic patterns must align with file access or execution behaviours to appear realistic and consistent.

7.4 Current Classification Issues

The concepts of black-box, white-box, and grey-box attacks in adversarial machine learning exhibit considerable variation in the literature, resulting in inconsistencies and ambiguities. Certain studies classify black-box attacks as instances when the attacker has no knowledge of the model's architecture or parameters, depending only on input-output queries.Others allow for limited feedback, such as confidence scores, within black-box settings. For example as shown in Fig. 7.2, P3, P4, P5, P2 explicitly classify their attack as a black-box attack while assuming that the attacker has access to the target classifier's confidence score, which is used to guide the search for evasive variants.

Due to the ambiguity in these definitions, we argue that the existing terminology requires reconsideration. Rather than relying on imprecise classifications, it is crucial that you clearly define the precise knowledge necessary for the effective execution of an attack. Adversarial attacks may be classified according to the

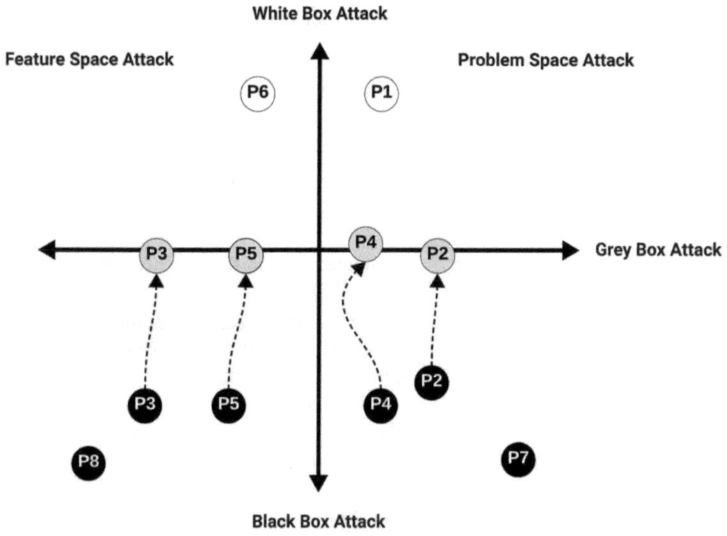

Fig. 7.2 Inconsistencies of the definitions of black-box, white-box, and grey-box attacks in the context of adversarial machine learning

problem space and feature space; the previous section examines the minimum knowledge required for each attack type.

7.5 Case Study of Malware Detection

We now classify the personas as per our taxonomy: (1) the attacker's knowledge level required, (2) the types of modifications utilized to bypass detection systems, and (3) the constraints used to preserve realism. These three essential components have to be carefully considered by an attacker to enable a successful and actionable adversarial attack against malware detection: Knowledge (K), Modifications (M), and Realism (R). To create such an attack, the attacker has to have knowledge about the target machine learning model, for instance, its structure, feature representation, and confidence scores. The modifications refer to the particular alteration to be applied to the malware samples, e.g., modifications to the API calls and injection of benign code to evade detection. At the same time, realism guarantees that the modified malware maintains its functionality, plausibility, and semantic coherence and successfully evades detection mechanisms. In Table 7.2 will complete the running example by showcasing the usage of the taxonomy on the personified papers.

As can be seen this allows a more fine grained classification of the problem. This is necessary to allow for, (1) finding gaps in the literature that requires further investigation, (2) understanding how and where defenses need to be placed

Table 7.2 Summary of adversarial attacks on malware detection

Paper	Knowledge (K)	Modifications (M)	Realism (R)
P1	The dataset, The classifier, The feature representation	Extract slices of byte code from benign donor apps	Semantic preservation, plausibility, preprocessing
P2	Labels and confidence scores	Changing manifest, injecting benign code	Semantic preservation, plausibility, preprocessing
P3	Labels and confidence scores	Manipulates static features of PDFs	Semantic preservation
P4	Confidence score	Modify API calls randomly or generative adversarial learning-based methods	Semantic preservation, plausibility
P5	Some prior knowledge about how ML-based malware detection works	Altering features that a machine learning model uses for detection	Semantic preservation, plausibility
P6	Classifier's architecture, weights, and decision rules, The feature representation	Adds benign-looking permissions or intents to deceive ML models	Semantic preservation
P7	Binary decision output (malware or benign)	Uses genetic programming (GP) to modify malware samples until they evade detection	Semantic preservation, plausibility
P8	Binary decision output (malware or benign)	The attack makes random modifications using a morpher	Semantic preservation, plausibility

within your system, and (3) discovering paper tigers, as attacks based on unrealistic assumptions are nor of practical relevance to real world scenarios.

7.6 Conclusion

In this paper, a taxonomy for adversarial attacks against machine learning-based malware detection systems has been proposed. Our framework argues that to properly classify these attacks you need to clarify three components, (1) Knowledge of the attacker, (2) Modifications made, and (3) Realism. To show this taxonomy in practice, eight personas have been defined that cover a variety of adversarial techniques. These personas were each based on currently published papers from top cybersecurity conferences, and yet, we show that some were incorrectly classified leading to confusions as to what they were doing. Our taxonomy provides a deeper understanding of these attack models beyond theoretical weaknesses and offers practical insights into real threats faced in practice. Results obtained from our study stress the need to understand attacks in a real-world setting, which forms a vital part in constructing effective defense schemes.

References

1. Goodfellow, I.J., Shlens, J., Szegedy, C.: Explaining and harnessing adversarial examples. In: Proceedings of the International Conference on Machine Learning (ICML) (2015) [Online]. Available https://arxiv.org/abs/1412.6572
2. Grosse, K., Papernot, N., Manoharan, P., Backes, M., McDaniel, P.: Adversarial perturbations against deep neural networks for malware classification. Preprint. arXiv:1606.04435 (2016)
3. Carlini, N., Wagner, D.: Towards evaluating the robustness of neural networks. In: 2017 IEEE Symposium on Security and Privacy (SP), pp. 39–57. IEEE, Piscataway (2017)
4. Biggio, B., Roli, F.: Wild patterns: ten years after the rise of adversarial machine learning. In: Proceedings of the 2018 ACM SIGSAC Conference on Computer and Communications Security, pp. 2154–2156 (2018)
5. Weilin, X., Yanjun, Q., Evans, D.: Automatically evading classifiers. In: Proceedings of the 2016 Network and Distributed Systems Symposium, vol. 10 (2016)
6. Xiao, C., Li, B., Zhu, J.-Y., He, W., Liu, M., Song, D.: Generating adversarial examples with adversarial networks. Preprint. arXiv:1801.02610 (2018)
7. Brendel, W., Rauber, J., Bethge, M.: Decision-based adversarial attacks: Reliable attacks against black-box machine learning models. Preprint. arXiv:1712.04248 (2017)
8. Aryal, K., Gupta, M., Abdelsalam, M., Kunwar, P., Thuraisingham, B.: A survey on adversarial attacks for malware analysis. IEEE Access **13**(2), 428–459 (2025)
9. Demetrio, L., Biggio, B., Lagorio, G., Roli, F., Armando, A.: Functionality-preserving black-box optimization of adversarial windows malware. IEEE Trans. Inf. Forens. Secur. **16**, 3469–3478 (2021)
10. Huang, L., Joseph, A.D., Nelson, B., Rubinstein, B.I., Tygar, J.D.: Adversarial machine learning. In: Proceedings of the 4th ACM Workshop on Security and Artificial Intelligence, pp. 43–58 (2011)
11. Yan, S., Ren, J., Wang, W., Sun, L., Zhang, W., Yu, Q.: A survey of adversarial attack and defense methods for malware classification in cyber security. IEEE Commun. Surv. Tutorials **25**(1), 467–496 (2022)
12. Anderson, H.S., Kharkar, A., Filar, B., Roth, P.: Evading machine learning malware detection. black Hat **2017**, 1–6 (2017)
13. Pierazzi, F., Pendlebury, F., Cortellazzi, J., Cavallaro, L.: Intriguing properties of adversarial ML attacks in the problem space. In: 2020 IEEE Symposium on Security and Privacy (SP), pp. 1332–1349. IEEE, Piscataway (2020)
14. Castro, R.L., Schmitt, C., Dreo, G.: Aimed: evolving malware with genetic programming to evade detection. In: 2019 18th IEEE International Conference on Trust, Security and Privacy in Computing and Communications/13th IEEE International Conference on Big Data Science and Engineering (TrustCom/BigDataSE), pp. 240–247. IEEE, Piscataway (2019)
15. Zhao, K., Zhou, H., Zhu, Y., Zhan, X., Zhou, K., Li, J., Yu, L., Yuan, W., Luo, X.: Structural attack against graph based android malware detection. In: Proceedings of the 2021 ACM SIGSAC Conference on Computer and Communications Security, pp. 3218–3235 (2021)
16. Song, W., Li, X., Afroz, S., Garg, D., Kuznetsov, D., Yin, H.: MAB-malware: a reinforcement learning framework for attacking static malware classifiers. Preprint. arXiv:2003.03100 (2020)
17. Li, H., Cheng, Z., Wu, B., Yuan, L., Gao, C., Yuan, W., Luo, X.: Black-box adversarial example attack towards {FCG} based android malware detection under incomplete feature information. In: 32nd USENIX Security Symposium (USENIX Security 23), pp. 1181–1198 (2023)

18. Lucas, K., Sharif, M., Bauer, L., Reiter, M.K., Shintre, S.: Malware makeover: Breaking ML-based static analysis by modifying executable bytes. In: Proceedings of the 2021 ACM Asia Conference on Computer and Communications Security, pp. 744–758 (2021)
19. Rosenberg, I., Shabtai, A., Elovici, Y., Rokach, L.: Query-efficient black-box attack against sequence-based malware classifiers. In: Proceedings of the 36th Annual Computer Security Applications Conference, pp. 611–626 (2020)
20. He, P., Xia, Y., Zhang, X., Ji, S.: Efficient query-based attack against ML-based android malware detection under zero knowledge setting. In: Proceedings of the 2023 ACM SIGSAC Conference on Computer and Communications Security, pp. 90–104 (2023)
21. Severi, G., Meyer, J., Coull, S., Oprea, A.: {Explanation-Guided} backdoor poisoning attacks against malware classifiers. In: 30th USENIX Security Symposium (USENIX security 21), pp. 1487–1504 (2021)
22. D'Onghia, M., Di Cesare, F., Gallo, L., Carminati, M., Polino, M., Zanero, S.: Lookin'out my backdoor! Investigating backdooring attacks against DL-driven malware detectors. In: Proceedings of the 16th ACM Workshop on Artificial Intelligence and Security, pp. 209–220 (2023)
23. Grosse, K., Papernot, N., Manoharan, P., Backes, M., McDaniel, P.: Adversarial examples for malware detection. In: Computer Security–ESORICS 2017: 22nd European Symposium on Research in Computer Security, Oslo, Norway, September 11–15, 2017, Proceedings, Part II 22, pp. 62–79. Springer, Cham (2017)
24. Smutz, C., Stavrou, A.: When a tree falls: using diversity in ensemble classifiers to identify evasion in malware detectors. In: NDSS (2016)
25. Gibert, D., Planes, J., Le, Q., Zizzo, G.: A wolf in sheep's clothing: Query-free evasion attacks against machine learning-based malware detectors with generative adversarial networks. In: 2023 IEEE European Symposium on Security and Privacy Workshops (EuroS&PW), pp. 415–426. IEEE, Piscataway (2023)
26. Dang, H., Huang, Y., Chang, E.-C.: Evading classifiers by morphing in the dark. In: Proceedings of the 2017 ACM SIGSAC Conference on Computer and Communications Security, pp. 119–133 (2017)
27. Yang, W., Kong, D., Xie, T., Gunter, C.A.: Malware detection in adversarial settings: exploiting feature evolutions and confusions in android apps. In: Proceedings of the 33rd Annual Computer Security Applications Conference, pp. 288–302 (2017)
28. Rosenberg, I., Shabtai, A., Rokach, L., Elovici, Y.: Generic black-box end-to-end attack against state of the art API call based malware classifiers. In: Research in Attacks, Intrusions, and Defenses: 21st International Symposium, RAID 2018, Heraklion, Crete, Greece, September 10–12 (2018), Proceedings 21, pp. 490–510. Springer, Cham (2018)
29. Fass, A., Backes, M., Stock, B.: HideNoSeek: camouflaging malicious JavaScript in benign ASTs. In: Proceedings of the 2019 ACM SIGSAC Conference on Computer and Communications Security, pp. 1899–1913 (2019)
30. Demetrio, L., Biggio, B., Lagorio, G., Armando, A., Roli, F.: Adversarial examples: functionality-preserving optimization of adversarial windows malware. In: ICML 2021 Workshop on Adversarial Machine Learning (2021)

Chapter 8
Adversarial Threats to Digital Twin Technology: A Taxonomy of Vulnerabilities and Attack Surfaces

Nasreen Anjum, Mahreen Kiran, Abu Alam, Javed Ali Khan, Katerina Kanta, Mo Adda, and Tamer Elboghdadly

8.1 Introduction

Digital Twin (DT) technology has emerged as one of the most transformative innovations of the twenty-first century, with applications spanning a multitude of domains. By creating virtual replicas of physical assets, systems, or processes, DTs enable real-time simulation, monitoring, and optimization [1, 2]. DT technology is revolutionising industries with its innovative and transformative technologies. For instance, in healthcare, DTs enable personalised diagnosis and treatment, improving outcomes [3]. Aerospace giants like Rolls-Royce use DTs to optimise engine performance, cutting maintenance costs and boosting safety [4]. In manufacturing, smart factories leverage DTs for real-time simulations, enhancing efficiency and reducing downtime, as seen in Siemens' advancements [5]. The energy sector uses DTs to optimise smart grids and renewable systems, with General Electric improving wind farm management [6].

Such widespread adoption and the immense value generated by DTs have made this technology a global trend. According to market studies, the DT market was

N. Anjum (✉) · A. Kanta · M. Adda · T. Elboghdadly
Faculty of Technology, School of Computing, University of Portsmouth, Portsmouth, UK
e-mail: nasreen.anjum@port.ac.uk; katerina.kanta1@port.ac.uk; mo.adda@port.ac.uk; tamer.elboghdadly@port.ac.uk

M. Kiran
Department of Health and Medicine, Anglia Ruskin University, Chelmsford, UK
e-mail: mk1476@pgr.aru.ac.uk

A. Alam
School of Computing & Engineering, University of Gloucestershire, Cheltenham, UK

J. A. Khan
Cybersecurity and Computing Research Group, Department of Computer Science, University of Hertfordshire, Hertfordshire, UK

E. Nowroozi et al. (eds.), *Adversarial Example Detection and Mitigation Using Machine Learning*, https://doi.org/10.1007/978-3-031-99447-0_8

valued at approximately USD 12.91 billion in 2023 and is projected to grow at a compound annual growth rate (CAGR) of 39.8%, reaching USD 259.32 billion by 2032 [7]. Industry analysts, including Gartner, have recognized the significance of DTs, ranking them among the top ten strategic technologies in recent years [8].

Despite the remarkable advancements and diverse applications of DT technology, its integration also presents significant cyber security challenges. DTs rely on a complex combination of emerging technologies, including Artificial Intelligence (AI), the Internet of Things (IoT), edge computing, cloud computing, big data analytics, and blockchain, to deliver their full potential. However, this interconnected ecosystem is inherently vulnerable to a range of cybersecurity threats. Among these, adversarial attacks have emerged as a critical concern, yet they are not explored adequately within the context of DTs [1, 2].

An adversarial attack on a DT is an intentional and malicious attempt to alter or disrupt its functioning and output. To execute such attacks, cybercriminals exploit vulnerabilities within the functional and operational layers of a DT by targeting its underlying core technologies, including AI, ML, communication technologies (e.g., 5G, WiFi), and IoT. For example, in a white-box adversarial attack, cybercriminals gain full access to the ML model, including its training dataset, features, model parameters, and computational processes. This deep level of control allows attackers to manipulate the outputs generated by DT processes and further develop strong adversarial examples [2]. Similarly, IoT, a core component of DT systems, is susceptible to various adversarial attacks, such as unauthorised access, tampering, and damage. These vulnerabilities further create entry points or backdoors for launching and propagating additional adversarial attacks within the DT ecosystem [9].

Although extensive work has been undertaken by the research community to develop and advance the DT technology [10]. However, little attention has been paid to explore the adversarial attacks and their impact on the integrity and reliability of DT processes and output. DT is a critical and emerging technology, it is essential to ensure its security, reliability, and sustainability. Research community should focus on understanding and exploring these security threads in DT. This chapter aims to bridge this gap. To the best of our knowledge, this is the first study that explores the adversarial attacks into the DT. To achieve this, we proposed a taxonomy based on functional and operational layer of the DT. Then, we explored and investigated vulnerabilities at each layer to identify adversarial attacks and cyber threats within each layer.

8.1.1 Organization of Chapter

This chapter is organised as follows: Sect. 8.2 provides an introduction to DT technology, covering its core components and functional layers. Section 8.3 discusses the proposed taxonomy of adversarial attacks on DT technology. Finally, Sect. 8.3 concludes the overall findings.

8.2 Background Introduction to Digital Twin Technology

A DT is a virtual representation of a physical object, system, or process, created using technologies such as sensors, IoT devices [11], AI, data analytics, and software tools. The concept traces its origins back to NASA's Apollo 13 mission in the 1970s, where engineers relied on a virtual model of the spacecraft to diagnose issues following an oxygen tank explosion. By simulating real-time conditions, they were able to devise a strategy to bring the astronauts safely back to Earth. The modern definition of DT technology was later formalized by Dr. Michael Grieves in 2002 during a presentation on Product Lifecycle Management (PLM) at the University of Michigan [12, 13].

8.2.1 Core Components of DT Technology [12]

DT technology is comprised of the following core components:

- **Physical Entity**: A real-world object, system, or process (e.g., machines, biological systems, industrial systems) replicated to create its DT.
- **Sensors and Actuators**: Sensors collect real-time data (e.g., temperature, pressure), while actuators enable feedback from the DT to control the physical entity (e.g., shutting down overheating machines).
- **Communication Networks**: Networks like 5G, LPWAN, and Wi-Fi facilitate real-time data transfer between the physical entity and its DT.
- **Data Acquisition, Storage and Management System**: Systems (e.g., cloud networks, edge computing, distributed databases) store and preprocess collected data for analysis.
- **AI and ML**: AI and ML processes data to predict outcomes, identify patterns, and recommend actions.
- **Visualisation and Interface Tools:** These tools, including dashboards, augmented reality (AR), and virtual reality (VR) allow users to monitor, analyze, and interact with the DT for informed decision-making.

8.3 Taxonomy of Adversarial Attacks on DT Technology

DT technology operates across multiple interconnected layers, integrating advanced technologies such as **IoT, AI, cloud computing, and communication networks** to enable real-time monitoring, simulation, and decision-making. However, each layer presents potential **vulnerabilities** that adversaries can exploit, posing risks to the **integrity, reliability, and security** of DT systems. A cyberattack on any layer could result in **data corruption, misinterpretation of results, and flawed decision-making**, ultimately affecting the physical system it represents [2].

Table 8.1 Vulnerabilities and attacks on each operational layer of DT technology

Operational layer	Vulnerabilities	Adversarial attacks
Physical layer	Weak authentication, unsecured IoT devices, hardware manipulation, firmware vulnerabilities	Sensor spoofing, physical tampering, device impersonation, IoT hijacking, signal interception, environmental interference, firmware tampering
Communication layer	Insecure handshake protocols, weak encryption, poor authentication, unprotected transmission	Eavesdropping, man-in-the-middle (MitM), signal jamming, replay attacks, protocol exploitation, DoS/DDoS attacks, unauthorized data interception
Data layer	Unprotected data ingestion, weak access controls, insecure storage, lack of anomaly detection	Data poisoning, unauthorized data access, ransomware, SQL injection, data manipulation, data exfiltration, side-channel attacks
Analytics and intelligence layer	AI/ML susceptibility to adversarial inputs, model vulnerabilities, weak anomaly detection	Data poisoning, adversarial example attacks, model inversion, model stealing, backdoor attacks, anomaly injection, targeted/untargeted attacks
Interface layer	Weak authentication, input validation flaws, unprotected dashboards and AR/VR interfaces	Interface manipulation, phishing, DoS attacks, command injection, privilege escalation, misinformation attacks, AR/VR exploitation
Feedback loop layer	Weak encryption in control signals, lack of sequence validation, unsecured actuators	Command injection, feedback misinformation, signal interception, privilege escalation, actuator hijacking, denial-of-service (DoS) attacks

To systematically address these **security concerns**, our proposed **taxonomy** examines vulnerabilities within each operational layer of the **DT framework**, identifying potential **adversarial threats** that could compromise its functionality. By categorising these risks, the taxonomy provides a **structured approach** to understanding **attack surfaces** and the cascading effects of security breaches across layers. Table 8.1 summarises key aspects of the DT, including its overview, vulnerabilities, and potential adversarial attacks. In the following sections, we provide a detailed examination of each layer, discussing its function, vulnerabilities, and the adversarial attacks that can be launched against it.

8.3.1 Physical Layer

The physical layer involves the collection of real-time data along with the execution of feedback actions. Sensors collect the real-time data, such as temperature and pressure, and transmit it to the gateways or edge servers for further processing and analysis. While actuators, such as robotic arms and motors, execute the feedback received from the digital counterpart of the DT. Various communication technologies, including 5G, Wi-Fi, LPWAN, and Zigbee modules, connect these components with the IoT network. 5G enables high-speed and efficient communication, Zigbee module facilitates energy-efficient, short-distance connectivity, suitable for smart industrial automation and smart city infrastructure; and Wi-Fi provides reliable, high-speed communication for local data transfer.

Vulnerabilities and Adversarial Attacks on Physical Layer

Adversarial attacks on the physical layer of DT technology exploit vulnerabilities in the hardware devices, sensors, actuators, and supporting technologies that connect the physical system to the DT. Below are some types of adversarial attacks specific to the physical layer:

- **Sensor spoofing attacks**: These attacks inject false or misleading data into the DT through spoofing or tampering with the sensors [14]. To execute these attacks, adversaries exploit the vulnerabilities of sensors such as weak authentication, passwords, and unencrypted communication. As a result, this attack distorts DT operations, leading to inaccurate simulations or predictions.
- **Hardware Compromise Attack**: The physical components of the physical layer, including their embedded firmware, are equally at risk. Firmware in sensors and actuators is responsible for various tasks, including data collection and transmission. However, these critical components are vulnerable to various cyber attack:
  - **Firmware Tampering**: This attack involves modifying the firmware to alter the device's behavior or introduce malicious functionality [15]. An attacker could inject malicious code into the firmware of a sensor, causing it to alter data collection, thus sabotaging the DT's processes.
  - **Backdoor Installation**: A malicious backdoor can be inserted into the firmware, allowing attackers to gain unauthorized access to the device at any time. If a backdoor is installed in the firmware of medical sensor equipment, an attacker could remotely access and manipulate diagnostic tools, potentially leading to misdiagnoses [16].
  - **Firmware Downgrade**: In this attack, devices are forced to downgrade to older and less secure versions of firmware that contain known vulnerabilities [17]. An attacker might force a lab instrument to revert to outdated firmware that has unpatched security flaws, which could then be exploited to gain control over the instrument.
  - **Permanent Denial of Service (PDoS)**: known as "bricking," this attack renders a device permanently inoperable by corrupting its firmware [18]. Deliberate corruption of the firmware on critical laboratory equipment could halt research and cause significant financial loss.
  - **Supply Chain Attacks:** Compromising firmware before it even reaches the user, typically during manufacturing or distribution. Malicious actors could

tamper with firmware at the source, inserting vulnerabilities or malware before the devices are even shipped to users [19].

- **Device impersonation and injection Attacks**: To carry out this attack, attackers introduce unauthorized devices into an IoT network. Hackers hack into the device via injecting malware or exploiting weak authentication protocols [20]. This allow attackers to feed malicious data into the DT system, ultimately leading to inaccurate assessments and flawed decision-making.
- **IoT Device Hijacking**: Cybercriminals can gain control over IoT devices by exploiting security loopholes such as easily guessable passwords, outdated or vulnerable firmware, and improperly secured network ports. Once penetrated, these hijacked devices can be used to interrupt the data collection process or carry out malicious operations, putting the DT's reliability at risk [21].
- **Signal Interception and Jamming**: In these attacks, adversaries interfere with the communication channels that connect physical and digital counterparts. The main goal is to halt the transmission or distort the signals, which eventually leads to infiltrate the DT with outdated information [22]. This causes to affect DT's performance and decision-making capabilities.

8.3.2 Communication Layer (CL)

The CL enables the communication between the physical layer and its virtual replica. This layer is also known as the backbone of the DT. Therefore, it relies on many advanced communication protocols such as TCP/IP, MQTT, and CoAP, as well as communication technologies like Wi-Fi, Bluetooth, and Zigbee.

Vulnerabilities and Adversarial Attacks on Communication Layer

The CL is vulnerable to various cybersecurity threats due to its reliance on communication technologies and protocols that can jeopardize both the integrity and confidentiality of sensitive data.

- **Eavesdropping and Data Interception:** Attackers exploit weak encryption protocols and insecure authentication mechanisms to intercept communication between the physical and its virtual replica. This allows attackers to eavesdrop on or modify the exchanged data, potentially leading to unauthorised access and data breaches [23].
- **Man-in-the-Middle (MitM) attacks**: These attacks exploit security gaps in handshake protocols, weak authentication, and unencrypted communication channels to intercept and modify transmitted data. By manipulating information, attackers can mislead the DT, leading to faulty predictions and incorrect decisions.

- **DoS Attack:** DoS attacks involve flooding servers or network infrastructure with an overwhelming amount of illegitimate traffic or requests. This is done to exhaust resources and bandwidth, thereby rendering the services inaccessible to legitimate users [24]. The primary impact of a DoS attack on DT processes includes delays in processes, and potential data loss.
- **Distributed Denial-of-Service (DDoS) Attack:** A DDoS attack is a more severe form of DoS that utilizes a network of compromised computers (a botnet) to launch a massive flood of traffic towards the target server or network [25]. In the context of DT, since DTs rely on continuous data exchange between physical and virtual components, a DDoS attack can overload the network and computational resources, leading to system failure or degraded performance.
- **Signal Jamming Attacks**: These attacks disrupt or block data exchange between physical entity and its virtual replica via flooding the network with excessive noise signals or interference. By blocking data transmission, jamming attacks delay or prevent real-time updates.
- **Replay Attacks**: Replay attacks involve capturing legitimate data packets during transmission and resending them later to disrupt operations or manipulate system behaviour. Exploiting communication protocols lacking security features such as timestamping and sequence validation, attackers inject outdated or unauthorised data into the DT, leading to flawed predictions and system disruptions.

8.3.3 Data Layer (DL)

The Data Layer in DT technology is responsible for collecting, storing, processing, and analyzing data from physical systems, ensuring accurate simulations and decision-making. It aggregates real-time data from IoT sensors, actuators, and gateways, processes it through frameworks like Apache Kafka and Spark, and stores it in cloud databases, edge devices, or data lakes. Secure data transfer is facilitated by protocols such as MQTT, HTTP/HTTPS, and OPC UA, while storage systems like relational (PostgreSQL) and NoSQL (MongoDB) databases manage structured and unstructured data. This layer ensures seamless interoperability, scalability for handling vast data volumes, and high availability for real-time applications [2].

Vulnerabilities and Adversarial Attacks on Data Layer

The Data Layer plays a crucial role in aggregating, processing, and storing information from various sources, making it a prime target for attacks aimed at compromising data integrity, confidentiality, and availability.

- **Data Poisoning Attacks**: This attack occurs when adversaries inject falsified or misleading data during collection or preprocessing. Exploiting weak validation mechanisms and unsecured data ingestion points, attackers introduce false information that corrupts ML models and analytics [26]. For example, in a smart manufacturing DT, an attacker could alter sensor data to trigger unnecessary maintenance procedures, disrupting production schedules.

- **Data Manipulation Attacks**: This attack occurs when attackers alter stored data to mislead DT operations. This type of attack targets insecure databases or storage systems that lack cryptographic integrity checks. By modifying historical records or operational metrics, attackers can skew analytics and cause incorrect system behavior [2].
- **Unauthorized Data Access**: Attackers exploit weak access controls or poorly configured databases to gain unauthorized access to sensitive data. This can result in breaches of confidentiality, theft of intellectual property, or further exploitation. In a healthcare DT, for example, unauthorized access to patient records could lead to privacy violations and regulatory penalties.
- **Ransomware Attacks:** In ransomware attacks, attackers encrypt or lock critical data in the DT's storage systems and demand payment for restoration. Such attacks exploit inadequate backup systems and endpoint security. For example, encrypting operational data in an industrial DT could halt production until the ransom is paid, causing financial and reputational damage [27].
- **Data Exfiltration:** Data exfiltration involves the unauthorized extraction of sensitive data from the DT's storage systems. Attackers exploit unencrypted storage or transmission channels, or insufficient network monitoring, to steal proprietary information. In an industrial DT, stolen operational data could provide competitors with critical insights, undermining a company's market position.
- **Data Injection Attacks:** Data injection attacks occur when adversaries insert malicious inputs or commands into the data streams. These attacks exploit inadequate input sanitization and validation mechanisms. Injected data can lead to incorrect analytics or decision-making. For example, falsifying vibration data in a turbine DT could trigger unnecessary shutdowns or maintenance actions.
- **SQL Injection Attacks:** SQL injection is a technique where attackers inject malicious SQL queries into database systems to manipulate or extract data. This attack targets insecure query handling and input validation. For instance, a SQL injection in a smart city DT could disable traffic management systems or expose sensitive urban planning data [28].
- **Side-Channel Attacks:** Side-channel attacks exploit indirect information leakage from data processing systems, such as timing, power usage, or electromagnetic emissions, to infer sensitive data. For example, an attacker could deduce encryption keys used in a financial DT system, leading to further exploitation [29].
- **Data Availability Attacks:** Data availability attacks overload storage systems or disrupt data retrieval processes, denying access to critical information. These attacks exploit limited storage resources or redundancy in data management systems. For instance, overwhelming a smart grid DT's storage system could delay real-time energy distribution adjustments, causing inefficiencies or outages

8.3.4 *Analytics and Intelligence Layer (AIL)*

Serving as the intelligence hub of DT technology, AIL leverages AI and ML algorithms to transform raw and structured data into valuable insights. It facilitates real-time analytics, predictive modeling, and scenario-based decision making, enabling operators to anticipate and respond to potential system changes efficiently.

Vulnerabilities and Adversarial Attacks on Analytics and Intelligence Layer

This layer is a key target by the cybercriminals for adversarial attacks due to its strong reliance on the ML and AL models. The attackers utilise various tricks to sabotage the true functionalities of Al and ML models, leading to incorrect predictions [30, 31].

- **Data poisoning attacks**: These attacks compromise the training phase of AI and ML models. Attackers execute this attack by injecting falsified data into datasets. This manipulation degrades the accuracy of AI models, leading to unreliable future predictions [32].
- **Adversarial Example Attacks**: These attacks exploit the susceptibility of ML algorithms to small, targeted modifications. Upon completion of training, such poisoned ML models produce flawed conclusions and decisions based on erroneous information [33]. For example, in a DT-enabled autonomous vehicle, an adversarially altered image of a stop sign might be misclassified as a speed limit sign, leading to dangerous driving decisions.
- **White Box Attack**: In this type of attack, cybercriminals gain full access to the ML model including its training dataset, features, ML model, and computational measures. This strong control allows the attacker to manipulate the output generated by the DT processes and design a strong adversary example attack [30, 31].
- **Black Box Attack**: In contrast to white box attacks, where attackers possess extensive knowledge about the ML model and internal data, attackers in black box attacks rely entirely on the output generated by the ML model. Then attackers infer the ML behaviours based on the input queries [30, 31].
- **Model Stealing Attacks**: In contrast to a white-box attack where the attacker has full knowledge about the ML model. The attacker in this type of attack attempts to gain knowledge about the targetted ML model without accessing it directly. Instead, attackers utilise extensive queries and use the responses to reconstruct a similar model.
- **Model Inversion Attacks**: In model inversion attacks, attackers infect the dataset with false information via manipulating it without interfering with the ML model. This unauthorised manipulation causes the ML model to generate incorrect predictions. Further, attackers can also retrieve sensitive information. For instance patience's health history record [30, 31].

- **Targeted Attacks**: In this type of attack, the attacker controls the output generated by the ML model by manipulating the classification model. For example, cybercriminals target the specific image of a cat so that ML models misclassify it as a dog [30, 31].
- **Untargeted Attacks**: in contrast to targeted attacks, in untargetted attacks, attackers do not control the output of the ML model. Instead, attackers make sure that the ML model performs incorrect classification and generate incorrect results [30, 31].
- **Anomaly Injection Attacks**: These attacks involve feeding artificial anomalies into DT data streams, confusing AI-driven monitoring, and triggering unnecessary interventions. A smart city DT, for example, could be deceived by false temperature spikes, shutting down cooling systems unnecessarily. Such attacks exploit ineffective validation measures, diverting resources to handling phantom issues [34].

8.3.5 Interface Layer (IL)

The Interface Layer acts as the bridge between users and the Digital Twin, enabling interaction through dashboards, AR/VR systems, and user interfaces. It translates complex data into actionable insights and allows operators to monitor, analyze, and control DT operations effectively.

Vulnerabilities and Adversarial Attacks on Analytics and Intelligence Layer

1. **Interface Manipulation Attacks**: The main objective of this attack is to mislead the users via tempering and altering the visual outputs on dashboards or user interfaces. For instance, modifying an industrial DT's user interface to trick the user into misinterpreting a normal operation into a faulty operation could delay the critical tasks [35].
2. **Phishing and Credential Theft**: In these attacks, cybercriminals may impersonate legitimate DT tools, creating deceptive websites or emails that mimic authentic platforms. Unsuspecting users may be tricked into providing their login credentials or downloading malicious software, compromising the security of their accounts or systems. Once obtained, these credentials can be exploited to gain unauthorised access to critical private data or to launch further cyberattacks within DT system [36].
3. **DoS Attacks**: This attack makes the user interfaces unavailable to the authentic users via flooding the IL with unlimited requests.
4. **Command Injection Attacks**: Attackers insert malicious scripts or commands into APIs that interact with databases or backend systems. This can result in unauthorised access to data, data corruption, or even full system control [37].

If not properly secured, these APIs can serve as gateways for further attacks, enabling unauthorized manipulation and access to services.

5. **Data Misinformation Attacks**: These attacks manipulate the interface layer by injecting false data, leading to misleading alerts. These attacks exploit weak anomaly detection and the lack of cross-layer validation in data streams. For example, in a smart city DT, falsified air quality readings could trigger unnecessary evacuations or improper policy decisions [38].
6. **Privilege Escalation Attacks**: Privilege escalation attacks exploit weaknesses in access control systems. This enables attackers to obtain administrative privileges and have control on restricted interface functions [39].
7. **AR and VR Exploitation**: Attackers exploit the vulnerabilities in AR/VR and weak authentication mechanisms to take control over the user interfaces. For example, in a smart factory DT, an attacker could alter VR simulations to hide equipment failures, preventing timely maintenance and increasing operational risks [40, 41].

8.3.6 Feedback Loop Layer (FLL)

The FLL facilitates bidirectional communication between physical and digital entities, enabling real-time updates, synchronization, and optimization. It integrates IoT sensors, actuators, communication protocols (e.g., MQTT, OPC UA), edge computing, and AI/ML algorithms for data collection, analysis, and control.

Adversarial Attacks on the Feedback Loop Layer

1. **Command Injection Attacks**: Exploiting insecure APIs or input validation, attackers inject malicious commands, causing unauthorized or harmful actions in the physical system.
2. **Data Manipulation Attacks**: Attackers alter real-time data to mislead the DT, leading to incorrect decisions. For example, falsifying temperature data could trigger unnecessary cooling processes, increasing energy consumption.
3. **Signal Interception and Replay Attacks**: Attackers capture and replay valid signals, disrupting system operations by introducing outdated or repeated commands.
4. **DoS Attacks**: This attack involves overwhelming the feedback loop with countless queries and injection of heavy data, thereby causing the disruption of feedback loop and real-time operations.
5. **Sensor and Actuator Hijacking**: This attack exploits weak authentication protocols or outdated firmware to temper the IoT devices, thereby disrupting the data collection mechanism during the feedback loop operations.

6. **Feedback Misinformation Attacks**: In this attack, attackers inject falsified feedback output and signals into the feedback loop, thereby causing to mislead the physical system of DT system.
7. **Privilege Escalation in Control Systems:** Attackers misuse control rights via unauthorized access to the feedback loop system and introduce system changes that may sabotage the overall functionality of the DT system.

8.4 Conclusion

In this study, we provide a structured approach to understanding adversarial threats in DT systems by categorizing vulnerabilities and attack vectors across multiple operational layers. As DT technology continues to evolve, implementing robust security measures such as encryption, anomaly detection, and authentication protocols is crucial to mitigating these risks and ensuring system integrity and resilience.

References

1. Wu, Y., Zhang, K., Zhang, Y.: Digital twin networks: a survey. IEEE Internet Things J. **8**(18), 13789–13804 (2021)
2. Alcaraz, C., Lopez, J.: Digital twin: a comprehensive survey of security threats. IEEE Commun. Surv. Tutorials **24**(3), 1475–1503 (2022)
3. Erol, T., Mendi, A.F., Doğan, D.: The digital twin revolution in healthcare. In 2020 4th International Symposium on Multidisciplinary Studies and Innovative Technologies (ISMSIT), pp. 1–7. IEEE, Piscataway (2020)
4. Li, L., Aslam, S., Wileman, A., Perinpanayagam, S.: Digital twin in aerospace industry: a gentle introduction. IEEE Access **10**, 9543–9562 (2021)
5. Kritzinger, W., Karner, M., Traar, G., Henjes, J., Sihn, W.: Digital twin in manufacturing: a categorical literature review and classification. Ifac-PapersOnline **51**(11), 1016–1022 (2018)
6. Ghenai, C., Husein, L.A., Al Nahlawi, M., Hamid, A.K., Bettayeb, M.: Recent trends of digital twin technologies in the energy sector: a comprehensive review. Sustain. Energy Technol. Assess. **54**, 102837 (2022)
7. Fortune Business Insights: Latest digital twin market trends (2024). Accessed 22 Dec 2024
8. Panetta, K.: Gartner top 10 strategic technology trends for 2019 (2025). Accessed 22 Dec 2024
9. Qiu, H., Dong, T., Zhang, T., Lu, J., Memmi, G., Qiu, M.: Adversarial attacks against network intrusion detection in IoT systems. IEEE Internet Things J. **8**(13), 10327–10335 (2020)
10. Attaran, M., Celik, B.G.: Digital twin: benefits, use cases, challenges, and opportunities. Decis. Anal. J. **6**, 100165 (2023)
11. Imran, K., Anjum, N., Alghamdi, A., Shaikh, A., Hamdi, M., Mahfooz, S.: A secure and efficient cluster-based authentication scheme for Internet of Things (IoTs). Comput. Mater. Continua **70**(1), 1033–1052 (2022)
12. Liu, M., Fang, S., Dong, H., Xu, C.: Review of digital twin about concepts, technologies, and industrial applications. J. Manuf. Syst. **58**, 346–361 (2021)
13. Lo, C.K., Chen, C.H., Zhong, R.Y.: A review of digital twin in product design and development. Adv. Eng. Inform. **48**, 101297 (2021)

14. Westerhof, W.: Pentester's view on "digital twins technology" (2024). Accessed 22 Dec 2024
15. Zhang, Y., Li, Y., Li, Z.: Aye: a trusted forensic method for firmware tampering attacks. Symmetry **15**(1), 145 (2023)
16. backdoor: FDA, CISA warn of backdoor in popular patient monitor used by US hospitals (2024). Accessed 22 Dec 2024
17. Wu, Y., Wang, J., Wang, Y., Zhai, S., Li, Z., He, Y., Sun, K., Li, Q., Zhang, N.: Your firmware has arrived: a study of firmware update vulnerabilities. In: USENIX Security Symposium (2023)
18. Abaimov, S.: Understanding and classifying permanent denial-of-service attacks. J. Cybersecur. Priv. **4**(2), 324–339 (2024)
19. Khadam, N., Anjum, N., Alam, A., Mirza, Q.A., Assam, M., Ismail, E.A.A., Abonazel, M.R.: How to punish cyber criminals: a study to investigate the target and consequence based punishments for malware attacks in UK, USA, China, Ethiopia & Pakistan. Heliyon **9**(12), e22823 (2023)
20. Gaber, T., El-Ghamry, A., Hassanien, A.E.: Injection attack detection using machine learning for smart IoT applications. Phys. Commun. **52**, 101685 (2022)
21. Yaacoub, J.P.A., Noura, H.N., Salman, O., Chehab, A.: Ethical hacking for IoT: security issues, challenges, solutions and recommendations. Internet Things Cyber-Phys. Syst. **3**, 280–308 (2023)
22. Sciancalepore, S., Oligeri, G., Di Pietro, R.: Strength of crowd (SoC)–defeating a reactive jammer in IoT with decoy messages. Sensors **18**(10), 3492 (2018)
23. He, C., Luan, T.H., Lu, R., Su, Z., Dong, M.: Security and privacy in vehicular digital twin networks: challenges and solutions. IEEE Wirel. Commun. **30**(4), 154–160 (2022)
24. de Neira, A.B., Kantarci, B., Nogueira, M.: Distributed denial of service attack prediction: challenges, open issues and opportunities. Comput. Netw. **222**, 109553 (2023)
25. Shukla, P., Rama Krishna, C., Patil, N.V.: IoT traffic-based DDoS attacks detection mechanisms: A comprehensive review. J. Supercomput. **80**, 1–58 (2023)
26. Zhang, Z., Fang, M., Chen, M., Li, G., Lin, X., Liu, Y.: Securing distributed network digital twin systems against model poisoning attacks. IEEE Internet Things **11**(21), 34312–34324 (2024)
27. Nam, S.M., Lee, S.M., Park, Y.S.: Development of information security practice contents for ransomware attacks in digital twin-based smart factories. J. Korea Inst. Inf. Secur. Cryptol. **31**(5), 1001–1010 (2021)
28. Shaikh, E., Al-Ali, A.R., Muhammad, S., Mohammad, N., Aloul, F.: Security analysis of a digital twin framework using probabilistic model checking. IEEE Access **11**, 26358–26374 (2023)
29. Benelhaouare, A.Z., Mellal, I., Oumlaz, M., Lakhssassi, A.: Mitigating thermal side-channel vulnerabilities in FPGA-based sip systems through advanced thermal management and security integration using thermal digital twin (TDT) technology. Electronics **13**(21), 4176 (2024)
30. Ozdag, M.: Adversarial attacks and defenses against deep neural networks: a survey. Procedia Comput. Sci. **140**, 152–161 (2018)
31. Deng, Y., Zheng, X., Zhang, T., Chen, C., Lou, G., Kim, M.: An analysis of adversarial attacks and defenses on autonomous driving models. In 2020 IEEE International Conference on Pervasive Computing and Communications (PerCom), pp. 1–10. IEEE, Piscataway (2020)
32. Tolpegin, V., Truex, S., Gursoy, M.E., Liu, L.: Data poisoning attacks against federated learning systems. In: Computer Security–ESORICs 2020: 25th European Symposium on Research in Computer Security, ESORICs 2020, Guildford, UK, September 14–18, 2020, Proceedings, part I 25, pp. 480–501. Springer, Cham (2020)
33. Yuan, X., He, P., Zhu, Q., Li, X.: Adversarial examples: attacks and defenses for deep learning. IEEE Trans. Neural Netw. Learn. Syst. **30**(9), 2805–2824 (2019)
34. Yan, Y., Kunhui, Y.: Novel cyber-physical architecture for optimal operation of renewable-based smart city considering false data injection attacks: digital twin technologies for smart city infrastructure management. Sustain. Energy Technol. Assess. **65**, 103733 (2024)

35. Jeremiah, S.R., El Azzaoui, A., Xiong, N.N., Park, J.H.: A comprehensive survey of digital twins: applications, technologies and security challenges. J. Syst. Architect. **151**, 103120 (2024)
36. Alkhalil, Z., Hewage, C., Nawaf, L., Khan, I.: Phishing attacks: a recent comprehensive study and a new anatomy. Front. Comput. Sci. **3**, 563060 (2021)
37. Ariffin, M.A.M., Ibrahim, M.F., Kasiran, Z.: API vulnerabilities in cloud computing platform: attack and detection. Int. J. Eng. Trends Technol. **1**, 8–14 (2020)
38. Suhail, S., Jurdak, R., Hussain, R.: Security attacks and solutions for digital twins. Preprint. arXiv:2202.12501 (2022)
39. Mehmood, M., Amin, R., Muslam, M.M.A., Xie, J., Aldabbas, H.: Privilege escalation attack detection and mitigation in cloud using machine learning. IEEE Access **11**, 46561–46576 (2023)
40. Lechachenko, T., Kozak, R., Skorenkyy, Y., Kramar, O., Karelina, O.: Cybersecurity aspects of smart manufacturing transition to industry 5.0 model. In: ITTAP, pp. 416–424 (2023)
41. Imran, K., Anjum, N.: Cluster-based group mobility support for smart IoT. Comput. Mater. Continua **68**(2), 491–507 (2021)

Chapter 9
Quantum Adversarial Artificial Intelligence in Secure Internet of Things Networks

Ilgın Şafak, Tapio Frantti, Fatih Alagöz, Tommi Mikkonen, and Majid Haghparast

9.1 Introduction

The attack surface and the possibility of a cyber-attacks have increased with the explosion of the Internet of Things (IoT). Adversarial attacks target IoT devices by changing inputs, outputs, or inner parameters for artificial intelligence (AI) systems to cause misclassifications, degrade performance, or extract information from these connected networks. Well-known examples include evasion, poisoning, and inference attacks that help to expose weaknesses in the AI models and get incorrect or undesired outputs. Such attacks may result in unauthorized access, data leaks, or service downtime.

IoT networks are heterogeneous by nature as they contain a multitude of devices, communications, and security protocols, which makes it even harder to implement standardized security approaches. Adversarial AI is both the creation of adversarial attacks and the implementation of counter-adversarial attacks. Adversarial AI focuses on discovering weaknesses in AI systems, creating security policies to prevent attacks, and solidifying models with the help of adversarial training,

I. Şafak (✉)
Fibabanka R&D Center, Istanbul, Turkey

University of Jyväskylä, IT Faculty, Jyväskylä, Finland
e-mail: ilgin.safak@fibabanka.com.tr; ilgin.i.safak@jyu.fi

T. Frantti · T. Mikkonen · M. Haghparast
University of Jyväskylä, IT Faculty, Jyväskylä, Finland
e-mail: tapio.k.frantti@jyu.fi; tommi.j.mikkonen@jyu.fi; majid.m.haghparast@jyu.fi

F. Alagöz
Computer Engineering Department, Bogazici University, Istanbul, Turkey
e-mail: fatih.alagoz@boun.edu.tr

E. Nowroozi et al. (eds.), *Adversarial Example Detection and Mitigation Using Machine Learning*, https://doi.org/10.1007/978-3-031-99447-0_9

differential privacy, and robust optimization. These are tactics within strategic adversarial AI, addressing offensive and defensive approaches to protect connected technologies from significant threats, such as potential device failures, service interruptions, and data breaches within interconnected networks, such as IoT [8]. Adversarial AI is gaining traction, as AI systems are built increasingly into critical infrastructure and world-facing applications. Adversarial AI is especially relevant in cybersecurity because it can circumvent conventional security mechanisms, resulting in unauthorized access, data breaches, and other forms of malicious conduct. This is what makes adversarial AI a dangerous attack, since they can change and evolve, making it necessary to define defenses against these types of attacks. Adversarial AI is the deliberate and minute perturbation of input data to manipulate an AI model. This may be the incorporation of noise at an undetectable level, resulting in mispredictions or misclassifications by the model. In the machine learning (ML) context, a model is a mathematical framework or program that evaluates the incoming data and generates outputs, predictions, or classes that it has learned during the training phase. Adversarial attacks are a particular class of attack in ML models that use IoT networks for different security functions such as intrusion detection systems, malware detection systems, and device identification systems [5].

The rise of quantum computing is expected to worsen the cybersecurity threat landscape of IoT networks. Quantum computing can enhance the effectiveness of adversarial AI attacks by utilizing quantum physics. Qubits are the primary units of quantum information and can be utilized for executing intricate computations. Unlike classical bits (which can be either 0 or 1 in traditional computing), the qubit exists in a superposition of states, which allows them to be both 0 and 1 at the same time. Typically, qubits are realized via subatomic particles, like electrons or photons, that carry quantum states with features such as spin or polarization. The power of qubits comes from the quantum phenomena of superposition and entanglement. Quantum computing allows you to perform complex calculations far beyond those of classical computers, such as in cryptography, optimization, pharmaceutical discovery, and complex simulations. The use of quantum computing speeds up the calculations from 2^n to $2^{\sqrt{n}}$, compared to classical computing, where number of bits involved in the computation. Quantum algorithms such as Shor's method for integer factorization and Grover's algorithm for searching unsorted databases, leverage quantum physics to obtain exponential computational speedups. Quantum computers can break encryption methods used in contemporary communication and increase the efficiency of complex operations using quantum algorithms. Quantum calculation is performed using quantum gates, which change the states of quantum bits through operations based on the principles of superposition and entanglement. Quantum computers perform calculations via quantum gates, which manipulate qubits through operations. Despite its potential, quantum computing has serious challenges, which include maintaining qubit coherence, executing error correction, and scaling the number of qubits. Research efforts can make usable quantum computing a reality in the near future, as most quantum computers available today are largely experimental [7].

9.2 Quantum Adversarial AI

9.2.1 Quantum Adversarial Attack Methods

Quantum adversarial attacks can be categorized as causative, exploratory, and inference (see Fig. 9.1):

Causative Attacks

- Quantum Data Poisoning: Intentionally inserting misleading or fake data in the training data during the quantum AI (QAI) model learning process, leading to misclassification or misprediction [6].
- Quantum State Perturbation (QSP): The modification of quantum states of a quantum system. It may make QAI models dysfunctional and yield a long-term deterioration of the performance of the model [3].
- Quantum Backdoor Attacks: Implanting secretive triggers into QAI models that make the system behave maliciously, despite appearing normal [6].

Exploratory Attacks

- Quantum Model Extraction implies replicating a QAI model using quantum algorithms. The attacker poses queries to the model, then uses the responses to replicate its parameters and structure [6].
- Quantum State Perturbation misleads the model by perturbing quantum states during inference without changing its fundamental structure. For example, the Quantum Fast Gradient Sign Method (QFGSM) [2] modifies quantum input states by utilizing gradient information from the quantum classifier to generate

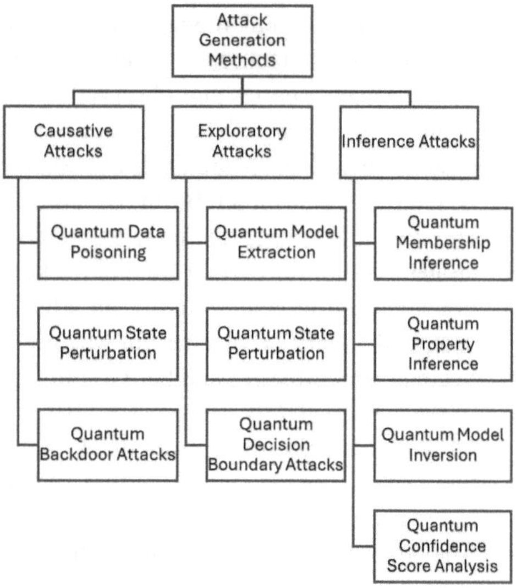

Fig. 9.1 Quantum adversarial attack methods

adversarial quantum states to optimize misclassification. It calculates the gradient of the loss function concerning the quantum state and introduces a minor perturbation in the direction that amplifies the classification error, rendering it the quantum equivalent of the classical FGSM attack.
- Quantum Decision Boundary Attacks aim to modify input data to take advantage of QAI models' decision boundaries, to push inputs over these boundaries and lead to incorrect input classification [6].

Inference Attacks
- Quantum Membership Inference: When an attacker can determine whether an attacker's data points belong to the training dataset of a QAI model, thus compromising the data confidentiality [1].
- Quantum Property Inference: When an attacker can infer features of the training data used in a QAI model, in turn potentially leaking sensitive information regarding the dataset [4].
- Quantum Model Inversion: Reconstructs sensitive attributes of the training data using correlations between the input data and the outputs of the model [1].
- Quantum Confidence Score Analysis: When an adversary uses the confidence scores of the predictions made by a QAI model to find and exploit vulnerabilities of the model, which may lead to stronger adversarial attacks [4].

Attacks can also be categorized depending on the access level of the attacker on the target model as follows:

- Black-box attacks: When an adversary has access only to the input and output of the model they want to attack and no information about the model itself, its architecture, parameters, or weights. The attacker, with no access to the exact details of the target model, can only learn about the model through carefully designed inputs and model outputs. These attacks are expensive to compute, requiring the attacker to reverse-engineer the behavior of the target model through many queries and observations [10].
- White-box attacks: When an attacker has complete details about the target system, including its architecture, source code, and internal parameters. White-box attacks can either be indiscriminate or targeted. Indiscriminate attackers aim to indiscriminately misclassify data by the classifier. The attacker may strategically train certain quantum classifiers based on a dataset to maximize efficiency. In targeted attacks, an attacker is interested in misclassifying some sample data labeled in a certain way preferred by the attacker. White-box attacks become very effective and hard to mitigate as a result of the complete knowledge of the system and its weaknesses [10]
- Gray-box attacks: In between black-box or white-box attacks, where the attacker has only a limited understanding of the system, such as restricted access to internal data or a fundamental understanding of the architecture of the system. The attacker may be an outsider who has acquired specific information about the system, or an insider with restricted access [9].

These attacks may comprise full, partial, or indirect manipulation of the data and quantum states, the model's architecture, the knowledge of the models' output, or intermediate circuits, qubits, layers, quantum states or decision boundaries.

The quantum adversarial attack methods (see Fig. 9.1) are summarized in Table 9.1 based on access levels.

Potential defense strategies for each attack type categorized by target and access level are provided in Table 9.2.

9.2.2 *Quantum Adversarial AI for the Defense of IoT Networks*

Quantum adversarial AI can be used to safeguard AI systems against quantum adversarial attacks [10]:

- Quantum-Secure Neural Networks are engineered to withstand quantum adversarial attacks. They employ post-quantum cryptography (PQC) and quantum-resistant architectures to guarantee security.
- Quantum Data Anonymization employs quantum algorithms to anonymize data, thwarting adversaries from obtaining sensitive information.
- Quantum Encryption techniques, including quantum key distribution (QKD), offer secure communication channels impervious to eavesdropping and interference by quantum adversaries.
- Adaptive Defense Mechanisms dynamically respond to the changing threat environment presented by quantum adversaries. To prevent adversarial attacks from modifying AI models, techniques like quantum noise injection are used.

Quantum adversarial AI for the defense of IoT networks can be categorized by attack surface, target domain, and defense technique as shown in Table 9.3.

9.2.3 *Reference Architecture for Quantum Adversarial Attack Resistance*

As no system is completely immune to all types of quantum adversarial attacks, quantum adversarial defenses are typically designed to counteract specific attack vectors, such as poisoning, inference, or evasion, rather than providing all-encompassing security. The probabilistic characteristics of quantum computing, coupled with difficulties in noise control, QEC, and adversarial resilience, render developing a comprehensive defense system a highly complex and unresolved challenge. A true quantum computer can only demonstrate genuine robustness against quantum-motivated attacks since quantum-specific defenses such as QEC, quantum cryptography, and entanglement-enabled security can be effectively utilized, something no classical system is able to institute inherently. At present,

Table 9.1 Quantum adversarial attacks for three types of access levels

Attack type	White-box	Gray-box	Black-box
Quantum data poisoning	Alters training quantum states with full access to the quantum data encoding and measurement process	Modifies partial qubits or measurement configurations of the training data without full access	Perturbs quantum inputs indirectly (e.g., via noise in the environment or corrupted quantum sensors)
Quantum membership inference	Exploits access to quantum gradients and amplitudes to determine if a quantum state was in training	Uses partial quantum circuit knowledge or intermediate results to infer training membership	Analyzes quantum output probabilities or measurement results to infer training membership
Quantum model extraction	Directly accesses quantum gates, unitary matrices, and amplitudes to replicate the model	Combines partial access to the quantum circuit structure and output states to approximate the model	Mimics the quantum model using queries of input-output pairs and measurement results
Quantum property inference	Analyzes full quantum circuit and state evolution to deduce training data properties	Uses limited access to specific layers or amplitudes to infer global data characteristics	Probes the quantum model with crafted states to deduce dataset-level properties via measurement results
Quantum state perturbation	Applies targeted quantum noise or perturbations to qubits during training or inference	Introduces environmental noise or decoherence into selected quantum states without full system access	Introduces random noise indirectly to influence quantum state evolution during inference
Quantum decision boundary attacks	Explores quantum model's decision boundary using full gradient and amplitude knowledge	Combines partial amplitude and measurement data to refine boundary exploration	Iteratively queries the quantum model with crafted quantum states to approximate decision boundaries
Quantum model inversion	Reconstructs input quantum states using gradients, amplitudes, or circuit parameters	Reconstructs partial quantum inputs via intermediate amplitudes or partial circuit information	Uses query outputs to approximate quantum states corresponding to given outputs
Quantum backdoor attacks	Injects specific triggers into training quantum states, exploiting superposition or entanglement	Modifies partial qubit states during training to introduce misclassification	Indirectly influences the quantum training environment to embed triggers without direct circuit access
Quantum confidence score analysis	Directly analyzes measurement probabilities and amplitudes to infer sensitive quantum data	Analyzes partially accessible output probabilities or circuit configurations for sensitive inference	Exploits output probabilities or measured states to infer sensitive quantum data or model weaknesses

Table 9.2 Defense strategies for quantum adversarial AI attacks

Attack type	Access level	Description of attack	Defense strategies
Quantum data poisoning (Data attacks)	White-box	Attacker manipulates quantum training data or states directly	Quantum data validation and sanitization. Train on multiple data distributions
	Gray-box	Attacker partially alters quantum states or training samples	Use quantum generative models to detect anomalies. Enforce data constraints using QEC
	Black-box	Indirectly corrupts input data, e.g., by introducing sensor noise	Robust adversarial training techniques. Use noise-aware quantum learning algorithms
Quantum backdoor attacks (Data attacks)	White-box	Attacker embeds triggers into quantum states during training	(1) Use trigger-detection protocols (e.g., random encoding checks). (2) Perform quantum circuit auditing
	Gray-box	Partially modifies training states to create backdoor vulnerabilities	Train with the random encoding of quantum states to mitigate malicious data manipulations or model extraction
	Black-box	Indirectly manipulates the environment to embed backdoor triggers	Periodically verify the integrity of quantum training data
Quantum model extraction (Model attacks)	White-box	Attacker accesses and duplicates quantum gates or parameters	Encrypt quantum circuit parameters. Use differential privacy to prevent data leakage
	Gray-box	Attacker uses partial circuit knowledge to approximate the model	Restrict access to intermediate qubits or gate configurations to prevent reverse-engineering
	Black-box	Queries the model's outputs to reconstruct its behavior	Limit query rates. Add randomness to output probabilities
Quantum membership inference (Output attacks)	White-box	Exploits quantum gradients or amplitudes to infer training membership	Use differential privacy to mask quantum training data contributions
	Gray-box	Partially analyzes intermediate quantum states or outputs for inference	Randomize intermediate states using quantum scrambling techniques
	Black-box	Observes output probabilities to deduce training data membership	Apply output obfuscation techniques (e.g., randomized measurement)

(continued)

Table 9.2 (continued)

Attack type	Access level	Description of attack	Defense strategies
Quantum confidence score analysis (Output attacks)	White-box	Analyzes full output amplitudes for sensitive data inference	Apply quantum dropout mechanisms during measurement
	Gray-box	Partially analyzes confidence scores or output probabilities	Introduce controlled decoherence in the measurement process
	Black-box	Exploits output labels or probabilities to infer sensitive data	Use noise-aware quantum classifiers to minimize exploitable patterns in outputs
Quantum decision boundary attacks (Intermediate attacks)	White-box	Explores decision boundaries using full gradient and amplitude knowledge	Regularize decision boundaries using adversarial training
	Gray-box	Refines boundary exploration with partial quantum state knowledge	Smooth decision boundaries with randomized encoding or adaptive learning
	Black-box	Examines the model using designed states to estimate decision boundaries	Apply smooth decision boundaries to make the quantum model more resistant to adversarial perturbations
Quantum state perturbation (Intermediate attacks)	White-box	Injects noise into qubits to degrade learning	Use QEC codes during training and inference
	Gray-box	Introduces decoherence to selected quantum states during inference	Use noise-resilient quantum architectures
	Black-box	Randomly perturbs input states to affect outputs	Apply quantum adversarial training to mitigate attacks, and to improve robustness against input noise

Table 9.3 Categorization of quantum adversarial AI for defense of IoT networks

Level	Type	Examples
Attack surface	Data, model, output	Data poisoning, model extraction, membership inference
Target domain	Quantum network security	QKD, securing quantum IoT networks
	Federated QAI	Gradient poisoning, federated update integrity
	Quantum sensing and IoT	Adversarial manipulation detection in quantum sensors
Defense techniques	Adversarial training	Robust quantum datasets, adversarial exposure in training
	QEC	Mitigating noise, using topological codes
	Randomized quantum encodings	Obfuscating input states, randomized intermediate computations
	Robust architectures	Defensive distillation, hybrid quantum-classical defenses

however, there are no known architectures that can resist quantum adversarial attacks. As a result, it is critically necessary to develop new architectures that achieve uniform fault tolerance, enhancement of performance, and security across all up-and-coming quantum hardware platforms to avoid fragmentation, which must naturally be involved in any sound QAI defense mechanism.

A six-layer reference quantum computing architecture for quantum adversarial attack resistance including the hybrid computing, quantum application, software, quantum network, quantum processing & control and physical layers is presented in Fig. 9.2. By organizing quantum hardware, control, computation, networking, and user access into distinct layers, we anticipate enhancing the fault tolerance of quantum systems through QEC and improved qubit control. Additionally, we expect to boost performance by optimizing quantum workloads, hybrid processing, and distributed computing. We can expect better defenses against quantum adversarial attacks due to quantum and post-quantum crypto improvements, better access controls, better access control mechanisms, and adaptive adversarial AI defenses. The Physical Layer includes the native quantum hardware such as qubits, quantum gates, cryogenic systems, measurement devices, and quantum control circuits in charge of managing qubits and physically carrying out gates utilizing quantum hardware, as well as quantum encoding to represent classical bits in qubits. QEC, quantum gate management, and quantum state management are handled by the Quantum Processing & Control Layer. Quantum state management refers to the preparation, storage, and manipulation of the quantum states to perform computational tasks wherein the decoherence reduction and feedback loops play a crucial role in maintaining the quantum states coherent and stable during the computations. The loss of quantum behavior is called quantum decoherence, where a quantum system loses its quantum properties by interacting with its environment.

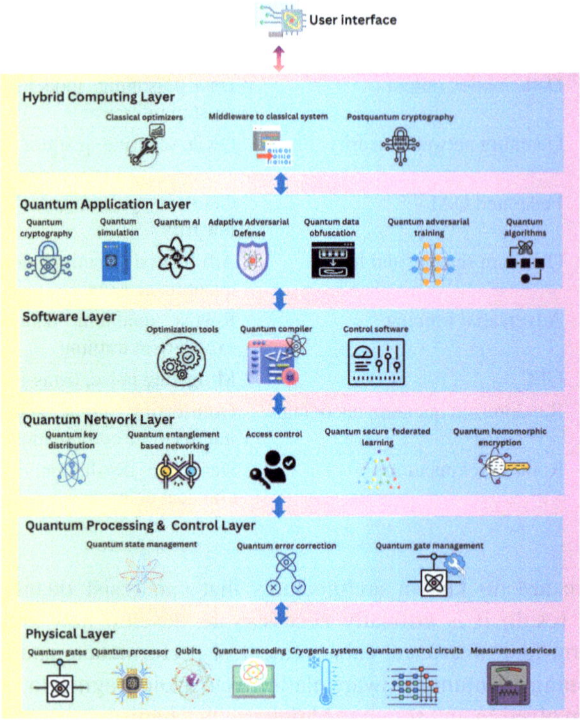

Fig. 9.2 Reference quantum computing architecture resilient to quantum adversarial attacks

Feedback control in quantum systems entails employing measurements and control operations to stabilize and regulate quantum states. Quantum gate management encompasses optimizing quantum gate operations at the software level, which may involve gate decomposition, breaking down complex gates into fundamental operations, and pulse shaping to enhance performance. Pulse shaping is the process of customizing control pulse temporal profile to maximize quantum operation performance by minimizing mistakes and raising quantum gate's fidelity. The Quantum Network Layer performs quantum secure communications and distributed secure quantum computing with QKD, access control, quantum entanglement-based networking, quantum secure federated quantum learning, and quantum homomorphic encryption. QKD is used to secure quantum model parameters and gradients for quantum secure federated learning. The Software Layer provides higher-level abstractions for quantum programming using quantum compilers, optimization tools and control software. The Quantum Application Layer supports quantum secure applications with quantum cryptography, quantum simulation, QAI, quantum data obfuscation, adaptive adversarial defense and quantum algorithms. The Hybrid Computing Layer executes hybrid quantum-classical algorithms that accelerate AI, optimization, and quantum-enhanced simulations. The User Interface provides the

API for end-users to quantum systems, as well as the programming environments, data visualization, and ICT security protocols for quantum application execution.

9.3 Conclusions

Quantum adversarial AI is an interdisciplinary field concerning offensive techniques that leverage weaknesses in AI systems, as well as defensive approaches aimed at addressing these threats using quantum computing and adversarial AI methods. Attacks in the quantum domain follow at multiple levels, such as quantum data poisoning, model extraction, decision boundary attacks, and inference attacks. Mitigating quantum adversarial attacks necessitates the utilization of both classical and quantum-specific methodologies. The primary defenses include QEC to reduce noise and decoherence, differential privacy to conceal sensitive information in gradients and outputs, and adversarial training to improve robustness by exposing quantum models to intentional perturbations. Randomized encoding enhances the security of quantum states by incorporating variability during training, whereas defensive distillation refines decision boundaries to mitigate susceptibility to adversarial examples. In contrast, defensive quantum adversarial AI strategies including quantum adversarial training, error correction, randomized encodings, and robust quantum architectures seek to improve the resilience of quantum adversarial AI models in applications such as IoT networks, quantum-enhanced sensing, and secure communications.

Both quantum technology and quantum adversarial AI have a long way to go before their promise can be realized, and this proof-of-concept offers insight into the potential danger facing IoT networks. Many quantum adversarial AI defense methods are still in their infancy and remain domain-specific, lacking the generality to be impactful across a wide spectrum of applications. Furthermore, a standardized framework is needed for assessing adversarial robustness in quantum systems, complicating the benchmarking of defense mechanisms' efficacy. The probabilistic nature of quantum systems poses unique challenges in generating adversarial examples and implementing effective countermeasures. A reference architecture proposed for quantum adversarial attack prevention is anticipated to enhance fault tolerance, performance, and security. The emergence of quantum adversarial AI creates significant challenges for securing the systems of critical infrastructure providers such as in healthcare, energy, transportation, banking and finance. The study of quantum adversarial AI presents a unique opportunity for the creative exploration of adversarial offense and defense from the perspective of emerging quantum technologies' security. Future research is anticipated to concentrate on scalable quantum adversarial training methodologies, hybrid quantum-classical defenses, the use of quantum cryptographic protocols in adversarial defense systems, and the influence of entanglement and superposition on the enhancement or detriment of robustness. This may include developing noise-resilient quantum architectures and integrated hybrid quantum-classical adversarial training frameworks, e.g., using topological qubits, which employ topological concepts to store and manipulate quantum data. Ongoing efforts focus on the development of quantum analogs of adversarial robustness measures, study of entanglement impact on attack resistance, and design of methods for quantum model verification and certification to enhance robust performance for relevant applications.

Acknowledgments This work was supported by Business Finland (BF) within the EUREKA CELTIC-NEXT project CISSAN (www.celticnext.eu) and the Securing the Quantum Software Stack (SeQuSoS) project (112/31/2024). The authors would like to thank Fibabanka, Boğaziçi University and University of Jyväskylä management for their support.

References

1. Akter, M.S., Shahriar, H., Iqbal, I., Hossain, M., Karim, M.A., Clincy, V., Voicu, R.: Exploring the vulnerabilities of machine learning and quantum machine learning to adversarial attacks using a malware dataset: a comparative analysis. In: 2023 IEEE International Conference on Software Services Engineering (SSE) (2023)
2. Akter, M.S., Shahriar, H., Cuzzocrea, A., Wu, F.: Quantum adversarial attacks: developing quantum FGSM algorithm. In: 2024 IEEE 48th Annual Computers, Software, and Applications Conference (COMPSAC), Osaka, Japan (2024)
3. Anil, G., Vinod, V., Narayan, A.: Generating universal adversarial perturbations for quantum classifiers. In: Proceedings of the AAAI Conference on Artificial Intelligence (2024)
4. Gong, W., Deng, D.L.: Universal adversarial examples and perturbations for quantum classifiers. Natl. Sci. Rev. **9**(6) (2022)
5. Khazane, H., Ridouani, M., Salahdine, F., Kaabouch, N.: A holistic review of machine learning adversarial attacks in IoT networks. Future Internet **16**(1), 32 (2024)
6. Lu, S., Duan, L.M., Deng, D.L.: Quantum adversarial machine learning. Phys. Rev. Res. **2**(3), 033212 (2020)
7. Nielsen, M.A., Chuang, I.L.: Quantum Computation and Quantum Information, 10th edn. Cambridge University Press, Cambridge, UK (2010)
8. Purushotham, C.S., Reese, J.S., Kirkwood, M.J., Kosec, K.: Adversarial machine learning: a taxonomy and terminology of attacks and mitigations. Tech. rep., National Institute of Standards and Technology (NIST) (2023)
9. Vivek, B.S., Mopuri, K.R., Babu, R.V.: Gray-box adversarial training. In: 15th European Conference on Computer Vision (ECCV) 2018, Munich, Germany (2018)
10. Yocam, E., Rizi, A., Kamepalli, M., Vaidyan, V., Wang, Y., Comert, G.: Quantum adversarial machine learning and defense strategies: challenges and opportunities (2024). arXiv:2412.12373 [quant-ph]

Part III
Defense Techniques and Robustness Strategies

Chapter 10
Detecting and Mitigating Adversarial Examples in Neural Networks: An Enhanced PGD Approach

Oras Baker, Kasthuri Subaramaniam, Sellappan Palaniappan, Ehsan Nowroozi, and Yoosef Habibi

10.1 Introduction

Neural networks are found to be effective hard problem solvers in a number of fields, from image recognition to natural language and autonomous vehicles. Their capacity for learning intricate patterns from large volumes of data has established them as an essential instrument in the field of modern artificial intelligence. But while they provide outstanding performance, neural networks are still vulnerable to adversarial examples—inputs with carefully introduced minute distortions to deceive models into making incorrect predictions [9, 22]. These vulnerabilities represent real problems, particularly in safety-critical applications such as autonomous driving, medicine, and cyber security.

Deep neural networks (DNNs) have performed very well on nearly any learning problem from computer vision, natural language processing, and speech recognition [11]. Experiments have, however, shown that DNNs are susceptible to adversarial

O. Baker
Computer Science, Ravensbourne University London, London, UK
e-mail: O.Alhassani@rave.ac.uk

K. Subaramaniam
Faculty of Business and Economics, Universiti Malaya, Kuala Lumpur, Malaysia
e-mail: s_kasthuri@um.edu.my

S. Palaniappan
Faculty of Computing and Digital Technology, Help University, Kuala Lumpur, Malaysia
e-mail: sellappan.p@help.edu.my

E. Nowroozi
University of Greenwich, London, UK
e-mail: E.Nowroozi@greenwich.ac.uk

Y. Habibi (✉)
University of Padova, Padova, Italy

E. Nowroozi et al. (eds.), *Adversarial Example Detection and Mitigation Using Machine Learning*, https://doi.org/10.1007/978-3-031-99447-0_10

examples–inputs that have been carefully crafted by adding imperceptibly small perturbations to correct examples. The perturbations try to mislead a DNN without being perceptibly noticeable to human observers [3, 8]. The attacks by the adversaries are of two kinds depending on the process by which adversarial perturbations are generated: single-step and multi-step attacks [18]. The adversarial example is generated through the direct conversion of the clean example in a single-step attack, and in a multi-step attack, the adversarial perturbation is optimized step-by-step. The one-step attack FGSM [9] calculates the adversarial perturbation with the scaled sign of the gradient, whose scaling factor is the attack strength. Unlike multi-step attacks like the Basic Iterative Method (BIM) [5] and Projected Gradient Descent (PGD) [19], which optimize the adversarial perturbation iteratively. The Carlini & Wagner (CW) attack [3] poses the generation of adversarial perturbation as an optimisation problem that is solved through iterative refinement.

Adversarial training and adversarial example detection are two powerful techniques for improving the robustness of deep learning (DL) models. Adversarial training is used to make the model robust by adding the adversarial examples into the training set and aims at aligning the hidden-layer representations or outputs of legitimate samples with the corresponding adversarial samples with regard to the target model. Conversely, adversarial example detection methods focus on preventing adversarial examples from reaching the target model. Adversarial detection [10] enhances the robustness of DL systems by intercepting adversarial examples before they can affect the target model. Recent defense techniques against attack by adversaries are of three types: discriminative, generative, and metric-based. Discriminative models learn a single or several classifiers to differentiate adversarial and normal examples. Generative methods [7] attempt to reconstruct the input to determine whether a given example is adversarial. Metric-based methods [15] detect adversarial examples using predefined metrics such as mean, standard deviation, kernel density, and various invariants.

The vulnerability of neural networks to adversarial examples can result in serious outcomes. For example, adversarial attacks on healthcare systems can misclassify medical images, leading to misdiagnosis [22]. Alternatively, in cyberspace, adversarial examples can bypass intrusion detection mechanisms, outsmarting sensitive information [14]. Although significant progress has been made in understanding and addressing these vulnerabilities [15, 21], critical gaps remain in evaluating robust defences, particularly under diverse attack scenarios.

This study seeks to address the following questions:

- How effective is the enhanced Projected Gradient Descent (PGD) method in exposing vulnerabilities in neural networks?
- Can a simple perturbation-based detection mechanism reliably identify adversarial examples in real-time?
- What is the impact of adversarial training on improving model robustness across clean and adversarial inputs?

The primary objectives of this study are to evaluate the performance of enhanced PGD attacks in degrading model accuracy, develop and validate a perturbation-based detection mechanism that efficiently identifies adversarial inputs, and explore the effectiveness of adversarial training in restoring model performance against adversarial examples.

This paper makes three significant contributions:

1. Systematically examines the effect of heightened PGD attacks on neural network performance, and sheds light on the insufficiency of current defenses [15].
2. Introduces a computationally efficient perturbation-based detection mechanism and validates its effectiveness in diverse scenarios [16, 23].
3. Demonstrates the robustness of adversarial training as a mitigation strategy, balancing performance on both clean and adversarial datasets [13, 14].

The rest of this paper is structured as follows: Sect. 10.2 summarizes the state of the art on adversarial examples, detection mechanisms, and mitigation methods. Section 10.3 outlines the methodology, including dataset generation, model training, and defense mechanisms. Section 10.4 contains the implementation details and results. Section 10.5 documents the findings, highlighting the effect of adversarial examples and how effective mitigation is strategies. Lastly, Sect. 10.6 summarizes the study and offers guidelines for future studies.

10.2 Literature Review

10.2.1 Adversarial Examples and Their Impact

Adversarial examples, which were found by [9], indicate inherent weaknesses in machine learning (ML) models. Inputs of this type are designed to take advantage of the vulnerabilities of decision neural networks, usually in the form of humongous misclassification [9]. proposed the Fast Gradient Sign Method (FGSM), one of the first adversarial attack methods, illustrating how small perturbations would mislead even best-performing models [9]. Subsequent work by [22] and [14] explained such attacks, highlighting their significance in fields such as autonomous systems, medicine, and computer security.

The effects of adversarial examples are grave. For example, in autonomous systems, imperceptibly tampered road signs may lead to catastrophic navigation [8]. Similarly, neural diagnostic task-based medical systems can commit deleterious misclassifications under adversarial input [22]. These vulnerabilities emphasize the critical need for resilient defenses against malicious attacks.

The attackers generate the adversarial inputs using their expertise in the target model, such as model parameter training or data. The inputs appear real but contain perturbations that are carefully designed to exploit the vulnerabilities of the model [3]. Most adversarial example generation techniques were originally developed for images. The Fast Gradient Sign Method (FGSM) [9] is among the most referenced methods, serving as the basis for following methods [5, 18]. In an FGSM attack, the perturbation η is computed as in the following equation:

$$\eta = \epsilon \cdot \text{sign}(\nabla_x J(\theta, x, y)), \quad (10.1)$$

where ϵ controls the perturbation magnitude, x is the input, y is the corresponding output, θ represents the model weights, and $J(\cdot)$ is the loss function. The adversarial sample is generated by adding η to x. FGSM is computationally efficient, requiring only the gradient sign. Although designed for image perturbations, Santana et al. [19] demonstrated its effectiveness in degrading the performance of DL models in photovoltaic (PV) power generation forecasting.

10.2.2 Adversarial Detection Mechanisms

Detecting adversarial examples is a critical step in mitigating their impact. Early detection techniques relied on feature-space transformations or statistical analysis of input distributions [17]. More recent approaches, such as those proposed by [10], employ adaptive denoising techniques and neural estimation methods to improve detection accuracy. These methods focus on identifying perturbation patterns or inconsistencies in network responses.

Despite their promise, many detection mechanisms are computationally expensive or vulnerable to bypass strategies [4, 17]. Yuan et al. [22] highlighted the difficulty of distinguishing minimal perturbations, emphasizing the need for scalable and generalizable detection frameworks. This study builds on these insights by proposing a perturbation-based detection method that balances computational efficiency with high detection rates.

10.2.3 Mitigation Strategies

Mitigation strategies aim to enhance the robustness of neural networks against adversarial attacks. Among these, adversarial training has emerged as one of the most effective approaches [15]. By augmenting the training dataset with adversarial examples, models can learn more robust decision boundaries. However, adversarial training often increases computational overhead and may reduce generalisability to unseen attack types [14].

Other notable mitigation techniques include defensive distillation [17], feature squeezing [20], and hybrid methods combining multiple defences [22]. Recent studies by [13] and [21] propose novel strategies, such as conditional diffusion models and label recovery mechanisms, to improve model resilience. This paper evaluates the effectiveness of adversarial training in restoring model performance across diverse datasets.

ML algorithms have shown promise in detecting attacks in traditional computer systems and smart grids [13, 23]. These approaches gather data sensitive to attacks and use them to train ML-based classifiers, which distinguish between legitimate and malicious instances [1, 6]. For example, network-based attacks are detected by analysing traffic variables, while false data injection in smart grids is identified through power grid state measurements. Chen et al. [4] computed the influence of adversarial examples in power quality classification and load forecasting for feed-forward neural networks (NNs) and recurrent neural networks (RNNs) on the necessity of enhanced robustness of the power system model.

10.2.4 Enhanced Adversarial Attacks

Enhanced adversarial attacks such as Projected Gradient Descent (PGD) are excellent improvements in adversarial work. Madry et al. [15] introduced PGD as a powerful iterative attack with infinitesimal perturbations within a bounded region. Random initialization to it further strengthens it [8].

Past efforts have included transfer-based and black-box attacks, avoiding gradient-based defenses by taking advantage of model generalization properties [4, 22]. These efforts help emphasize defenses that are immune to present and forthcoming attack strategies.

10.3 Methodology

10.3.1 Dataset Generation

In order to experiment with adversarial robustness in neural networks, a synthetic data set was established through the use of Scikit-learn's make_classification feature. The data set contains 1000 instances and 20 features of which 15 are useful and 5 redundant. That the distribution is class-balanced provides testability of unbiased model performance. The data set was split into the training and test sets at 70:30 ratios, and feature values were normalised by applying standard normalisation techniques in an attempt to include uniformity and prevent magnitudes with dominant effects from interfering with the learning process [22].

Synthetic data generation provides an environment for controlled experiments, eliminating confounding variables and producing reproducible outcomes [2, 12]. It is at the cost of having poor generalisability to real data. Elimination of this restriction is one of the areas of future adversarial work of paramount significance [8].

10.3.2 Baseline Neural Network Model

The baseline model employed in this research is a Multi-Layer Perceptron (MLP) instantiated with Scikitlearn's MLPClassifier. The model has two hidden layers of 64 and 32 neurons, respectively. The ReLU activation function is used for the hidden

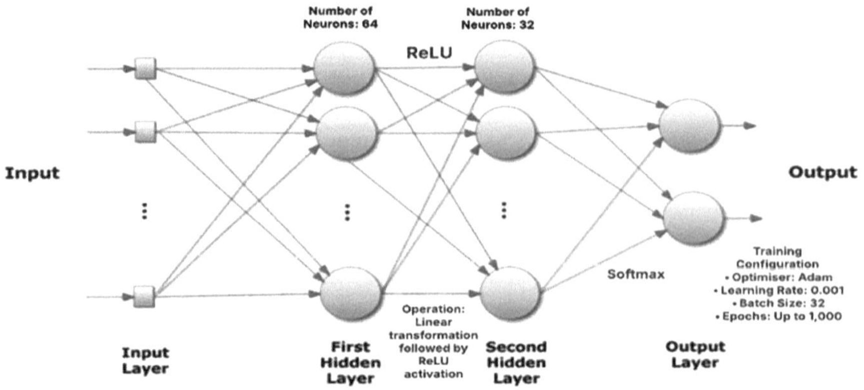

Fig. 10.1 Baseline MLP model architecture

layers, and the output layer uses a softmax activation function for probabilistic classification. The model was trained with the Adam optimiser and learning rate of 0.001 and a batch size of 32, training up to 1000 epochs, as indicated in Fig. 10.1.

Baseline neural networks are necessary to serve as a baseline to compare the effect of adversarial attacks and test the efficiency of defence methods [9]. The use of MLP, a common feedforward architecture, facilitates the simple demonstration of adversarial weaknesses and their mitigation.

10.3.3 Adversarial Example Generation

Adversarial examples were created with a stronger Projected Gradient Descent (PGD) attack. The iterative attack makes minor perturbations to input data, with a limit on the size of the perturbation (ϵ). ϵ was used as a value of 1.0 with step size (α) set to 0.2 and 20 iterations in this work. Random initialization was used to enhance adversarial example diversity and prevent gradient masking defences [15].

PGD attack magnifies is an effective approach to making model vulnerability observable, as Yuan et al. [22] and Gilmer et al. [8] argue. Dynamic improvement of perturbations guarantees the attack model is properly fooled by adversarial examples and thus proving the necessity to be resilient to defence.

10.3.4 Detection Mechanism

The detection mechanism used in this research approximates perturbation magnitude of inputs given. Particularly, the Euclidean distance between clean and adversarial inputs was approximated. Inputs with a perturbation magnitude greater than 0.5 were classified as adversarial. This straightforward yet effective approach aligns with the findings of Zarpelão et al. [23], who demonstrated the effectiveness of perturbation-based measurements in detecting adversarial attacks.

Although this technique has the potential to have high detection rates, its low perturbation robustness is a limitation, as attested by Carlini and Wagner [3]. Adaptive threshold techniques or feature-space mapping [10] might further evolve this in the future.

10.3.5 Mitigation Strategy

Adversarial training with the inclusion of adversarial examples in the training data was employed as the principal mitigation approach. The dataset augmentation helped the model to learn stable and robust decision boundaries that are not easily subject to adversarial perturbations [15]. The model was subsequently tested after mitigation on both the clean test set and the adversarial test set to quantify its performance.

Adversarial training has long been a go-to as a remedy to thumping adversarial attacks [9, 14]. It, however, incorporates computational expense and is prone to overfit on particular forms of attack, loss of generalisability. Hybrid defence methods involving the use of adversarial training coupled with other defence mechanisms, including feature squeezing or statistical preprocessing, are future research directions [13, 20].

10.4 Implementation and Outputs

This section describes the implementation of the above process, with pseudocode for the main processes and description of code. Outputs, plots, and figures have been added to clarify. The entire code for the implementation has been presented in the appendix as a reference.

10.4.1 Dataset Generation

Below is the pseudocode for building the dataset for this research:

1. Import required libraries (e.g., NumPy, Scikit-learn).
2. Generate a synthetic dataset using make_classification:
   - Specify 1000 samples and 20 features.
   - Set 15 informative features and 5 redundant features.
   - Ensure balanced class distribution.

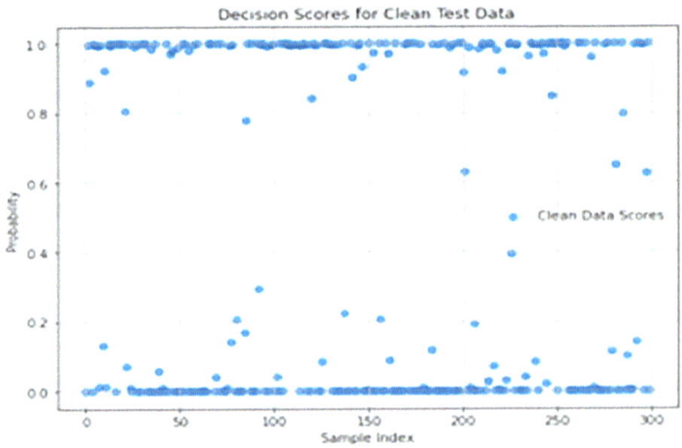

Fig. 10.2 Decision scores for clean test data

3. Split the dataset into training (70%) and testing (30%) sets.
4. Scale the feature values for consistency.

Explanation The dataset simulates a binary classification problem with well-defined informative features and noise. Scaling ensures gradient-based methods are effective across all features, as highlighted in Yuan et al. [22]. Synthetic datasets, such as those generated here, provide controlled environments for experimentation [15].

Output Figure 10.2 visualises the model's prediction probabilities for clean data, serving as a baseline reference.

10.4.2 Baseline Neural Network Training

The neural network serves as the baseline model for comparison. It is trained to classify clean data with high accuracy, forming the foundation for evaluating adversarial examples and mitigation techniques [9]. The architecture and training process align with best practices for feedforward neural networks, ensuring reliable performance. The pseudocode below illustrates the steps to generate the output:

1. Initialize the neural network using `MLPClassifier`:
   - Use two hidden layers with 64 and 32 neurons, respectively.
   - Employ ReLU activation for the hidden layers.
   - Use Softmax for the output layer.
2. Train the model using the Adam optimizer.
3. Evaluate the model on the test dataset to calculate baseline accuracy.
4. Save the trained model for further evaluation.

Figure 10.3 shows the baseline model decision boundary. It illustrates the decision-making capability of the neural network when predicting on clean test data.

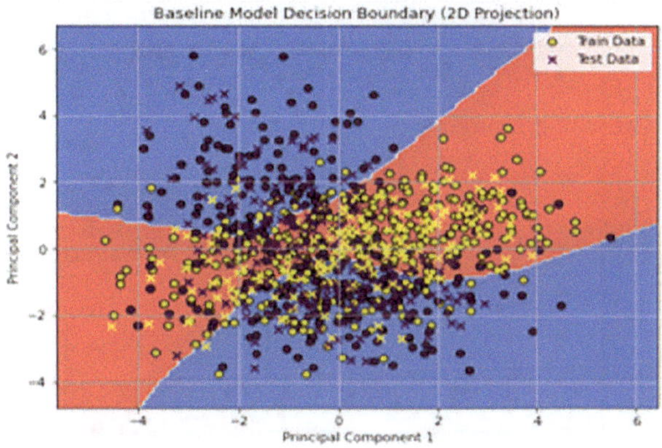

Fig. 10.3 Baseline model decision boundary

10.4.3 Adversarial Example Generation

The enhanced PGD method iteratively applies perturbations to test inputs, generating adversarial examples designed to deceive the model [8, 15]. Random initialisation ensures diversity, making the attack robust against defences. The effectiveness of the attack is assessed by evaluating the model's performance on adversarial examples. The pseudocode below illustrates the steps to generate the adversarial example:

1. Initialize adversarial examples with random noise bounded by ϵ.
2. For `num_iter` iterations:
   - Compute gradients of the loss with respect to inputs.
   - Add scaled perturbations (α) to inputs.
   - Clip values to stay within valid feature ranges.
3. Evaluate the model on adversarial inputs to compute adversarial accuracy.

Figure 10.4 shows the comparison of decision scores for clean and adversarial data. It highlights the impact of adversarial perturbations.

10.4.4 Detection of Adversarial Examples

The detection mechanism identifies adversarial examples based on perturbation magnitudes. Specifically, the Euclidean distance between clean and adversarial inputs is computed. Inputs exceeding a defined threshold are classified as adversarial [20, 22]. This method effectively separates clean and adversarial samples in most scenarios. Figure 10.5 shows the histogram of perturbation magnitudes with the

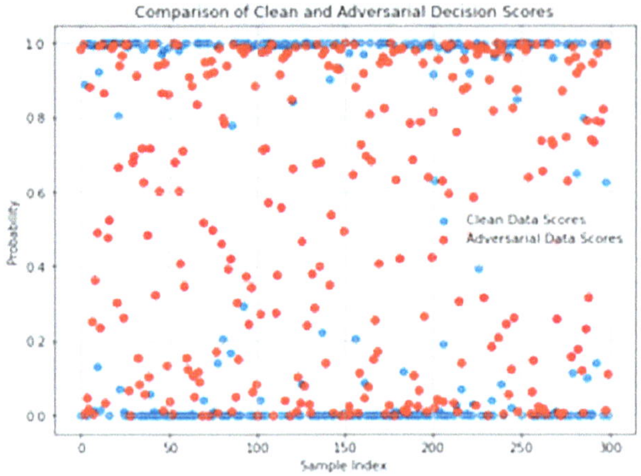

Fig. 10.4 Comparison of decision scores for clean and adversarial data

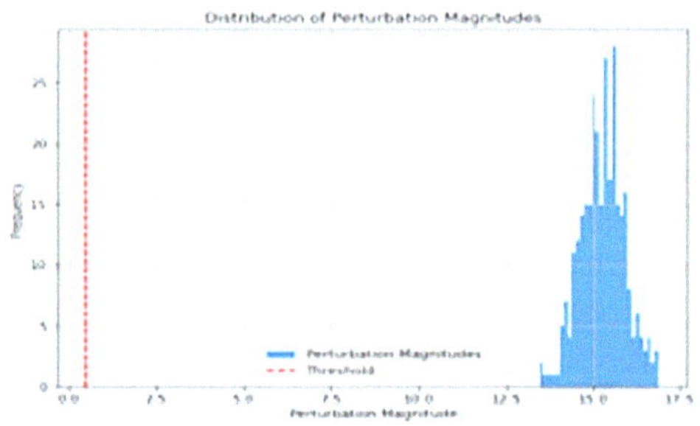

Fig. 10.5 Histogram of perturbation magnitudes with detection threshold

detection threshold. This histogram visualizes the distribution of perturbation magnitudes and highlights the detection threshold.

10.5 Results and Discussion

10.5.1 Baseline Model Performance

The baseline neural network achieved a classification accuracy of 96.33% on the clean test dataset, demonstrating its effectiveness under standard conditions. As demonstrated in Table 10.1, the clean data decision scores represent the model's performance in classifying unperturbed inputs. The baseline performance serves as a baseline for measuring the effect of adversarial examples, consistent with previous results that neural networks perform well on clean data [9]. The robust baseline provides a solid basis for measuring the degradation caused by adversarial attacks.

10.5.2 Impact of Adversarial Examples

When attempted on adversarial examples generated based on the enhanced PGD method, the model's accuracy dropped drastically to 62.33%, which indicates that neural networks are vulnerable to carefully crafted perturbations. This is consistent with Madry et al. [15] and Yuan et al. [22], who had already established that even slight perturbations could heavily undermine model performance. Figure 10.3 is the decision scores for adversarial and clean data, which shows how adversarial examples bias the model's predictions and undermine its decision boundaries. This sudden decrease in accuracy emphasizes the necessity for robust defenses against attack by malicious parties, particularly for safety-critical domains such as autonomous systems and health care [8]. The more efficient PGD attack's iterative mode ensures that the perturbations are optimized to effectively take advantage of the weaknesses of the model.

10.5.3 Detection of Adversarial Examples

The mechanism based on perturbation detection achieved 100% accuracy in identifying adversarial inputs. By calculating the Euclidean distance between clean and adversarial samples, the mechanism effectively distinguished the two classes, as illustrated in Fig. 10.4. The distribution of perturbation magnitudes clearly demonstrates the separation between normal and adversarial samples, thereby justifying the selected detection threshold of $\delta = 0.5$. These findings align with those of Xu et al. [20] and Kherchouche et al. [10], who emphasized the effectiveness of perturbation-based approaches in adversarial detection. However, as noted by Carlini and Wagner [3], detection methods may struggle with minimal perturbations, indicating the need for further refinement to address more sophisticated attack scenarios.

10.5.4 Mitigation Through Adversarial Training

Adversarial training significantly improved model robustness. The model trained had a 94.33% classification accuracy on clean test data and 95.33% classification accuracy on adversarial examples. These results confirm that adversarial training can restore the performance of a model against an adversarial attack, as the authors had discovered empirically by Madry et al. [15] and Macas et al. [14].

Table 10.1 provides the pre- and post-mitigation performance measurements for showing the robustness of adversarial training in having good accuracy even on adversarial data. Adversarial as well as clean examples in richer training sets

Table 10.1 Summary of model performance metrics

Metric	Value
Baseline accuracy	96.33%
Adversarial accuracy	62.33%
Detection rate	100.00%
Clean accuracy after mitigation	94.33%
Adversarial accuracy after mitigation	95.33%

allowed the model to build strong decision boundaries that could tolerate adversarial perturbations.

Adversarial training is effective but incurs computational overhead and overfits to certain types of attacks [9]. Hybrid methods that incorporate adversarial training with feature-space mappings or statistical preprocessing methods are potential avenues for future work [13, 20].

10.6 Conclusion and Future Directions

In this research, it has been established that neural networks can be attacked with adversarial examples and methods for detection and mitigation proposed by us proved effective. The performance of the baseline model was 96.33% against clean data but plummeted by 62.33% when the adversarial examples created through the improved Projected Gradient Descent (PGD) attack were used. This huge disparity highlights the urgent need for having strong defence algorithms in ML systems. Through combining a stronger PGD attack, perturbation-based detection approach, and adversarial training, the present work offered a general framework to robustify models against adversarial attacks. The detection based on the introduced approach was as good as 100% when detecting adversarial samples, and utilizing adversarial training, the performance restored to 94.33% on test and 95.33% on adversarial samples, depicting efficacy against such attacks. These findings are in agreement with existing research, e.g., Madry et al. [15] and Xu et al. [20], that identify the importance of adversarial training and detection techniques in securing ML models.

The overall contributions of this research are threefold:

- For starters, the paper offers a rigorous analysis of PGD attacks amplified, with the ability to make model predictions unstable and uncover weaknesses of neural networks.
- Second, a computationally effective perturbation-based detection algorithm was presented with 100% detection rate and showing its resistance to various attack vectors.
- Third, the study provided evidence for adversarial training efficacy in restoring robustness to the model to allow high accuracy for both clean and adversarial dataset. The above contributions as a whole enhance adversarial ML through

defence bridging and the provision of practical solutions to reinforce model resilience.

The importance of this work is that it unifies detection and mitigation techniques under a sound framework that can be tailored to address real-world problems, including healthcare, autonomous systems, and cybersecurity.

There are some areas worth investigation in the future based on this research. Scalability of detection mechanisms, especially for real-time, is one of the most hopeful directions. Improving with adaptive or dynamic thresholds may be useful for increasing capability in responding to more complex adversarial attacks and maintaining high performance under changing conditions. Another direction to continue working on is transferability of mitigation strategies. Their performance against unseen types of attacks and datasets would give significant insights into their generalisability to a variety of domains to make sure that defences are applicable in changing threat scenarios.

Another line of research is hybrid defence strategy design. Merging different mitigation strategies, for instance, feature squeezing and ensemble models, would negate the limitations of a singular approach and provide more resilient systems. Another line is applying the presented framework to application domains, i.e., healthcare, autonomous systems, and cybersecurity, and thus making it applicable to practical issues and addressing domain-specific issues. This would not only make ML systems stronger in high-risk use cases but also offer customized solutions to expert-level adversarial attacks.

Finally, upcoming research would need to confront novel threats of adversarial ML. Investigating new frontiers of attack vectors like conditional adversarial generation or transfer-based attacks would pave the way for forward-looking development of strong defences. By controlling the innovation race against competitors, researchers can construct adaptive models that foresee and conquer novel threats. These avenues in aggregate seek to increase the applicability and applications coverage of the current research and to provide resilient and adaptive ML systems for applications from various varieties.

References

1. Anthi, E., Williams, L., Javed, A., Burnap, P.: Hardening machine learning denial of service (DoS) defences against adversarial attacks in IoT smart home networks. Comput. Secur. **108**, 102352 (2021)
2. Birchfield, A.B., Schweitzer, E., Athari, M.H., Xu, T., Overbye, T.J., Scaglione, A., Wang, Z.: A metric-based validation process to assess the realism of synthetic power grids. Energies **10**(8), 1233 (2017)
3. Carlini, N., Wagner, D.A.: Towards evaluating the robustness of neural networks. In: 2017 IEEE Symposium on Security and Privacy (SP), San Jose, CA, USA, May 22–26 2017, pp. 39–57 (2017)
4. Chen, Y., Tan, Y., Deka, D.: Is machine learning in power systems vulnerable? In: 2018 IEEE International Conference on Communications, Control, and Computing Technologies for Smart Grids (SmartGridComm), Aalborg, Denmark, Oct 29–31 2018, pp. 1–6 (2018)

5. Dong, Y., Liao, F., Pang, T., Su, H., Zhu, J., Hu, X., Li, J.: Boosting adversarial attacks with momentum. In: IEEE Conference on Computer Vision and Pattern Recognition (CVPR), Salt Lake City, UT, USA, June 18–23 2018, pp. 9185–9193 (2018)
6. Duy, P.T., Khoa, N.H., Nguyen, A.G.T., Pham, V.H.: DIGFuPAS: Deceive IDS with GAN and function-preserving on adversarial samples in SDN-enabled networks. Comput. Secur. **109**, 102367 (2021)
7. Gao, J., Wang, H., Chen, Y.: Adversarial attacks and defences in deep learning: a survey. IEEE Trans. Pattern Anal. Mach. Intell. **45**(3), 1234–1245 (2023)
8. Gilmer, J., Metz, L., Faghri, F., Schoenholz, S., Raghu, M., Wattenberg, M.: Adversarial examples are a natural consequence of test error in noise. In: 35th International Conference on Machine Learning (ICML), Stockholm, Sweden, Jul 10–15 2018 (2018)
9. Goodfellow, I.J., Shlens, J., Szegedy, C.: Explaining and harnessing adversarial examples. In: 3rd International Conference on Learning Representations (ICLR), (May 7–9 2015 (2015)
10. Kherchouche, A., Fezza, S.A., Hamidouche, W.: Detect and defense against adversarial examples in deep learning using natural scene statistics and adaptive denoising (2021). https://doi.org/10.48550/arXiv.2107.05780
11. Krizhevsky, A.: Learning multiple layers of features from tiny images. Technical Report (2009)
12. Li, H., Wert, J.L., Birchfield, A.B., Overbye, T.J., San Roman, T.G., Domingo, C.M., Marcos, F.E.P., Martinez, P.D., Elgindy, T., Palmintier, B.: Building highly detailed synthetic electric grid data sets for combined transmission and distribution systems. IEEE Open Access J. Power Energy **7**, 478–488 (2020)
13. Li, J., Yang, Y., Sun, J.S., Tomsovic, K., Qi, H.: ConAML: Constrained adversarial machine learning for cyber-physical systems. In: 2021 ACM Asia Conference on Computer and Communications Security (AsiaCCS), pp. 52–66. Hong Kong, China, Jun 7–11 2021 (2021)
14. Macas, M., Wu, C., Fu, W.: Adversarial attacks and defences in cybersecurity: a survey. IEEE Access **12**, 12345–12367 (2024)
15. Madry, A., Makelov, A., Schmidt, L., Tsipras, D., Vladu, A.: Towards deep learning models resistant to adversarial attacks. In: 6th International Conference on Learning Representations (ICLR), Vancouver, BC, Canada (2017)
16. Martins, N., Cruz, J.M., Cruz, T., Abreu, P.H.: Adversarial machine learning applied to intrusion and malware scenarios: a systematic review. IEEE Access **8**, 35403–35419 (2020)
17. Papernot, N., McDaniel, P., Wu, X., Jha, S., Swami, A.: Distillation as a defence to adversarial perturbations against deep neural networks. In: 2016 IEEE Symposium on Security and Privacy (SP), San Jose, CA, USA, May 22–26 2016, pp. 582–597 (2016)
18. Rozsa, A., Rudd, E.M., Boult, T.E.: Adversarial diversity and hard positive generation. In: IEEE Conference on Computer Vision and Pattern Recognition Workshops (CVPRW), Las Vegas, NV, USA, Jun 27–30 2016, pp. 25–32 (2016)
19. Santana, E.J., Silva, R.P., Zarpelão, B.B., Barbon Junior, S.: Detecting and mitigating adversarial examples in regression tasks: a photovoltaic power generation forecasting case study. Information **12**(10), 394 pp. (2021)
20. Xu, W., Evans, D., Qi, Y.: Feature squeezing: detecting adversarial examples in deep neural networks (2018). https://10.14722/ndss.2018.23210
21. Ye, J., Wang, H., Zhang, L.: Label recovery mechanisms for adversarial robustness. IEEE Trans. Neural Netw. Learn. Syst. **35**(2), 1234–1245 (2024)
22. Yuan, X., He, P., Zhu, Q., Li, X.: Adversarial examples: attacks and defences for deep learning. IEEE Trans. Neural Netw. Learn. Syst. **30**(9), 2805–2824 (2019)
23. Zarpelão, B.B., Barbon, S., Acarali, D., Rajarajan, M.: How machine learning can support cyberattack detection in smart grids. In: Artificial Intelligence Techniques for a Scalable Energy Transition, pp. 225–258. Springer, Cham (2020)

Chapter 11
The Role of Explainable AI (XAI) in Enhancing the Security of Machine Learning Systems Against Adversarial Attacks

Ali Akbar ForouzeshNejad, Farzad Arabikhan, and Ramin Taheri

11.1 Introduction

With the rapid advancement of machine learning and deep learning, these technologies have been integrated into critical domains such as transportation, healthcare, and finance. However, the inherent complexity of these models makes them vulnerable to adversarial attacks, which can significantly disrupt their performance and reliability. These attacks involve subtle yet deliberate modifications to the model's input, leading to misleading or incorrect outputs. Such vulnerabilities pose serious threats to the security and trustworthiness of machine learning-based systems. In this context, the need for methods that provide deeper insights into the decision-making processes of machine learning models has become increasingly critical. Explainable AI (XAI) has emerged as a promising approach that enhances transparency and interpretability, facilitating the analysis of model decisions and revealing their weaknesses against adversarial attacks. This chapter aims to explore the role of XAI in strengthening the security of machine learning systems against adversarial attacks. By introducing advanced interpretability techniques such as LIME, SHAP, and attention-based models, we demonstrate how these methods can uncover vulnerabilities in machine learning models and contribute to the development of more effective security strategies. Additionally, this chapter focuses on the empirical evaluation of XAI's impact on improving model resilience, highlighting that interpretability not only enhances transparency but also plays a crucial role in reducing the susceptibility of models to adversarial threats. Finally, we discuss the challenges and limitations associated with the application of XAI in

A. A. ForouzeshNejad (✉) · F. Arabikhan · R. Taheri
PAIDS Research Centre, School of Computing, University of Portsmouth, Portsmouth, England, UK
e-mail: Ali.Forouzesh@port.ac.uk; Farzad.arabikhan@port.ac.uk; Ramin.Taheri@myport.ac.uk
https://www.port.ac.uk/

E. Nowroozi et al. (eds.), *Adversarial Example Detection and Mitigation Using Machine Learning*, https://doi.org/10.1007/978-3-031-99447-0_11

this domain and propose future research directions to further advance the security of machine learning systems.

11.1.1 Definition of Explainable AI (XAI)

Explainable AI (XAI) refers to a set of methods and techniques aimed at enhancing the transparency and interpretability of machine learning models. Despite their high accuracy, deep learning models and other advanced algorithms are often considered "black boxes" since their decision-making processes remain opaque to users and even developers. XAI seeks to address this issue by making models more interpretable in terms of how they process data and generate outputs. This is particularly crucial in domains such as medicine, law, and cybersecurity, where decision-making must be fully explainable and justifiable to ensure trustworthiness in AI-based systems [20]. XAI techniques can be broadly categorized into two groups: intrinsic methods and post-hoc methods. Intrinsic methods involve designing models that are inherently interpretable, such as decision trees and rule-based models. In contrast, post-hoc methods are applied after training complex models to explain their outputs. Among the most widely used post-hoc techniques are LIME (Local Interpretable Model-agnostic Explanations) and SHAP (Shapley Additive Explanations), which help analyze the contribution of input features to the model's predictions. Additionally, attention-based models in deep neural networks provide insights into which parts of the input data the model focuses on during decision-making. By leveraging XAI, not only can greater trust in model decisions be established, but it also becomes possible to identify and mitigate security vulnerabilities in machine learning systems.

11.1.2 Security Challenges in Machine Learning Models

With the widespread adoption of machine learning in high-stakes domains such as cybersecurity, healthcare, and finance, security-related challenges associated with this technology have also intensified. Due to their high complexity and opaque behavior, machine learning models are vulnerable to various adversarial attacks, data manipulation, and information leakage. Identifying and understanding these threats is a crucial step toward developing robust and secure AI-based systems [7].

- Security challenges in machine learning models can be categorized into the following key areas [1, 9, 21]:
- Adversarial Attacks: Small but strategically crafted modifications in the input data can mislead a model into making incorrect decisions. For example, slight perturbations in an image can cause a misclassification, leading to critical security risks in autonomous systems and biometric authentication.

- Model Inversion Attacks: Attackers can recover sensitive information from a trained model, allowing them to reconstruct private training data from the model's output. This poses a major privacy threat, especially in domains involving medical records, financial transactions, and personal user data.
- Data Poisoning Attacks: In this attack, malicious data is injected into the training dataset to manipulate the model's learning process. This can lead to reduced accuracy, biased predictions, or intended misclassifications in security-sensitive applications.
- Model Stealing Attacks: Attackers attempt to reproduce or clone a machine learning model by sending multiple queries and analyzing the model's responses, even without direct access to the original model. This can result in intellectual property theft and compromised security of proprietary AI models.
- Membership Inference Attacks: These attacks determine whether a specific data sample was used in the model's training phase, leading to potential privacy leaks. This is particularly concerning in healthcare and personalized recommendation systems, where revealing a person's data inclusion could expose sensitive personal information.
- Model Update Vulnerabilities: In online learning models, attackers can exploit the update process by injecting manipulated data, causing the model to learn incorrect patterns. This can be especially dangerous in fraud detection systems, real-time cybersecurity defenses, and recommendation algorithms.
- Model Uncertainty & Errors: Machine learning models often lack proper uncertainty quantification, making it difficult to assess how confident the model is in its predictions. This can be problematic in critical applications such as medical diagnosis, autonomous driving, and risk assessment, where incorrect predictions can have severe consequences.
- Side-Channel Attacks: Attackers can extract sensitive information from a model by analyzing indirect factors such as power consumption, processing time, or other non-obvious attributes during execution. These attacks are particularly relevant in edge AI, embedded systems, and cloud-based ML applications.
- Performance Degradation Under Distribution Shifts: Machine learning models often fail to generalize well when faced with data distributions different from the training set. This can lead to performance degradation and incorrect decisions, especially in dynamic environments such as fraud detection, cybersecurity, and autonomous navigation.

The security of machine learning models is increasingly threatened by adversarial manipulations, data leaks, and model exploitation. Understanding these vulnerabilities is crucial to designing robust and resilient AI systems capable of withstanding malicious attacks while maintaining high accuracy and reliability in critical applications.

11.1.3 The Role of XAI in Enhancing Model Decision Transparency

Machine learning models, particularly deep learning models, are often regarded as "black boxes" because their decision-making processes are not easily interpretable by users or even developers. This lack of transparency can lead to several challenges, including reduced user trust, difficulty in error diagnosis, and increased security risks such as adversarial attacks [14].

Explainable AI (XAI) addresses these challenges by providing meaningful explanations of how models function, enabling users to understand why a model has made a particular decision. XAI employs techniques such as LIME, SHAP, and attention-based models to analyze the influence of input features on the model's output, helping to identify strengths and weaknesses in decision-making. This capability not only enhances trust and acceptance of AI models but also enables experts to detect malicious patterns, unfair biases, and security vulnerabilities [16]. For instance, in cybersecurity, XAI can explain why an intrusion detection model classifies a particular request as a cyberattack, aiding in the refinement of defensive strategies. Overall, interpretability through XAI not only increases transparency but also paves the way for the development of more intelligent, resilient, and secure AI systems.

11.2 Adversarial Attacks in Machine Learning

With the advancement of machine learning and its increasing application in critical domains such as cybersecurity, healthcare, and finance, new challenges have emerged regarding the safety and reliability of these models. One of the most significant challenges is adversarial attacks, which involve making subtle yet intentional modifications to a model's input to deceive the system and produce incorrect outputs. These modifications are often imperceptible to humans but can mislead machine learning models into making incorrect decisions. Adversarial attacks pose a serious threat to the security of AI systems as they can cause failures in facial recognition, security systems, malware detection, image processing, and even autonomous vehicles. For example, a minor perturbation in an image can cause a computer vision model to misclassify a stop sign as a speed limit sign, which could lead to a catastrophic accident in a self-driving car [12, 19]. In the following sections, we will explore different types of adversarial attacks, their impact on machine learning systems, and common techniques used to execute such attacks.

11.2.1 Definition and Types of Adversarial Attacks

Adversarial attacks refer to deliberate and strategically crafted modifications in the input of a machine learning model with the goal of deceiving the model and producing incorrect outputs. These modifications are often imperceptible to humans but can cause machine learning models to misclassify inputs [6]. Adversarial attacks pose a serious threat to the security and reliability of machine learning models, potentially compromising AI applications in cybersecurity, computer vision, and speech recognition. Adversarial attacks can be classified in various ways, but the most common classification is based on the attacker's access level to the model and the execution method [5, 8].

Classification Based on the Attacker's Access to the Model

- White-box Attacks: In this type of attack, the attacker has full access to the model's architecture, parameters, and weights. With this level of access, the attacker can use gradient-based techniques to generate precise perturbations that force the model to misclassify inputs.
- Black-box Attacks: In black-box attacks, the attacker has no knowledge of the model's architecture or parameters and can only observe the model's outputs based on given inputs. The attacker can generate adversarial examples through trial and error, leveraging model responses to create deceptive inputs.

Classification Based on the Attack Execution Method

- Offline (Non-targeted) Attacks: In this type of attack, the adversary's goal is not to misclassify inputs into a specific category, but rather to cause the model to make any incorrect classification. For instance, a picture of a cat may be misclassified as a dog.
- Targeted Attacks: In targeted attacks, the attacker intentionally manipulates the input so that the model classifies it into a specific, incorrect category. For example, an attacker may modify an image of one person so that a facial recognition system falsely identifies them as someone else.

Classification Based on the Attack Timing

- Training-time Attacks (Poisoning Attacks): In poisoning attacks, the attacker injects malicious data into the training dataset, causing the model to learn incorrect patterns. As a result, the model makes incorrect predictions when deployed.

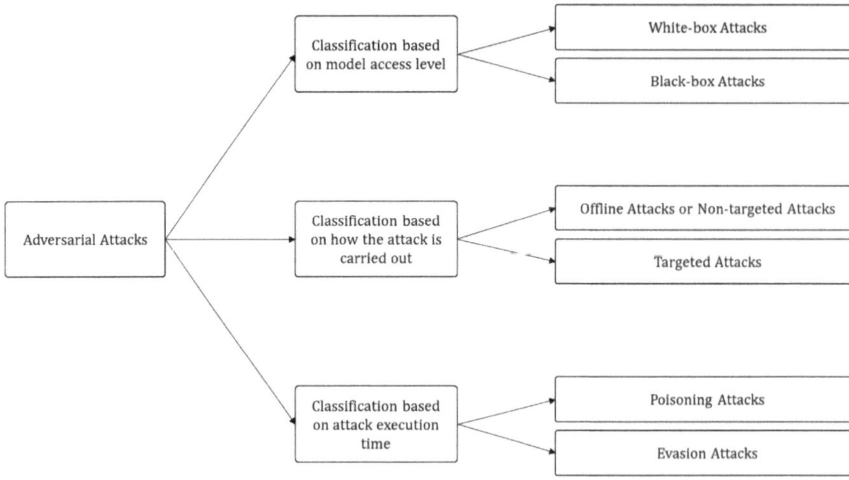

Fig. 11.1 Classification of adversarial attacks

– Inference-time Attacks (Evasion Attacks): Evasion attacks occur after the model has been trained, meaning the attacker modifies inputs at inference time to deceive the model. The attacker crafts adversarial inputs to exploit the model's weaknesses and manipulate its decisions.

Figure 11.1 provides a summary of different types of adversarial attacks.

11.2.2 The Impact of Adversarial Attacks on Machine Learning Systems

Adversarial attacks can have severe and widespread consequences for machine learning systems, particularly in high-stakes applications such as cybersecurity, autonomous vehicles, facial recognition, and medical diagnostics. These attacks not only degrade model accuracy but can also lead to dangerous decision-making, data breaches, and security vulnerabilities. The impact of adversarial attacks can be examined from multiple perspectives:

Reduction in Model Accuracy and Performance
Adversarial attacks can cause machine learning models to misclassify input data, leading to a significant drop in accuracy. This reduction can be especially harmful in computer vision, speech processing, and fraud detection systems.
Examples:

– In an image classification model, an image of a cat might be misclassified as a dog.

– In a facial recognition system, an unauthorized individual may be falsely identified as a legitimate user.

Cybersecurity Threats and Increased Vulnerabilities
Machine learning-based security systems, such as intrusion detection, malware detection, and intelligent firewalls, can be compromised by adversarial attacks. Attackers can subtly modify malicious code to bypass detection mechanisms, making malware appear benign.

Risks to Autonomous Vehicles and Smart Transportation
One of the most critical threats posed by adversarial attacks is their impact on self-driving cars and intelligent transportation systems. Even minor modifications in traffic signs can mislead autonomous systems, leading to potentially catastrophic consequences.
Example:

– A small perturbation in a stop sign might cause an autonomous vehicle to interpret it as a speed limit sign (e.g., 80 km/h), potentially leading to a fatal accident.

Disruptions in Medical Systems and Disease Diagnosis
Machine learning models are widely used in medical diagnostics for detecting diseases and analyzing medical images. However, adversarial attacks can manipulate these models to misclassify medical conditions, leading to serious health risks.
Example:

– A model may incorrectly classify a healthy MRI scan as cancerous or fail to detect a serious illness, leading to misdiagnosis and incorrect treatment plans.

Challenges in Financial Systems and Fraud Detection
Machine learning plays a crucial role in fraud detection and credit risk assessment in banks and financial institutions. Attackers can exploit adversarial techniques to bypass fraud detection models, allowing fraudulent transactions to be marked as legitimate.
Example:

– A fraudster can modify transaction details just enough to deceive a fraud detection system, enabling large-scale financial fraud.

Data Leakage and Privacy Attacks
Adversarial attacks can be exploited to extract sensitive information from machine learning models.
Example:

– Membership inference attacks enable attackers to determine whether a specific data sample was included in the training dataset, leading to privacy breaches and exposure of confidential data.

Adversarial attacks pose a serious threat to the security and reliability of machine learning systems. They can lead to reduced model accuracy, disruptions in critical applications, cybersecurity threats, and privacy breaches. Therefore, identifying and mitigating these attacks is crucial. In the next section, we will explore common techniques used to generate adversarial attacks and their underlying mechanisms.

11.3 Explainable AI: Concepts and Techniques

This section examines three important and widely used techniques: LIME, SHAP, and attention-based models. Each of these methods employs different approaches to explain machine learning model decisions and can be applied across various domains, including computer vision, natural language processing (NLP), and cybersecurity.

11.3.1 LIME (Local Interpretable Model-Agnostic Explanations)

LIME, short for Local Interpretable Model-agnostic Explanations, is a model-agnostic approach used to interpret the outputs of machine learning models. This method aims to explain the behavior of a complex model at a local level by constructing a simple, linear model around a specific instance of data [22]. LIME is particularly useful for highly complex models, such as deep neural networks and random forest-based models [17].

How LIME Works [22]

1. Generating Similar Samples: LIME creates multiple new samples by introducing small perturbations to the original input instance.
2. Predicting Using the Original Model: The perturbed samples are fed into the machine learning model, and their corresponding predictions are obtained.
3. Training a Simple Local Model: Based on the model's predictions and the features of the generated samples, LIME trains a simple, interpretable model (e.g., linear regression or decision tree) that approximates the behavior of the complex model only in the local vicinity of the given input.

Evaluating Feature Importance The local model identifies the most influential input features that contribute to the predictions of the original model.

Advantages of LIME [17]

- Model-agnostic—Can be applied to any machine learning model, regardless of its complexity.

- Local interpretability—Provides explanations for individual model predictions.
- Versatile—Applicable to image, text, and numerical data, making it useful across multiple domains.

Limitations of LIME [22]
- Provides only a local approximation—Cannot fully capture the overall behavior of the model.
- Sensitive to small data variations—May produce inconsistent explanations for slightly altered inputs.

LIME is widely used in computer vision, NLP, and cybersecurity to evaluate model performance and identify vulnerabilities in decision-making processes.

11.3.2 SHAP (Shapley Additive Explanations)

SHAP, short for Shapley Additive Explanations, is a game theory-based approach for interpreting machine learning models. The goal of SHAP is to quantify the contribution of each input feature to a model's prediction and assign a fair value to it [2]. This method is built upon the Shapley value, a concept from game theory used to fairly distribute rewards or costs among team members [11].

How SHAP Works [3]
- Generating all possible combinations of input features: SHAP creates various subsets of input features and evaluates the model's prediction for each subset.
- Calculating changes in model predictions: The impact of each feature is assessed by comparing the model's output with and without that feature.
- Computing the Shapley value for each feature: Using game-theoretic formulas, SHAP quantifies each feature's contribution to the final model prediction.
- Graphical and analytical representation of results: The computed Shapley values are visualized using intuitive plots such as Summary Plot, Force Plot, and Dependence Plot, making the model's decision process more understandable.

Execution Structure of SHAP [10]
- Input: A single data instance (or a dataset) for model interpretation.
- Process: The model's prediction variations are analyzed by computing Shapley values for all feature combinations.
- Output: SHAP values indicating the relative importance of each feature in the final model decision.

Advantages of SHAP [11]
- Fair and consistent explanations based on game theory principles.
- More precise and detailed interpretations compared to methods like LIME.
- Effective for complex models such as neural networks, XGBoost, Random Forest, and SVM.
- Advanced visualization tools for deeper analysis of model decisions.

Limitations of SHAP [10]

- Computationally expensive due to the need to evaluate multiple feature combinations.
- Time-consuming when applied to highly complex models or large datasets.

SHAP is widely used in financial data analysis, fraud detection, NLP, and cybersecurity to examine model decisions, identify weaknesses, and uncover biases in machine learning systems.

11.3.3 Attention-Based Models

Attention-based models are a type of deep learning architecture that enable models to focus on the most relevant parts of the input while extracting key information for decision-making. These models are particularly effective in natural language processing (NLP), computer vision, and pattern recognition. The core idea behind attention mechanisms is that not all input components hold equal importance, and the model should identify and emphasize the most critical features that influence the final output [13].

How Attention-Based Models Work [15]

- Computing attention scores: The model assigns an importance score (weight) to each input component based on its relevance to the output.
- Assigning weights to different input sections: Input segments with greater relevance receive higher attention weights.
- Integrating information based on attention: The model extracts the most critical information based on the computed weights and uses it for final predictions.

Execution Structure of Attention-Based Models [13]

- Input: Raw data such as text, images, or time-series data.
- Process: Compute attention scores for each input section $\rightarrow$ Integrate the most important features $\rightarrow$ Generate the model's final prediction.
- Output: A model prediction with weighted focus on the most critical parts of the input.

Types of Attention-Based Architectures [15]

- Self-Attention: This mechanism allows the model to analyze relationships between different parts of a single input. Commonly used in Transformer models for NLP (e.g., BERT, GPT).
- Soft Attention The model assigns a continuous probability distribution of weights over the entire input sequence.
- Hard Attention: The model focuses only on specific parts of the input while ignoring the rest, making it sparser and more computationally efficient.

Advantages of Attention-Based Models [13, 15]
- Improves deep learning model accuracy by prioritizing the most relevant input components.
- Enhances interpretability by highlighting the key input features that impact decisions the most.
- Well-suited for sequential data such as text, video, and time-series information.

Limitations of Attention-Based Models [15]
- Computationally expensive as they require processing the entire input simultaneously.
- Optimization challenges in complex models, requiring proper initialization techniques.

By improving model interpretability in complex decision-making, attention-based models play a crucial role in enhancing security and transparency in machine learning. They can also be used to detect adversarial attacks by identifying input patterns that may indicate malicious data manipulations.

11.4 The Role of XAI in Detecting and Mitigating Adversarial Attacks

With the growing security threats in machine learning, Explainable AI (XAI) has emerged as a powerful tool for identifying and mitigating model vulnerabilities against adversarial attacks. Machine learning models typically operate as black boxes, making their decision-making processes opaque. This lack of transparency allows attackers to exploit model weaknesses [4]. By providing detailed explanations of how models make decisions, XAI helps identify vulnerabilities and enables the precise analysis of adversarial attack impacts. Furthermore, interpretability techniques such as LIME, SHAP, and attention-based models can detect anomalous patterns and suspicious behaviors in model outputs, aiding in the detection of adversarial attacks.

This section explores how XAI can be effectively used to uncover model vulnerabilities, detect attack patterns, and propose defensive strategies against security threats.

Explainable AI techniques can reveal machine learning model vulnerabilities by analyzing model decisions, identifying sensitive features, and detecting unusual patterns in the data. These methods assess how input features influence model predictions, highlighting weaknesses that attackers may exploit. Below are some of the most important XAI methods for identifying model vulnerabilities:

Sensitivity Analysis of Models Using Feature Importance Technique
Techniques such as LIME (Local Interpretable Model-agnostic Explanations) and SHAP (Shapley Additive Explanations) help determine which input features con-

tribute most to the model's predictions. If a model is overly dependent on specific features, attackers can manipulate these features to launch targeted adversarial attacks.

Examples:

- In facial recognition models, if the system relies heavily on a few specific facial features, slight modifications to those features could cause misclassification.
- In malware detection models, if only a single key feature determines classification, attackers can subtly alter that feature to make malware appear as benign software.

Detecting Unusual and Anomalous Data Patterns

XAI algorithms can analyze model predictions to identify inputs with abnormal behavior. Attention-based models, such as Transformers, can pinpoint which parts of an input influence decision-making the most. If an irrelevant but manipulated feature significantly alters the output, XAI methods can detect such anomalous patterns during an adversarial attack.

Identifying Adversarial Attack Patterns Through Anomaly Detection

Using SHAP visualizations and other model analysis tools, it is possible to differentiate between normal model behavior and adversarial attack patterns. If a machine learning model produces highly unusual predictions under specific conditions, it may indicate the presence of an adversarial attack.

Examples in Security Systems

- In an intrusion detection model, a malicious request with minimal modifications might bypass security filters. SHAP can reveal how such slight changes influence the model's decision.
- In fraud detection models, adversarial techniques may enable fraudulent transactions to bypass standard detection methods. XAI can help uncover the hidden patterns behind these manipulations.

Analyzing Model Errors and Weaknesses Against Attacks

By evaluating model predictions under various conditions, it is possible to determine where and why the model fails. For example, in computer vision models, LIME can highlight specific regions of an image that should not be crucial for classification, yet still impact the model's decision. Such insights allow developers to identify model weaknesses and improve robustness.

Explainable AI techniques such as feature sensitivity analysis, anomaly detection, and adversarial pattern recognition can reveal weaknesses in machine learning models. These capabilities enable developers to design more robust, secure, and trustworthy models, reducing the risk of adversarial exploitation. By leveraging XAI, organizations can enhance model transparency, improve cybersecurity defenses, and build resilient AI systems capable of withstanding adversarial attacks.

Enhancing ML Model Resilience Using XAI

With the increasing adoption of machine learning in high-stakes domains such as cybersecurity, medical diagnosis, and autonomous transportation, challenges such as adversarial attacks, data leakage, and uncertainty in model predictions have gained significant importance. Complex machine learning models, especially deep neural networks, are often highly sensitive and vulnerable to even minor changes in their input data. This sensitivity means that even an imperceptible modification, which is undetectable to humans, can drastically alter the model's output. In such scenarios, enhancing model resilience—the ability of a model to maintain its performance despite threats and uncertainties—has become a key priority [4].

One of the most effective approaches to improving the resilience of machine learning models is the use of Explainable AI (XAI). By providing transparent explanations of model decision-making processes, XAI can help analyze and identify model weaknesses and suggest strategies to improve security and robustness. Techniques such as SHAP, LIME, and attention-based models can reveal which features the model relies on and how an adversarial attack or data perturbation can affect predictions. This information enables developers to design more resilient models that are less susceptible to adversarial attacks and data alterations. A major challenge in machine learning is ensuring model robustness against unforeseen changes, uncertain data, and adversarial threats. Deep learning models often exhibit hidden vulnerabilities that, when exploited by attackers, can lead to incorrect outputs and security breaches. Explainable AI (XAI) can play a crucial role in this regard by identifying key features, revealing weaknesses in the model, and assisting in the improvement of its design.

The role of XAI in identifying key features, uncovering model weaknesses, and contributing to model improvement can be outlined as follows [4, 18]:

Identifying and Reducing Overdependence on Sensitive Features

Many machine learning models may overly rely on specific features within the dataset, making them highly sensitive to slight modifications. A minor alteration in these features can cause unexpected changes in model predictions. XAI methods like SHAP and LIME can identify these high-sensitivity areas, revealing which features contribute the most to the model's decisions.

Practical Examples:

- In facial recognition models, if XAI indicates that the model is overly dependent on skin texture or lighting conditions, attackers could exploit this by subtly modifying these areas, leading to misclassification.
- In malware detection models, if XAI analysis reveals that the model relies on a few specific code patterns, attackers could slightly modify those patterns to bypass security filters.

Solution:

Once these high-sensitivity features are identified, the model should be redesigned to reduce its reliance on them. This can be achieved by:

- Augmenting the training dataset with diverse data to improve model generalization.
- Adjusting model weights to decrease dependence on vulnerable features.
- Implementing robust learning strategies, such as adversarial training, to enhance resistance to small perturbations.

Strengthening Model Robustness Against Adversarial Attacks Using XAI Analysis

Adversarial attacks typically exploit weaknesses in machine learning models by introducing subtle yet deceptive perturbations. XAI techniques can analyze these attacks and help developers design more resilient and stable model architectures.

Practical Examples:

- In computer vision, Grad-CAM and SHAP can highlight the regions of an image that the model focuses on the most. If the model primarily relies on image edges, attackers could slightly modify these edges to deceive the model.
- In speech analysis models, XAI can reveal the frequency bands that the model is most sensitive to. Attackers could inject imperceptible noise in those bands to confuse the model's predictions.

Solution:

- Adversarial Training: Exposing the model to adversarial examples during training to improve its resistance against such modifications.
- Regularization Techniques: Implementing dropout, weight decay, or feature smoothing to reduce dependency on specific features.
- Ensemble Learning: Combining multiple models with different structures to increase model robustness and reduce the impact of adversarial inputs.

Enhancing Decision Transparency and Correcting Model Errors

XAI techniques can pinpoint incorrect model decisions and explain why the model makes errors in certain cases. Such insights can guide model redesign and increase its resilience.

Practical Examples:

- In fraud detection systems, if a model flags a transaction as fraudulent without a clear reason, XAI can clarify the decision, helping financial analysts determine whether the classification is justified.
- In medical diagnosis, if an AI model incorrectly classifies a patient as having cancer, XAI can reveal which MRI regions or patient attributes contributed to the error.

Solution:

- Refining the dataset to remove or mitigate biases that lead to incorrect decisions.

– Revising the model architecture based on XAI insights to ensure more reliable predictions.

Explainable AI techniques such as SHAP, LIME, and attention-based models play a crucial role in enhancing the resilience of machine learning models by:

– Identifying and reducing reliance on highly sensitive features.
– Detecting and mitigating adversarial attack vulnerabilities.
– Providing transparency into decision-making to correct errors.

By integrating XAI-driven insights into model development, organizations can design machine learning models that are more robust, secure, and trustworthy, ensuring their effectiveness against adversarial attacks and real-world uncertainties.

11.5 Conclusion and Future Research Directions

The research presented in this chapter demonstrated that Explainable AI (XAI) plays a crucial role in enhancing the security and resilience of machine learning systems against adversarial attacks. While machine learning models, particularly deep neural networks, exhibit high accuracy, they often function as black boxes, meaning that their decision-making processes remain opaque to users and even developers. This lack of transparency not only raises trust issues but also increases vulnerability to adversarial attacks. In this context, XAI techniques such as SHAP, LIME, and attention-based models have proven to be effective tools for interpreting and analyzing machine learning model decisions, helping to identify weaknesses, strengthen security, and improve robustness against adversarial threats.

One section of this chapter explored XAI's ability to analyze model dependency on specific input features. By examining these dependencies, researchers can identify high-sensitivity areas in the model, which can be exploited by adversaries. For example, in computer vision systems, using Grad-CAM has shown that minor modifications in certain pixel regions can lead to incorrect predictions. These findings emphasize that to enhance security, it is not only necessary to develop robust models but also to employ techniques that explain why a model reaches a particular decision.

Additionally, the use of XAI in analyzing adversarial attacks has demonstrated that explainability techniques can detect attack patterns, enabling developers to design more effective defense mechanisms. Self-attention-based models, particularly in natural language processing (NLP) and financial transaction analysis, have shown high potential in detecting anomalies and strengthening model security. In the cybersecurity domain, XAI-based analyses have revealed that many threat detection models rely on specific network traffic patterns, which attackers can subtly manipulate to bypass security mechanisms. These findings highlight that integrating XAI with security frameworks can create more resilient systems against adversarial attacks. Despite the success of XAI in improving the security of machine learning systems, several challenges remain, which must be addressed in future research. Some of the most critical challenges and research directions include:

Improving XAI's Robustness Against Adversarial Attacks

- Just as machine learning models are vulnerable to adversarial attacks, XAI techniques themselves can be targeted and manipulated.
- Attackers can exploit methods such as Explanation Manipulation to alter XAI outputs, generating misleading explanations that conceal model weaknesses.
- Future research should focus on developing adversarially robust XAI methods that provide reliable and tamper-resistant explanations.

Integrating XAI with Robust Learning Techniques

- Most current research uses XAI as a post-hoc analysis tool, but future advancements could involve embedding XAI into the training process to automatically improve model resilience.
- Potential research directions include combining XAI with Reinforcement Learning (RL) and Transfer Learning, allowing models to learn adaptively while improving security and resilience.Integrating XAI with Robust Learning Techniques
- Most current research uses XAI as a post-hoc analysis tool, but future advancements could involve embedding XAI into the training process to automatically improve model resilience.
- Potential research directions include combining XAI with Reinforcement Learning (RL) and Transfer Learning, allowing models to learn adaptively while improving security and resilience.

Optimizing XAI for Large-Scale and Real-Time Processing

- Existing XAI methods, particularly SHAP and LIME, are computationally expensive, making them inefficient for large-scale models and real-time applications.
- Future research should focus on scalable and optimized XAI techniques that maintain high interpretability while reducing computational overhead, particularly for online learning environments and high-frequency data streams.

Leveraging XAI for Bias Detection in Machine Learning Models

- Many machine learning models exhibit undesirable biases, leading to unfair and discriminatory decisions.
- XAI has proven effective in identifying and mitigating biases, particularly in applications such as facial recognition, hiring systems, and financial decision-making.
- Future studies should focus on developing XAI-driven fairness evaluation frameworks to improve algorithmic accountability and fairness.

Adapting XAI for Emerging Deep Learning Architectures

- As Transformer models, Graph Neural Networks (GNNs), and Self-Supervised Learning architectures continue to evolve, conventional XAI methods may not be fully applicable to these frameworks.
- A key research challenge is to develop new XAI methodologies tailored for advanced deep learning architectures to maintain interpretability in cutting-edge AI systems.

Explainable AI (XAI) has emerged as a fundamental tool for improving the security, resilience, and trustworthiness of machine learning systems. By helping developers identify weaknesses, reduce vulnerabilities, and design more effective defense mechanisms, XAI plays an essential role in ensuring the safe deployment of AI technologies. However, several challenges remain, shaping the future research landscape in this field. The development of more robust, efficient, and scalable XAI methods will be critical to advancing machine learning security. Future research should focus on combining XAI with robust learning, cybersecurity defense mechanisms, and bias mitigation strategies to create safer and more accountable AI systems. Additionally, optimizing XAI for large-scale applications and integrating it with evolving deep learning architectures will be crucial in ensuring that AI remains both interpretable and secure in future applications.

References

1. Ahmad, R., Wazirali, R., Abu-Ain, T.: Machine learning for wireless sensor networks security: an overview of challenges and issues. Sensors **22**(13), 4730 (2022)
2. Antwarg, L., Miller, R.M., Shapira, B., Rokach, L.: Explaining anomalies detected by autoencoders using SHAP. Preprint. arXiv:1903.02407 (2019)
3. Antwarg, L., Miller, R.M., Shapira, B., Rokach, L.: Explaining anomalies detected by autoencoders using Shapley additive explanations. Expert Syst. Appl. **186**, 115736 (2021)
4. Baniecki, H., Biecek, P.: Adversarial attacks and defenses in explainable artificial intelligence: a survey. Inf. Fusion **107**, 102303 (2024)
5. Chakraborty, A., Alam, M., Dey, V., Chattopadhyay, A., Mukhopadhyay, D.: A survey on adversarial attacks and defences. CAAI Trans. Intell. Technol. **6**(1), 25–45 (2021)
6. ForouzeshNejad, A.A., Arabikhan, F., Aheleroff, S.: Optimizing project time and cost prediction using a hybrid XGBoost and simulated annealing algorithm. Machines **12**(12), 867 (2024)
7. ForouzeshNejad, A.A., Arabikhan, F., Williams, N., Gegov, A., Sari, O., Bader, M.: Agile project status prediction using interpretable machine learning. In: 2024 IEEE 12th International Conference on Intelligent Systems (IS), pp. 1–8. IEEE, Piscataway (2024)
8. Huang, S., Papernot, N., Goodfellow, I., Duan, Y., Abbeel, P.: Adversarial attacks on neural network policies. Preprint. arXiv:1702.02284 (2017)
9. Khan, M., Ghafoor, L.: Adversarial machine learning in the context of network security: challenges and solutions. J. Comput. Intell. Robot. **4**(1), 51–63 (2024)
10. Mangalathu, S., Hwang, S.H., Jeon, J.S.: Failure mode and effects analysis of RC members based on machine-learning-based Shapley additive explanations (SHAP) approach. Eng. Struct. **219**, 110927 (2020)
11. Nohara, Y., Matsumoto, K., Soejima, H., Nakashima, N.: Explanation of machine learning models using Shapley additive explanation and application for real data in hospital. Comput. Methods Program Biomed. **214**, 106584 (2022)

12. Nowroozi, E., Taheri, R., Hajizadeh, M., Bauschert, T.: Verifying the robustness of machine learning based intrusion detection against adversarial perturbation. In: 2024 IEEE International Conference on Cyber Security and Resilience (CSR), pp. 9–15. IEEE, Piscataway (2024)
13. Osman, A.A., Shalaby, M.A.W., Soliman, M.M., Elsayed, K.M.: A survey on attention-based models for image captioning. Int. J. Adv. Comput. Sci. Appl. **14**(2) (2023)
14. Rudin, C.: Stop explaining black box machine learning models for high stakes decisions and use interpretable models instead. Nat. Mach. Intell. **1**(5), 206–215 (2019)
15. Srivastava, P., Bej, S., Yordanova, K., Wolkenhauer, O.: Self-attention-based models for the extraction of molecular interactions from biological texts. Biomolecules **11**(11), 1591 (2021)
16. Tan, Z.: Exploring enterprise digital transformation: the perspective of explainable machine learning. In: 2024 6th International Conference on Machine Learning, Big Data and Business Intelligence (MLBDBI), pp. 1–5. IEEE, Piscatawy (2024)
17. Toğaçar, M., Muzoğlu, N., Ergen, B., Yarman, B.S.B., Halefoğlu, A.M.: Detection of covid-19 findings by the local interpretable model-agnostic explanations method of types-based activations extracted from CNNs. Biomed. Signal Process. Control **71**, 103128 (2022)
18. Vadillo, J., Santana, R., Lozano, J.A.: Adversarial attacks in explainable machine learning: a survey of threats against models and humans. Wiley Interdiscip. Rev. Data Min. Knowl. Discov. **15**(1), e1567 (2025)
19. Wang, Y., Wu, L., Jin, J., Wang, E., Zhang, Z., Zhao, Y.: An imperceptible adversarial attack against 3d object detectors in autonomous driving. IEEE Internet Things J. (2025)
20. Xu, F., Uszkoreit, H., Du, Y., Fan, W., Zhao, D., Zhu, J.: Explainable AI: a brief survey on history, research areas, approaches and challenges. In: Natural Language Processing and Chinese Computing: 8th CCF International Conference, NLPCC 2019, Dunhuang, China, October 9–14, 2019, Proceedings, Part II 8, pp. 563–574. Springer, Cham (2019)
21. Xue, M., Yuan, C., Wu, H., Zhang, Y., Liu, W.: Machine learning security: threats, countermeasures, and evaluations. IEEE Access **8**, 74720–74742 (2020)
22. Zafar, M.R., Khan, N.M.: DLIME: A deterministic local interpretable model-agnostic explanations approach for computer-aided diagnosis systems. Preprint. arXiv:1906.10263 (2019)

Chapter 12
Neurodevelopmental-Inspired Training Enhances Adversarial Robustness of a Primary Visual Cortex-Based Model

Ehsan Ur Rahman Mohammed, Elham Bagheri, Apurva Narayan, and Yalda Mohsenzadeh

12.1 Introduction

As artificial intelligence (AI) models are increasingly deployed in safety-critical applications, ensuring robustness against adversarial attacks has become a critical area of research. These attacks exploit model vulnerabilities by injecting imperceptible perturbations, leading to misclassifications that can compromise decision-making in domains such as autonomous vehicles, medical diagnosis, and cybersecurity [5, 10, 21, 22]. Despite ongoing research in adversarial defenses, most existing methods struggle to generalize across different types of attack, limiting their practical effectiveness in real-world scenarios. The human visual system demonstrates remarkable robustness against noisy and adversarial distortions in vision tasks. Unlike AI models, humans can filter out misleading perturbations and maintain stable recognition under varying conditions. Although studies show that humans can be fooled by adversarial attacks in time-limited settings [9, 24], these instances rarely lead to significant real-time behavioral problems [27]. This resilience is attributed to brain's neurodevelopmental processes, which shape visual representations through early-life learning experiences. Inspired by this phenomenon, researchers are exploring biologically motivated strategies to improve AI robustness. These strategies aim to overcome the limitations of conventional AI approaches through innovations in architecture, training, and regularization techniques.

Traditional convolutional neural networks (CNNs), like Alexnet [15] and ResNet [11], excel in object recognition when tested on data from the same distribution as the training data. However, studies have demonstrated their limited generalization

E. U. R. Mohammed (✉) · E. Bagheri · A. Narayan · Y. Mohsenzadeh
Western University, London, ON, Canada
e-mail: mrahm326@uwo.ca; ebagher5@uwo.ca; apurva.narayan@uwo.ca; ymohsenz@uwo.ca

E. Nowroozi et al. (eds.), *Adversarial Example Detection and Mitigation Using Machine Learning*, https://doi.org/10.1007/978-3-031-99447-0_12

ability, particularly their failure to perform well on unseen or out-of-distribution data [12], which makes them vulnerable to both adversarial and natural distribution shifts. They have demonstrated susceptibility to various sources of noise and common real-world variations, including weather conditions, lighting conditions, measurement errors [8], and occlusion [20]. While non-adversarial sources of noise have not been as extensively studied as adversarial noises, research has shown a correlation between robustness against both forms of noise [16].

VOneNet [6], a brain-inspired model designed to mimic the V1 area of the primate visual cortex, has shown increased adversarial robustness compared to conventional CNNs. The core component of VOneNet, the VOneBlock, incorporates biologically realistic Gabor filters that encode texture and frequency content in images, aligning with human early-stage visual processing mechanisms [19]. This biologically inspired structure improves resistance to high-frequency adversarial perturbations but does not fully eliminate model vulnerabilities. Beyond model architecture, brain-inspired AI also includes training mechanisms that mimic the brain-development cycle of primates and learning mechanisms.

Blur-training methods [14], including strong blur training (SBT), weak blur training (WBT), and blurry-to-clear (B2C) training [13, 14], simulate how humans refine visual perception over time. These methods gradually expose models to increasing levels of image detail, reducing over-reliance on high-frequency information while enhancing generalization to adversarial and natural distortions. In particular, strong blur training (SBT) and weak blur training (WBT) emphasize different levels of spectral bias, while B2C training progressively sharpens visual details over training iterations, mimicking human perceptual learning. However, the potential of these training methods to enhance the adversarial robustness of brain-inspired networks like VOneNet remains underexplored.

Recent surveys highlight the expanding landscape of adversarial attacks and defenses. Techniques such as Projected Gradient Descent (PGD) [18] and Carlini and Wagner [4] have been adapted across different domains. Further, advancements in explainable AI (XAI) have revealed that adversarial attacks can not only manipulate model predictions but also distort their interpretability, further emphasizing the need for more robust learning strategies [2].

In this study, we propose that human neurodevelopmental-inspired training methods, such as blur training and blurry-to-clear (B2C) training, can significantly boost the adversarial robustness of brain-inspired architectures like VOneNet. We also explore how these training regimes impact VOneNet's performance compared to traditional models like AlexNet and ResNet50.

12.2 Materials and Methods

The human visual system's ability to maintain robustness against adversarial distortions is closely tied to its hierarchical visual processing mechanisms, particularly its handling of blurred retinal inputs at different spatial resolutions. Whenever a visual scene is projected onto the retina, much of it appears blurred due to a combination of peripheral low-resolution vision and optical defocus in central vision. However, despite the frequent exposure to blur in natural vision, CNNs are typically trained on clear images, which may lead to an over-reliance on high spatial frequency details for object recognition [14]. The lack of blurry training inputs and dependence on fine-grained features could potentially result in systematic differences from biological vision.

A critical aspect of human vision is its developmental trajectory, where visual acuity starts off poor in infancy and gradually improves over the first few months of life. This developmental trajectory is essential for learning to interpret and refine blurry visual inputs into clearer, more detailed perceptions. Research by Jang and Tong [13] explored whether the inherent robustness to blurred images in humans could be linked to these early visual development stages. The study investigated the effects of such early experiences with blur in both face and object recognition settings, finding that CNNs trained with brain-inspired approaches, such as blur training and blurry-to-clear (B2C) training, maintained robustness to blur, aligning with findings by Vogelsang et al. [25].

Building on these insights, our study introduces a novel exploration of these biologically inspired training techniques—SBT, WBT, B2C—applied to a biologically inspired vision model: VOneNet. VOneNet, which explicitly models early-stage visual processing in the primate brain, employs Gabor filters that mimic the frequency and orientation selectivity of V1 neurons. This unique combination of blur-based training methods with a brain-inspired architecture like VOneNet represents a significant step forward in aligning machine learning models with the fundamental principles of human visual development, potentially leading to more robust and human-like visual processing in AI systems.

By applying these training methods to VOneNet and comparing their impact on robustness against traditional CNN architectures like AlexNet and ResNet50, we seek to determine whether neurodevelopmental-inspired learning improves adversarial robustness beyond standard architectural modifications.

Hypotheses Our primary hypothesis is that by manipulating the training data, we can reduce models' sensitivity to higher-frequency details like texture, thereby enhancing their robustness against adversarial attacks. We propose that training models with blurred images, which emphasize lower-frequency content, can lead to a more balanced model performance, particularly under adversarial conditions. Additionally, we hypothesize that these brain-inspired training methods, particularly SBT, WBT, and B2C training, may enhance VOneNet's performance more effectively than traditional models like AlexNet and ResNet50, as the integration of biologically inspired architectures and training methodologies may lead to greater adversarial robustness.

Model Development We began with normal clear training (CT), where the models were trained with the original, unaltered images. We then implemented two variants of blur training, as proposed by Jang et al. [13, 14], which include SBT, WBT, and B2C. Experiments are applied on MNIST [17], CIFAR10 [1], and ImageNet100 (a 100-class subset of ImageNet) [7]. After training, we evaluated the models' adversarial robustness using two widely recognized adversarial attacks: Projected Gradient Descent (PGD) and Carlini & Wagner (C&W). Both attacks are highly effective at generating adversarial examples, making them critical benchmarks for assessing model resilience.

PGD is an iterative method where small perturbations are added to the input image in the direction that maximizes the model's loss function. This perturbation is constrained within an ϵ-ball to ensure the altered image remains close to the original. In our experiments, ϵ was carefully tuned to balance the strength of the attack with the need to keep perturbations minimal. The step size α and the number of iterations k were adjusted to ensure convergence to a successful adversarial example while maintaining computational efficiency. The Carlini & Wagner (C&W) attack, known for finding minimal perturbations that lead to misclassification, was implemented with a focus on optimizing the confidence parameter c. This parameter controls the trade-off between minimizing the perturbation and ensuring that the model misclassifies the input. Higher c values increase the likelihood of a successful attack but result in larger perturbations.

Both attacks were applied in their untargeted forms, aiming to cause misclassification without forcing the model to predict a specific incorrect class. This approach provides a comprehensive measure of the model's robustness across different training methodologies. A wide range of perturbations was used for both attacks. To quantitatively measure adversarial robustness, we evaluated clean accuracy and adversarial accuracy. Clean accuracy reflects model performance on unaltered test data, while adversarial accuracy assesses performance when test images have been perturbed by PGD or C&W attacks.

Interpreting the Results For qualitative analysis of the model's performance, we used additional techniques including t-SNE visualization and ScoreCAM. t-SNE [23] is a dimensionality reduction technique used to project high-dimensional data into 2D/3D space, allowing for the visualization of class separations and decision boundaries. ScoreCAM [26] is an XAI tool that highlights the regions in an image most relevant to a model's prediction.

12.3 Results

MNIST: Insights from Simplified Data For the MNIST dataset, the top-1 accuracy values for all models were above 80% with tight confidence intervals, indicating high reliability (Fig. 12.1). VOneNet model with AlexNet backbone consistently outperformed the rest of the baseline models, including VOneNet with the ResNet50 backbone, especially when the perturbation budget is high ($\epsilon = 0.1$) under PGD attacks (Fig. 12.2). However, both SBT and WBT increased attack success rates among all models, suggesting potential drawbacks of increased sensitivity to texture rather than shape. This can be attributed to the MNIST

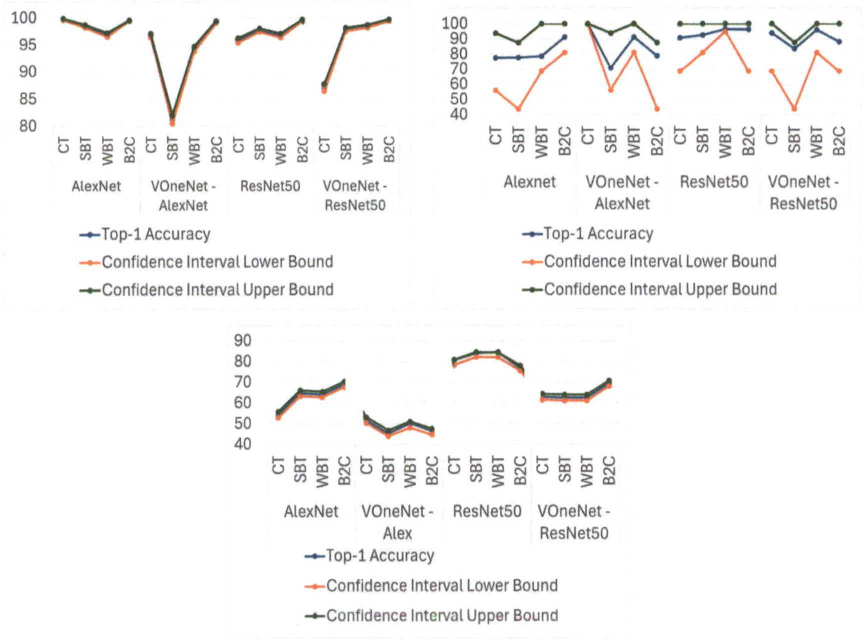

Fig. 12.1 Top-1 accuracy trends: Left to right top row: MNIST, CIFAR10; bottom row: ImageNet100

dataset's lack of texture information, thereby biasing the models towards texture-focused learning, which ultimately reduces the focus on shapes. B2C training models outperformed all the baseline models and other blur training models. This suggests that initially focusing on low-frequency details and gradually transitioning to high-frequency features enhances overall robustness. The additional t-SNE plots, presented in Fig. 12.5, further illustrate that B2C training leads to a clear separation between similar classes (e.g., '0' and '8'), particularly in the B2C models with the ResNet-50 backbone, indicating superior robustness. The t-SNE plots for the rest of the datasets, presented in the supplementary material, further illustrate that B2C training leads to better separation between similar classes (e.g., '0' and '8'), particularly in B2C models with the ResNet50 backbone, indicating superior robustness.

CIFAR10: Challenges with Low-Resolution Data The CIFAR10 dataset posed significant challenges for the VOneNet models. Due to low resolution, the lack of detailed edges and textures made it difficult for the VOneBlock filters to effectively extract meaningful features, leading to poorer performance, particularly with SBT, WBT, and B2C methods. In contrast, traditional models like AlexNet and ResNet50 performed better, illustrating the difficulty posed by the complexity of CIFAR10's images. Accuracy curves for CIFAR10 (Fig. 12.1) showed much wider confidence intervals than those observed in MNIST and ImageNet100, suggesting greater

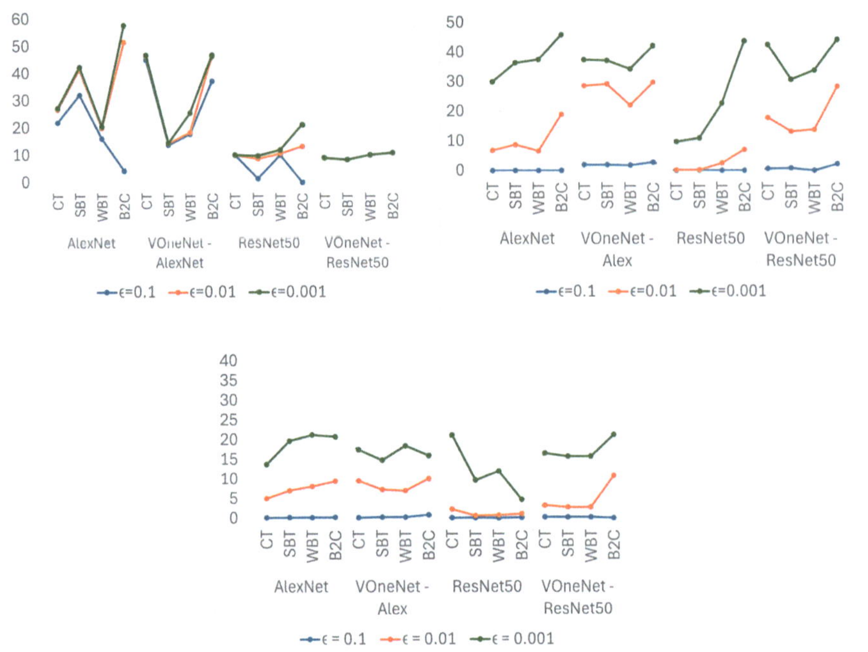

Fig. 12.2 Accuracy under PGD attack. Left to right top row: MNIST, CIFAR10; bottom row: ImageNet100

variability and less reliable parameter estimation. Among all models, VOneNet with the AlexNet backbone performed best, exhibiting the narrowest confidence intervals. B2C training improved performance over WBT, underscoring its effectiveness even with low-resolution data. However, the performance of VOneNet with the ResNet50 backbone remained suboptimal, with minimal drops in attack success rates even under increased attack strength, highlighting the unique challenges posed by CIFAR10.

In Fig. 12.6, the ScoreCAM heatmaps present an explanation on CIFAR10 dataset examples. Due to the wider confidence intervals owing to the low resolution of the images in the CIFAR10 dataset, the explanations tend to be generally on the poorer case and are pretty similar to one another. However, B2C training in some cases tends to present better explanations than the rest of the training methods, such as in the case of ResNet50 on ship and car images.

ImageNet100: B2C Training Reduces Dependence on Confounding Features ResNet50 outperformed all models in clean accuracy across all training mechanisms on the ImageNet100 dataset. B2C training also outperformed other training methods in clean accuracy, particularly for AlexNet, VOneNet with AlexNet backbone, and ResNet50 backbone, reflecting similar trends observed in the MNIST results. Under the PGD attack with a low perturbation budget, B2C-trained models consistently

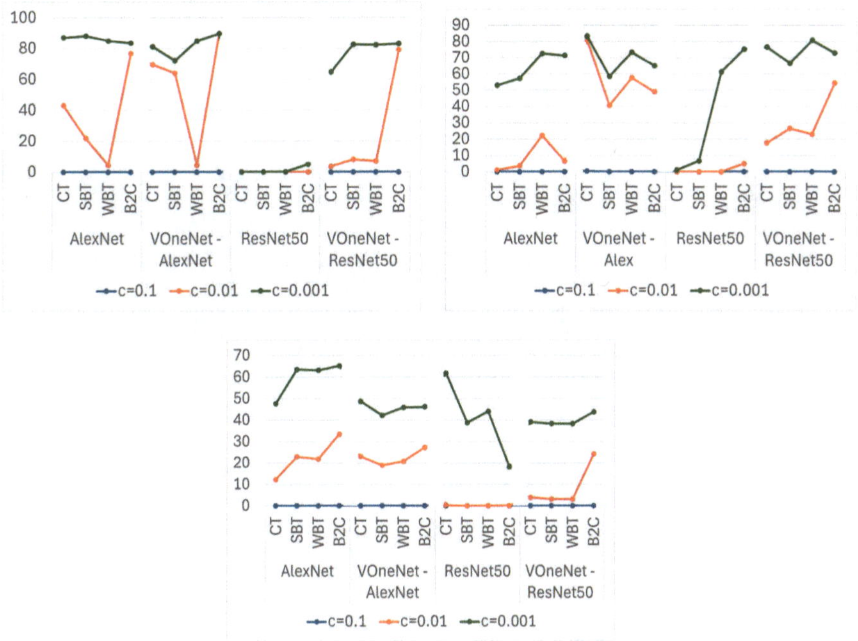

Fig. 12.3 Accuracy under C&W Attack. Left to right top row: MNIST, CIFAR10; bottom row: ImageNet100

outperformed others, particularly at a budget of (ϵ = 0.01 (Fig. 12.2). This trend persisted under C&W attacks, where B2C models showed greater robustness (Fig. 12.3). Interestingly, ResNet50 models performed the worst under these adversarial conditions, likely due to potential overfitting.

The t-SNE plots confirm these findings, showing improved class separation in WBT models, particularly in ResNet50 and VOneNet with ResNet50 backbones. This shows WBT's effectiveness in maintaining class distinctions. ScoreCAM images obtained for ImageNet100 in Fig. 12.4 also demonstrated that WBT and B2C models exhibit improved interpretability, focusing more effectively on prediction-relevant features, suggesting that B2C training not only enhances adversarial robustness but also improves model interpretability by directing attention toward critical image regions (Figs. 12.5 and 12.6).

VOneNet, especially with AlexNet backbone, consistently showed superior resistance to adversarial attacks compared to traditional models. This finding aligns with previous research by Dapello et al. [6], which reported a 30% higher adversarial accuracy for VOneNet with a ResNet50 backbone on ImageNet. The robustness of ResNet-based models on the ImageNet100 and MNIST datasets was not examined in their study. SBT, WBT, and B2C training significantly improved model robustness by increasing spectral bias [3] towards low-frequency information. B2C training, in particular, showed the best results across datasets. This study validates

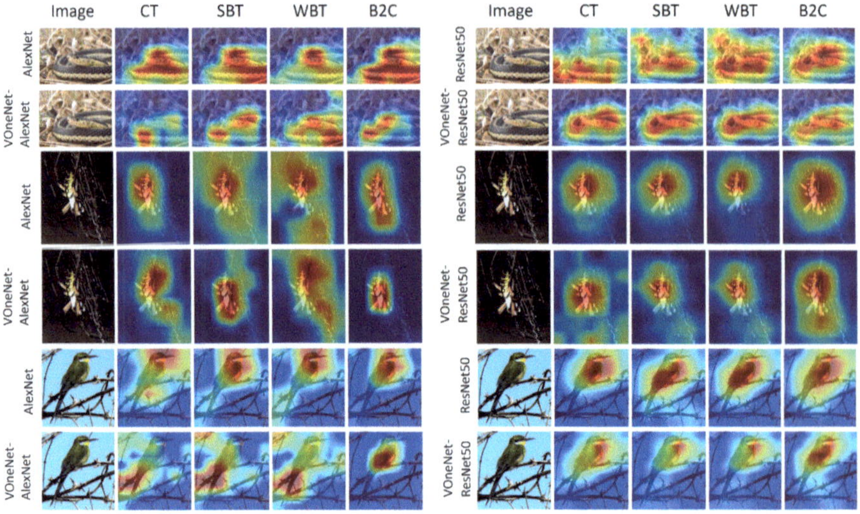

Fig. 12.4 ScoreCAM on ImageNet100. Brighter shades mean higher contribution

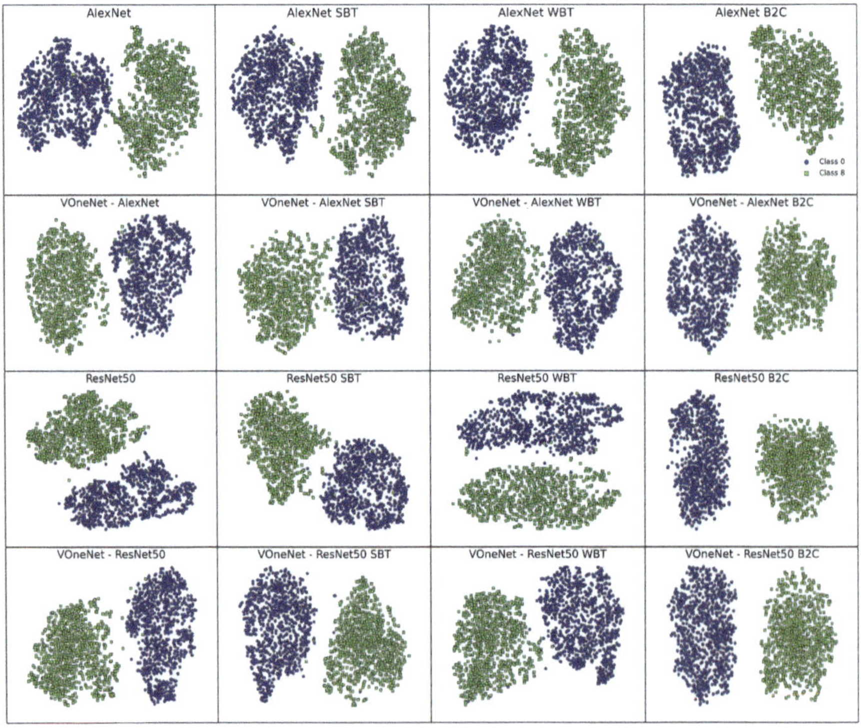

Fig. 12.5 Decision boundaries of test dataset of MNIST classes '0' and '8'

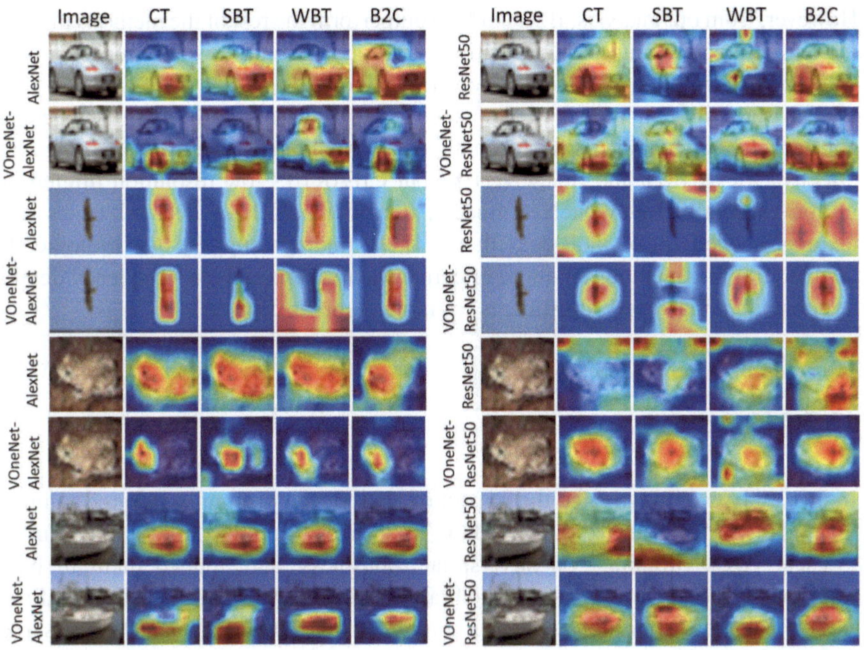

Fig. 12.6 ScoreCAM visual explanations on CIFAR10 images highlight which parts of the images contributed to the prediction and quantify their contribution where brighter shade means higher contribution

the findings of Jang et al. [14], who showed that training CNNs with blurred images improved adversarial robustness, as evidenced by their evaluation using PGD attacks with an L_∞ norm and $\epsilon = 0.001$.

12.4 Discussion

The results show that B2C training greatly enhances model robustness, especially in datasets like MNIST and ImageNet100 where shape and low-frequency information are key. VOneNet's strong performance, particularly with the AlexNet backbone, underscores the potential of brain-inspired architectures for enhancing robustness, aligning with findings by Dapello et al. [6]. t-SNE and ScoreCAM visualizations further validate B2C training's effectiveness, revealing clearer class separations and more focused attention on prediction-relevant features.

To the best of our knowledge, the adversarial robustness of B2C training [13] has not been demonstrated in the literature, making this work the first to present its merit. Jang and Tong [13] claimed that CNNs trained on objects under similar conditions did not retain the same robustness to blur as CNNs trained on faces.

However, from our analysis, B2C tends to outperform the rest of the methods in most model architecture-dataset combinations, calling for further exploration of model behavior under both traditional and adversarial noise.

The variability in model performance across datasets highlights the need to consider dataset characteristics when choosing training methods. Blur training was effective for MNIST and ImageNet100, but CIFAR10's low resolution limited its utility, especially for VOneNet models, due to the lack of detailed edges and textures needed for VOneBlock filters. These findings suggest that biologically inspired training methods are more effective where shape and low-frequency details dominate, but may need adjustments for more complex datasets.

12.5 Conclusions

In this study, we investigated the impact of neurodevelopmentally inspired training methods—Strong Blur Training (SBT), Weak Blur Training (WBT), and Blurry-to-Clear (B2C) training—on adversarial robustness, and applied them to both traditional CNN architectures and the brain-inspired VOneNet model. Our findings reveal that these training methods improve model resilience against adversarial attacks. B2C training particularly demonstrates the most substantial impact. By gradually transitioning from low-frequency to high-frequency content, B2C training reduces spectral bias, making models less reliant on high-frequency details. This approach aligns closely with human visual development, where visual acuity gradually improves over time.

We found that VOneNet, when trained with B2C, outperformed traditional CNN models in resisting adversarial attacks, especially on simpler datasets like MNIST. To the best of our knowledge, this study is the first to show that B2C training outperforms other methods across multiple model architecture-dataset combinations, including both traditional CNNs and brain-inspired models like VOneNet. This study highlights the potential of combining biologically inspired training methods with architectural designs that mimic the human visual system. Future work could explore expanding these methods to more complex vision tasks, further integrating biologically inspired mechanisms to close the gap between artificial and human-like perception.

References

1. Alex, K.: Learning multiple layers of features from tiny images (2009). https://www.cs.toronto.edu/kriz/learning-features-2009-TR.pdf
2. Baniecki, H., & Biecek, P.: Adversarial attacks and defenses in explainable artificial intelligence: a survey. Inf. Fusion **107**, 102303 (2024)
3. Cao, Y., Fang, Z., Wu, Y., Zhou, D.X., Gu, Q.: Towards understanding the spectral bias of deep learning. Preprint. arXiv:1912.01198 (2019)

4. Carlini, N., Wagner, D.: Towards evaluating the robustness of neural networks. In: 2017 IEEE Symposium on Security and Privacy (SP), pp. 39–57. IEEE, Piscataway (2017)
5. Chernikova, A., Oprea, A., Nita-Rotaru, C., Kim, B.G.: Are self-driving cars secure? Evasion attacks against deep neural networks for steering angle prediction. In: 2019 IEEE Security and Privacy Workshops (SPW), pp. 132–137. IEEE, Piscataway (2019)
6. Dapello, J., Marques, T., Schrimpf, M., Geiger, F., Cox, D., DiCarlo, J.J.: Simulating a primary visual cortex at the front of CNNs improves robustness to image perturbations. Adv. Neural Inf. Process. Syst. **33**, 13073–13087 (2020)
7. Deng, J., Dong, W., Socher, R., Li, L.J., Li, K., Fei-Fei, L.: Imagenet: a large-scale hierarchical image database. In: 2009 IEEE Conference on Computer Vision and Pattern Recognition, pp. 248–255. IEEE, Piscataway (2009)
8. Drenkow, N., Sani, N., Shpitser, I., Unberath, M.: A systematic review of robustness in deep learning for computer vision: mind the gap? Preprint. arXiv:2112.00639 (2021)
9. Elsayed, G., Shankar, S., Cheung, B., Papernot, N., Kurakin, A., Goodfellow, I., Sohl-Dickstein, J.: Adversarial examples that fool both computer vision and time-limited humans. Adv. Neural Inf. Process. Syst. **31**, 3914–3924 (2018)
10. Girdhar, M., Hong, J., Moore, J.: Cybersecurity of autonomous vehicles: a systematic literature review of adversarial attacks and defense models. IEEE Open J. Veh. Technol. **4**, 417–437 (2023)
11. He, K., Zhang, X., Ren, S., Sun, J.: Deep residual learning for image recognition. In: Proceedings of the IEEE Conference on Computer Vision and Pattern Recognition, pp. 770–778 (2016)
12. Hendrycks, D., Basart, S., Mu, N., Kadavath, S., Wang, F., Dorundo, E., Desai, R., Zhu, T., Parajuli, S., Guo, M., et al.: The many faces of robustness: a critical analysis of out-of-distribution generalization. In: Proceedings of the IEEE/CVF International Conference on Computer Vision, pp. 8340–8349 (2021)
13. Jang, H., Tong, F.: Convolutional neural networks trained with a developmental sequence of blurry to clear images reveal core differences between face and object processing. J. Vis. **21**(12), 6–6 (2021)
14. Jang, H., Tong, F.: Improved modeling of human vision by incorporating robustness to blur in convolutional neural networks. Nat. Commun. **15**(1), 1989 (2024)
15. Krizhevsky, A., Sutskever, I., Hinton, G.E.: Imagenet classification with deep convolutional neural networks. Adv. Neural Inf. Process. Syst. **25**, 1097–1105 (2012)
16. Laugros, A., Caplier, A., Ospici, M.: Are adversarial robustness and common perturbation robustness independent attributes? In: Proceedings of the IEEE/CVF International Conference on Computer Vision Workshops (2019)
17. LeCun, Y.: The MNIST database of handwritten digits (1998). http://yann.lecun.com/exdb/mnist/
18. Madry, A., Makelov, A., Schmidt, L., Tsipras, D., Vladu, A.: Towards deep learning models resistant to adversarial attacks. Preprint. arXiv:1706.06083 (2017)
19. Olshausen, B.A., Field, D.J.: Emergence of simple-cell receptive field properties by learning a sparse code for natural images. Nature **381**(6583), 607–609 (1996)
20. Rajaei, K., Mohsenzadeh, Y., Ebrahimpour, R., Khaligh-Razavi, S.M.: Beyond core object recognition: recurrent processes account for object recognition under occlusion. PLoS Comput. Biol. **15**(5), e1007001 (2019)
21. Szegedy, C., Zaremba, W., Sutskever, I., Bruna, J., Erhan, D., Goodfellow, I., Fergus, R.: Intriguing properties of neural networks. Preprint. arXiv:1312.6199 (2013)
22. Vakhshiteh, F., Nickabadi, A., Ramachandra, R.: Adversarial attacks against face recognition: a comprehensive study. IEEE Access **9**, 92735–92756 (2021)
23. Van der Maaten, L., Hinton, G.: Visualizing data using t-SNE. J. Mach. Learn. Res. **9**(11), 2579 (2008)
24. Veerabadran, V., Goldman, J., Shankar, S., Cheung, B., Papernot, N., Kurakin, A., Goodfellow, I., Shlens, J., Sohl-Dickstein, J., Mozer, M.C., et al.: Subtle adversarial image manipulations influence both human and machine perception. Nat. Commun. **14**(1), 4933 (2023)

25. Vogelsang, L., Gilad-Gutnick, S., Ehrenberg, E., Yonas, A., Diamond, S., Held, R., Sinha, P.: Potential downside of high initial visual acuity. Proc. Natl. Acad. Sci. **115**(44), 11333–11338 (2018)
26. Wang, H., Wang, Z., Du, M., Yang, F., Zhang, Z., Ding, S., Mardziel, P., Hu, X.: Score-CAM: score-weighted visual explanations for convolutional neural networks. In: Proceedings of the IEEE/CVF Conference on Computer Vision and Pattern Recognition Workshops, pp. 24–25 (2020)
27. Zhou, Z., Firestone, C.: Humans can decipher adversarial images. Nat. Commun. **10**(1), 1334 (2019)

Chapter 13
Evaluating and Defending Against Adversarial Attacks on LLM-Generated LSTM Models

Mani Ghahremani, Arsenii Podshyvalin, and Rahim Taheri

13.1 Introduction

Large Language Models (LLMs), including widely adopted systems like ChatGPT, LLaMa, Falcon, and PaLM, as well as newer entrants such as Databricks' Dolly 2.0, have redefined natural language processing by extending their capabilities beyond traditional text-based tasks [12, 27]. These models are now integral to applications in search engines [28], customer support [21], recruiting [10], healthcare [8, 16, 22], education [18], and finance [24], demonstrating their versatility and transformative potential across industries. For instance, ChatGPT alone serves over 180 million users globally [9], reflecting the widespread reliance on these systems. However, their increasing ubiquity also underscores the importance of addressing ethical and practical concerns, such as ensuring fairness, privacy, and mitigating biases as well as security issues [1].

The application of LLMs in code generation has received noteworthy attention in the literature, particularly for their potential to simplify complex tasks such as building AI models like Long Short-Term Memory (LSTM) networks. LLMs, including ChatGPT, PaLM, and LLaMa 2, have demonstrated the ability to generate executable code for LSTM models directly from natural language prompts [12, 19]. This capability is transformative for data analysts and software developers, enabling them to create sophisticated forecasting models without extensive programming expertise [17]. For example, ChatGPT has been extensively studied for its state-of-the-art performance in code generation, showing promising results on benchmarks like HumanEval and CodeContests [19].

M. Ghahremani (✉) · A. Podshyvalin · R. Taheri
School of Computing, University of Portsmouth, Portsmouth, UK
e-mail: mani.ghahremani@port.ac.uk

E. Nowroozi et al. (eds.), *Adversarial Example Detection and Mitigation Using Machine Learning*, https://doi.org/10.1007/978-3-031-99447-0_13

This paper aims to answer the question of how robust are LSTM models generated by LLMs to adversarial attacks, and how effective are existing defenses in mitigating these vulnerabilities. To address this, we adopt the methodology of Gopali et al. [12], which demonstrated that state-of-the-art LLMs such as ChatGPT, Falcon, LLaMa, and Mistral can generate well-performing LSTM models comparable to manually crafted architectures. While Gopali et al.'s work focused on evaluating the accuracy and performance of these models, we extend their research by systematically analyzing their robustness.

To this end, we recreate the experimental setup of Gopali et al. [12], where the following LLMs were used to generate LSTM models for time series forecasting: ChatGPT (GPT-3.5-Turbo), Falcon, LLaMa 2, and Mistral. In our work, we utilize four similar LLMs–Meta Llama 3 8B Instruct, Mistral 7B Instruct v0.3, Gemma 1.1 7B Instruct, and gpt-4o-mini (2024–07–18)–to generate LSTM models tailored for financial time series forecasting, utilizing the same dataset as Gopali et al. [12], sourced from Yahoo Finance [26]. We then evaluate the robustness of these models under adversarial conditions, focusing on training-time data poisoning attacks. Moreover, we implement and assess sanitization and anomaly detection defense mechanisms to mitigate these vulnerabilities. By employing these contemporary LLMs and incorporating adversarial testing, our work extends the study of Gopali et al. [12] by examining the functional accuracy of LLM-generated models and their resilience to security threats.

The remainder of the paper is organized as follows: In Sect. 13.2, we review relevant literature, including attacks and defense mechanisms. Section 13.3 describes our methodology, including the dataset and experimental setup. Section 13.4 presents our results, comparing the performance of LLM-generated LSTMs under different conditions. Finally, Sect. 13.5 discusses our findings' implications and suggests future research directions.

13.2 Literature Review

LLMs have demonstrated remarkable advancements in code generation, transforming the landscape of software development. Tools like GitHub Copilot, based on LLMs, have been extensively studied for their usability and impact on programming workflows. Vaithilingam et al. [23] found that while Copilot provided helpful starting points and reduced the need for web searches, users often encountered challenges in debugging the generated code. Similarly, Jiang et al. [14] highlighted the rapid evolution of LLMs in generating code from natural language prompts, showcasing their applicability in various domains.

Surveys by Jin et al. [15] and Chen et al. [5] emphasize the increasing reliance on LLMs for tasks like autonomous software design, test generation, and code maintenance. They note the critical role of benchmarks in evaluating code generation quality, focusing on metrics like correctness, readability, and efficiency. These works also underline ongoing challenges, such as ensuring consistency, mitigating errors, and refining evaluation methodologies to better align with real-world software development practices.

Our paper examines the security implications of LLM-generated models, particularly in the context of time series forecasting using LSTM networks. While LLMs have shown promise in generating accurate LSTM models for forecasting tasks, their susceptibility to adversarial attacks remains a critical concern. In the following sections, we review existing literature on adversarial attacks and defenses in machine learning models, focusing on LSTM architectures and time series forecasting applications. It is important to note that the attacks considered in this paper are solely those applied to the training dataset X, and correspondingly the defenses are designed to operate on X. We do not consider attacks that directly target the model or defenses that are applied at the model level.

The poisoning attacks and defenses examined in this paper operate on time series datasets formatted as input sequences to LSTM models. Each dataset is structured as a three-dimensional tensor X of shape $[N, T, V]$, where: N is the number of training samples in the dataset, T is the number of consecutive time steps used as input for predicting the next time step, and V is the number of features per time step. For the dataset used in our paper, the values of these values are as follows: $N = 78$, $T = 5$, and $V = 4$. See Sect. 13.3 for more details on the dataset used in our experiments.

We first consider three common adversarial attacks on LSTM models that are applied to the training dataset. We then explore defense strategies against these attacks. Each defense strategy discussed is designed to mitigate a specific class of attack: The Outlier Detection Defense is intended to counter the Backdoor Data Poisoning Attack, the Red Team/Blue Team Defense is designed to combat Arithmetic-Based Poisoning, and the Reverse-Engineering Defense is selected to defend against the Adversarial Perturbations Attacks (such as FGSM and BIM).

Later, in Sect. 13.3, we will discuss our methodology and implement these attacks and defenses to evaluate the robustness of LSTM models generated by the selected LLMs for time series forecasting. For a more comprehensive discussion of the attacks and defenses presented in this paper, we recommend the interested reader to visit survey papers such as [11, 13].

13.2.1 Arithmetic-Based Poisoning (Attack 1)

Chen et al. [6] proposed an arithmetic-based poisoning attack, which introduces structured perturbations that degrade model accuracy while remaining inconspicuous. We have summarized this in Algorithm 1.

Chen's results demonstrate that this methodology significantly increases the Root Mean Square Error (RMSE) of LSTM predictions. For instance, tailored attacks on road segments caused a dramatic increase in RMSE, with optimal strategies varying based on specific road segment characteristics.

Algorithm 1 Arithmetic-based poisoning attack

```
Require: Dataset X, arithmetic operation A, tampering range [a, b]
Ensure: Poisoned dataset X'
Select subset T from X
for each sample X[i] ∈ T do
    Sample Δ ~ U(a, b)
    Apply A(X[i], Δ)
end forreturn X'
```

Algorithm 2 Backdoor data poisoning attack

```
Require: Dataset X, perturbation magnitude σ, target class t*
Ensure: Poisoned dataset X', updated labels y'
Select subset from X
for each selected sample X[i] do
    Perturb features: X'[i] = X[i] + U(−σ, σ)
    Set label y'[i] = t*
end forreturn X', y'
```

13.2.2 Backdoor Data Poisoning Attacks (Attack 2)

Dai [7] and Schuessler [20] explore backdoor poisoning attacks on LSTM models. These attacks introduce systematic perturbations into training data, misleading models into misclassifying specific target classes (t^*) while maintaining high accuracy on unaltered data.

Schuessler's approach perturbs time-series data by introducing noise υ sampled from a uniform distribution $U(-\sigma, \sigma)$ across all time steps. The poisoned dataset closely resembles the original but contains systematic perturbations designed to mislead the model. This method is summarized in Algorithm 2.

Schuessler's experiments highlight a trade-off between perturbation magnitude σ and the effectiveness of the attack. Their best results achieved 92% poisoning accuracy while maintaining 92% accuracy on unaltered data.

13.2.3 Adversarial Perturbations (Attack 3)

Gradient-based adversarial attacks, such as the Fast Gradient Sign Method (FGSM) explored by Arslan [2], manipulate inputs to reduce the accuracy of LSTM models. These attacks perturb data points using gradients of the model's loss function with respect to the input. Algorithm 3 summarizes this approach:

The attack caused a significant drop in the R^2 score in both regression and classification tasks. In the latter, FGSM caused reductions from 81.8% to 64.2% in univariate data, and from 98.05 to 18.68% in multivariate contexts.

Algorithm 3 FGSM attack algorithm

Require: Input dataset X, perturbation magnitude ϵ
Ensure: Poisoned dataset X'
1: **for** each selected sample $X[i]$ **do**
2: Compute gradient $G = \nabla_{X[i]}\mathcal{L}(X[i])$
3: Perturb input: $X'[i] = X[i] + \epsilon \cdot sign(G)$
4: **end forreturn** X'

Algorithm 4 Red team/blue team defense

Require: Clean dataset X, attacked dataset X', detection threshold θ
Ensure: Restored dataset X''
1: Train classifier to detect tampered samples
2: Train regression model to restore tampered samples
3: **for** each sample $X'[i]$ **do**
4: **if** Classifier($X'[i]$) $> \theta$ **then**
5: Restore $X''[i]$ using regression model
6: **end if**
7: **end forreturn** X''

13.2.4 Red Team/Blue Team Defense (Defense 1)

To counter the arithmetic-based poisoning attack, Chen et al. [6] defined a Red Team/Blue Team defense strategy. This defense detects and restores arithmetic-poisoned data using classifiers and regression models trained on clean reference data. The simplified method is summarized in Algorithm 4:

However, reliance on the availability of clean reference data restricts its applicability in practical scenarios.

13.2.5 Outlier Detection Defense (Defense 2)

Outlier detection methods, such as Local Outlier Factor (LOF) [4], aim to identify and remove poisoned data points by assessing deviations in input space. Inputs that deviate significantly from their class centroids or exhibit anomalously high local outlier factors are flagged as potential attack samples. Methods based on robust mean estimation and K-Nearest Neighbors (KNN) have been adapted for this purpose. For further details on outlier detection, the reader is referred to [3]. A typical implementation is provided in Algorithm 5.

A key limitation is that removing outliers from time-series datasets may introduce gaps, potentially reducing the predictive accuracy of LSTM models.

Algorithm 5 Outlier detection defense

Require: Dataset X, outlier threshold τ
Ensure: Filtered dataset X'
1: Compute LOF scores for samples in X
2: **for** each sample $X[i]$ **do**
3: **if** LOF($X[i]$) $> \tau$ **then**
4: Remove $X[i]$ from dataset
5: **end if**
6: **end forreturn** X'

Algorithm 6 Reverse-engineering defense

Require: Clean dataset X, attacked dataset X', classifier C, threshold τ
Ensure: Restored dataset X''
1: Compute perturbation vectors v between X and X'
2: Identify minimal perturbation vector v^*
3: **for** each sample $X'[i]$ misclassified by C **do**
4: **if** perturbation norm of $X'[i] \leq \tau$ **then**
5: Restore $X''[i] = X'[i] - v^*$
6: **end if**
7: **end forreturn** X''

13.2.6 Reverse-Engineering Defense (Defense 3)

Xiang et al. [25] proposed a reverse-engineering defense against gradient-based adversarial perturbations. The simplified approach involves estimating minimal perturbations to reverse attacks, as summarized in Algorithm 6.

This defense, although effective, has high computational overhead and assumes access to clean data, which may not always be available.

13.3 Methodology

We utilize a dataset sourced from Yahoo Finance [26], namely the Apple stock ticker (AAPL) within the Technology sector in the USA. We point out that this dataset was originally used in [12], and consisted of 77 data points spanning the period from 2022–01–01 to 2022–04–23.

To generate our LSTM models, we used the four LLMs, Meta Llama 3 8B Instruct, Mistral 7B Instruct v0.3, Gemma 1.1 7B Instruct, and gpt-4o-mini (2024–07–18). Our LLMs were run on Azure's OpenAI service but due to API limitations on OpenAI, we had to access them via different API services: OpenAI API for gpt-4o-mini, Hugging Face Serverless API for Gemma and Mistral, and Google Vertex AI for Llama 3.

We used the highest-ranking prompt from [12] to generate the LSTM models:

Can you assist me in creating a comprehensive Python script to build an LSTM architecture using the time series dataset enclosed within double backticks "{data}"? My objective is to execute steps such as preprocessing, splitting, building, compiling, training, and evaluating models using RMSE.

To assess the robustness of LLM-generated LSTM models, we implemented the three presented in Sect. 13.2, namely Backdoor Data Poisoning, Arithmetic-Based Poisoning, and Adversarial Perturbations. We then applied our implementation of these attacks to the dataset. These attacks introduced perturbations to the data, manipulated numerical values, and used gradient-based techniques to degrade model performance, respectively. The attacks resulted in three more poisoned datasets (Attack 1, Attack 2, and Attack 3).

To mitigate the impact of these attacks, we implemented the three defense strategies discussed in Sect. 13.2, specifically Outlier Detection Defense, Red Team/Blue Team Defense, and Reverse-Engineering Defense. Our implementations of these defense strategies were respectively applied to restore the poisoned datasets Attack 1, Attack 2, and Attack 3. The resulting (restored) datasets were named Defense 1, Defense 2, and Defense 3 respectively.

This adds up to a total of seven datasets for each of the four models iterated 10 times, resulting in 280 experiments, which we will discuss in Sect. 13.4

13.4 Results

The performance of the LSTM models generated by the LLMs was measured using Root Mean Squared Error (RMSE). RMSE quantifies the average prediction error of the model by computing the square root of the mean of the squared differences between actual and predicted stock prices. We also calculated the Mean Absolute Percentage Error (MAPE) to evaluate the models' performance. MAPE measures the average percentage error of the model predictions relative to the actual stock prices. A lower MAPE and RMSE indicate predictive accuracy and generalization of the LSTM models.

In our experiments, we evaluated model accuracy using RMSE and MAPE during training, recording validation losses at each epoch. Additionally, we measured RMSE and MAPE on poisoned datasets to assess model performance under adversarial conditions, as well as after applying defense strategies. While validation losses were tracked throughout training, they were not directly analyzed in our results; future work may explore their potential insights (Table 13.1).

The RMSE values for each of the four LLM-generated LSTM models were computed across the seven datasets (clean, three poisoned datasets, and their corresponding restored versions). The RMSE values are presented below:

To further illustrate the variations in RMSE across datasets and models, Fig. 13.1 presents a visual representation.

The results indicate that the GPT and Llama-generated LSTMs consistently exhibited the highest RMSE values, signifying the worst performances. Conversely,

Table 13.1 RMSE values across different datasets

Dataset	Gemma	GPT	Llama	Mistral
Clean	0.6438	4.5877	6.0337	3.5794
attack 1 (Arithmetic-based poisoning)	8.7368	5.0094	4.3263	4.5330
Attack 2 (Backdoor data poisoning attacks)	8.5176	4.9116	6.0797	4.9625
Attack 3 (Adversarial perturbations)	8.5738	4.7449	4.7490	4.0186
Defense 1 (Red team/blue team defense)	7.8033	4.1392	4.4932	3.7786
Defense 2 (Outlier detection defense)	7.6936	6.1182	7.2474	2.5447
Defense 3 (Reverse-engineering defense)	7.9483	4.1737	4.5666	3.8151

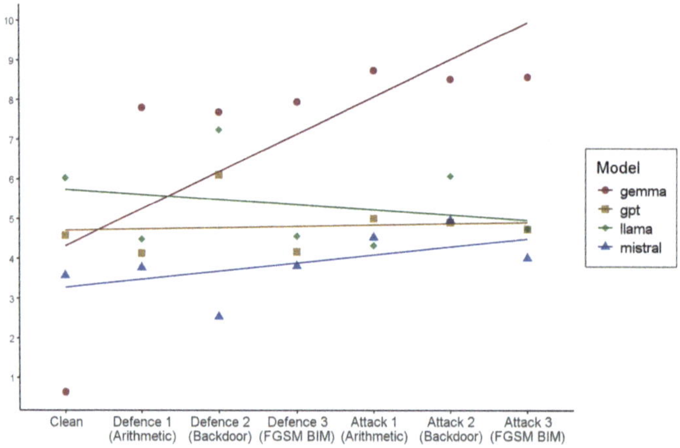

Fig. 13.1 RMSE values for each LSTM model across the seven datasets. A lower RMSE indicates better model performance and accuracy

the Gemma-generated LSTM achieved the lowest RMSE across most datasets, making it the most effective in terms of predictive accuracy.

As expected, models trained on the clean dataset generally yielded lower RMSE values compared to those trained on poisoned datasets. The exception was observed in Mistral's model under the Backdoor Data Poisoning Attack, where the RMSE was slightly lower than that of the clean dataset. This anomaly suggests that the attack might have induced a form of regularization effect.

The application of defense mechanisms showed a positive impact in reducing RMSE values. Notably, the Outlier Detection Defense (Defense 2) and Reverse-Engineering Defense (Defense 3) achieved RMSE values lower than even the clean dataset in some instances (e.g., GPT and Gemma models). This outcome suggests that the defenses, while effective at mitigating attack effects, may have also introduced some overcorrections, leading to artificially low RMSE scores.

We need to point out a distinction between self-accuracy–measured using RMSE and MAPE–which quantifies how well the model predicts values relative to the given dataset, and the deviation between model predictions and the clean dataset values, which is influenced by poisoning. A highly accurate model in a clean setting

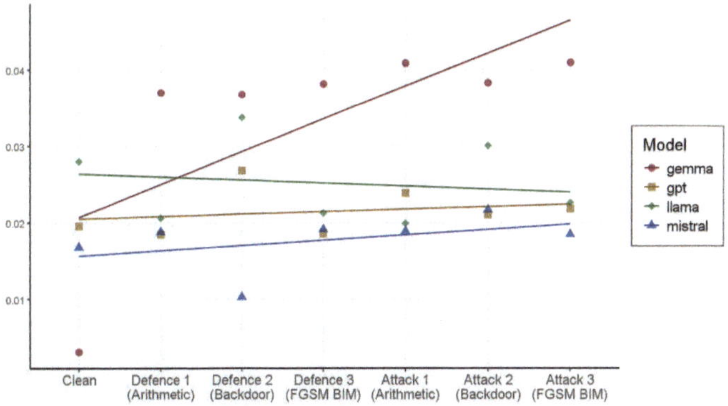

Fig. 13.2 MAPE values averaged over all models

may be more susceptible to poisoning, as it learns dataset-specific patterns more effectively. Conversely, models with inherently lower self-accuracy (such as GPT) show less deviation when poisoned because their errors are already high.

This trend is reflected in Fig. 13.1, where models with high self-accuracy exhibit steeper RMSE increases when poisoned, while lower-accuracy models display a more gradual shift. Additionally, models that performed well in clean datasets (e.g., Mistral and Gemma) also demonstrated better recovery when restored, highlighting the effectiveness of defenses for high-accuracy models.

To decrease the impact of outliers, we introduce Fig. 13.2, which uses MAPE instead of RMSE. Percentage-based error is often more interpretable, especially in financial forecasting.

The main limitation in these experiments stems from the small dataset size (77 data points). Given that LSTM models typically require large datasets for effective learning, the observed RMSE instability is likely a result of overfitting. The complex architectures of the AI-generated LSTMs struggle to generalize patterns effectively, resulting in inconsistent performance across datasets.

Figure 13.1 represents the deviation of predictions across our experiments. The slope of the line depicts the steepness of the trend, indicating the impact of poisoning on model performance. The steeper the line (with a positive slope), the more significant the effect of poisoning on the model. The results suggest that models with higher self-accuracy (e.g., Mistral and Gemma) are more susceptible to poisoning, as evidenced by the steeper trend lines. Moreover, these models perform better with restored datasets than with poisoned ones. Conversely, models with a lower self-accuracy (e.g., GPT) exhibit less sensitivity to poisoning, as indicated by the flatter (or inverted) trend line.

This trend is also supported by the MAPE values in Fig. 13.2. We can summarize by ranking models in terms of self-accuracy from most accurate to least: Gemma, Mistral, Llama, and GPT.

Figure 13.3 shows the average RMSE values for the datasets in all models. The clean dataset consistently yielded the lowest RMSE values, while the poisoned

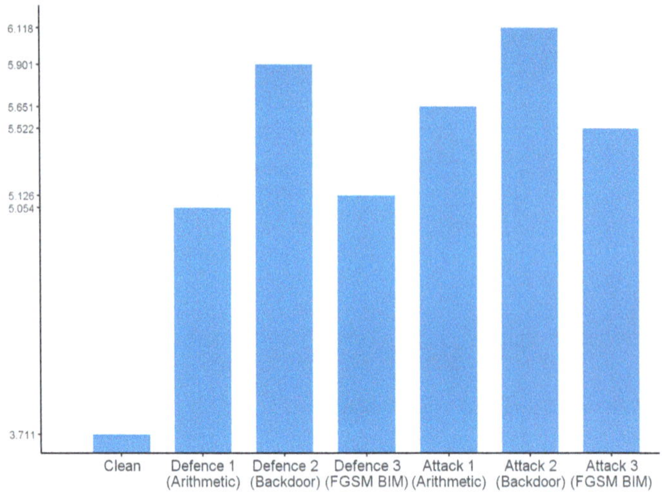

Fig. 13.3 RMSE values averaged over all models

datasets resulted in higher RMSE values and the restored datasets showed a decrease in RMSE values compared to the poisoned counterparts.

13.5 Conclusion and Future Work

This study builds upon earlier research by Gopali et al. in [12]. In our paper, we simulated their experiment to a degree but then evaluated the robustness of LSTM models generated by LLMs against adversarial attacks and assessed the effectiveness of various defense strategies. Our results demonstrate that while AI-generated LSTMs can achieve reasonable predictive accuracy, they lack inherent security mechanisms, making them susceptible to data poisoning and adversarial perturbations.

The RMSE values across different datasets highlight the impact of adversarial attacks on model performance, with poisoned datasets leading to significantly higher error rates. Although the implemented defenses mitigated some of these effects, certain defense strategies produced RMSE values lower than the clean dataset, indicating potential overcorrection or data leakage issues. The instability of results across models further emphasizes the challenges posed by using complex LSTM architectures on small datasets.

Another conclusion of our work is that in models in which MAPE is bigger than 2%, defense mechanisms have little impact on prediction accuracy. This is because the base deviation is already too high. On the positive side, the poison attacks did not influence predictions in such a model either. Simply, defense mechanisms are effective only when MAPE is below a certain level, in our experiments this is under 2–1.6%.

Future work will explore refining the prompt engineering process to generate simpler and more robust LSTMs that align better with the dataset size. In addition, investigating alternative defense mechanisms, such as adversarial training and data augmentation, could provide more stable and effective mitigation strategies. Extending the study to larger datasets with more comprehensive feature sets will also help determine whether the observed vulnerabilities persist in more realistic forecasting scenarios. Also, to deeper analyze data poisoning effects, future work could use validation losses collected during model training.

Furthermore, in this paper, we only used one of the five datasets from Gopali et al. [12]. Future work could involve replicating the experiments on the remaining datasets as well as other financial time series datasets to assess the generalizability of our findings. Also, as part of our experiment, we did not perform any pre-processing on the dataset. Future work could involve pre-processing, such as harmonizing missing values, to ensure consistency across time intervals and improve model performance.

Finally, integrating automated adversarial robustness evaluations within the LLM-generated model pipeline could offer a proactive approach to improving the security and reliability of these models, reducing the risk of deploying vulnerable architectures in real-world applications.

References

1. Abdali, S., He, J., Barberan, C., Anarfi, R.: Can llms be fooled? Investigating vulnerabilities in LLMS (2024). https://arxiv.org/abs/2407.20529
2. Arslan, A.: Afraarslan/adversarial-attacks: adversarial attacks - time-series data - LSTM - regression - classification. https://github.com/afraarslan/adversarial-attacks
3. Boukerche, A., Zheng, L., Alfandi, O.: Outlier detection: methods, models, and classification. ACM Comput. Surv. **53**(3), 1–37 (2020). https://doi.org/10.1145/3381028
4. Breunig, M.M., Kriegel, H.P., Ng, R.T., Sander, J.: LOF: identifying density-based local outliers. SIGMOD Record **29**(2), 93–104 (2000). https://doi.org/10.1145/335191.335388
5. Chen, L., Guo, Q., Jia, H., Zeng, Z., Wang, X., Xu, Y., Wu, J., Wang, Y., Gao, Q., Wang, J., et al.: A survey on evaluating large language models in code generation tasks (2024). arXiv:2408.16498
6. Chen, Y.Y., Hung, H.N., Yang, S.R., Yen, C.C., Lin, P.: On poisoning attacks and defenses for LSTM time series prediction models: speed prediction as an example. In: 2024 International Wireless Communications and Mobile Computing (IWCMC), pp. 610–615 (2024). https://doi.org/10.1109/IWCMC61514.2024.10592413
7. Dai, J., Chen, C., Li, Y.: A backdoor attack against LSTM-based text classification systems. IEEE Access **7**, 138872–138878 (2019). https://doi.org/10.1109/ACCESS.2019.2941376. https://ieeexplore.ieee.org/document/8836465
8. Das, A., Tariq, A., Batalini, F., Dhara, B., Banerjee, I.: Exposing vulnerabilities in clinical LLMS through data poisoning attacks: case study in breast cancer. medRxiv (2024). https://doi.org/10.1101/2024.03.20.24304627
9. Duarte, F.: Number of chatgpt users (2023). Explod. Topics **16**, 2023 (2023)

10. Gan, C., Zhang, Q., Mori, T.: Application of LLM agents in recruitment: a novel framework for resume screening (2024). https://arxiv.org/abs/2401.08315
11. Goldblum, M., Tsipras, D., Xie, C., Chen, X., Schwarzschild, A., Song, D., Mądry, A., Li, B., Goldstein, T.: Dataset security for machine learning: data poisoning, backdoor attacks, and defenses. IEEE Trans. Pattern Anal. Mach. Intell. **45**(2), 1563–1580 (2023). https://doi.org/10.1109/TPAMI.2022.3162397
12. Gopali, S., Siami-namini, S., Abri, F., Siami Namin, A.: The performance of the LSTM-based code generated by large language models (LLMS) in forecasting time series data. Nat. Lang. Process. J. **9**, 100120 (2024). https://doi.org/10.1016/j.nlp.2024.100120. https://www.sciencedirect.com/science/article/pii/S2949719124000682
13. Guo, W., Tondi, B., Barni, M.: An overview of backdoor attacks against deep neural networks and possible defences. IEEE Open J. Signal Process. **3**, 261–287 (2022). https://doi.org/10.1109/OJSP.2022.3190213. https://ieeexplore.ieee.org/document/9827581
14. Jiang, J., Wang, F., Shen, J., Kim, S., Kim, S.: A survey on large language models for code generation (2024). arXiv:2406.00515
15. Jin, H., Huang, L., Cai, H., Yan, J., Li, B., Chen, H.: From LLMS to LLM-based agents for software engineering: a survey of current, challenges and future (2024). arXiv:2408.02479
16. Li, J., Dada, A., Puladi, B., Kleesiek, J., Egger, J.: Chatgpt in healthcare: a taxonomy and systematic review. Comput. Methods Program Biomed. **245**, 108013 (2024). https://doi.org/10.1016/j.cmpb.2024.108013. https://www.sciencedirect.com/science/article/pii/S0169260724000087
17. Lin, F., Kim, D.J., et al.: When LLM-based code generation meets the software development process (2024). arXiv:2403.15852
18. Mbakwe, A.B., Lourentzou, I., Celi, L.A., Mechanic, O.J., Dagan, A.: Chatgpt passing usmle shines a spotlight on the flaws of medical education. PLOS Digital Health **2**(2), 1–3 (2023). https://doi.org/10.1371/journal.pdig.0000205
19. Ouyang, S., Zhang, J.M., Harman, M., Wang, M.: An empirical study of the non-determinism of chatgpt in code generation. ACM Trans. Softw. Eng. Methodol. 34, 1–28 (2024). https://doi.org/10.1145/3697010
20. Schuessler, J.: Adversarial attacks and defense in long short-term memory recurrent neural networks. Master's Thesis, Pennsylvania State University (2021). https://etda.libraries.psu.edu/catalog/12345678. Graduate Program: Electrical Engineering, Master of Science
21. Spatharioti, S.E., Rothschild, D.M., Goldstein, D.G., Hofman, J.M.: Comparing traditional and LLM-based search for consumer choice: a randomized experiment (2023). arXiv:2307.03744
22. Thirunavukarasu, A.J., Ting, D.S.J., Elangovan, K., Gutierrez, L., Tan, T.F., Ting, D.S.W.: Large language models in medicine. Nat. Med. **29**(8), 1930–1940 (2023)
23. Vaithilingam, P., Zhang, T., Glassman, E.L.: Expectation vs. experience: evaluating the usability of code generation tools powered by large language models. In: CHI Conference on Human Factors in Computing Systems Extended Abstracts, pp. 1–7 (2022)
24. Wu, S., Irsoy, O., Lu, S., Dabravolski, V., Dredze, M., Gehrmann, S., Kambadur, P., Rosenberg, D., Mann, G.: Bloomberggpt: a large language model for finance (2023). arXiv:2303.17564
25. Xiang, Z., Miller, D.J., Kesidis, G.: Detection of backdoors in trained classifiers without access to the training set. IEEE Trans. Neural Netw. Learn. Syst. **33**(3), 1177–1191 (2022). https://doi.org/10.1109/TNNLS.2020.3041202. https://ieeexplore.ieee.org/document/9296553
26. Yahoo! finance. https://finance.yahoo.com/
27. Yao, B., Jiang, M., Yang, D., Hu, J.: Benchmarking LLM-based machine translation on cultural awareness (2023). arXiv:2305.14328
28. Ziems, N., Yu, W., Zhang, Z., Jiang, M.: Large language models are built-in autoregressive search engines (2023). arXiv:2305.09612

Chapter 14
Statistical Feature-Based Detection of Adversarial Noise and Patch Attacks in Image and Deepfake Analysis

Niklas Bunzel, Raphael Antonius Frick, Lukas Graner, Nicolas Göller, and Martin Steinebach

14.1 Introduction

Neural networks are increasingly deployed in safety-critical applications such as autonomous vehicles [4, 5], drones [18], robotics [23, 35], and malware detection [41, 58]. Their widespread use has exposed critical vulnerabilities, making it essential to understand and mitigate security threats. Since 2013, a variety of adversarial attacks have emerged [2, 29, 36, 40, 48], where imperceptible modifications to input images can drastically alter a model's predictions. Defense strategies against these attacks include preprocessing techniques [16, 20], robust network architectures [53, 55], and notably, Adversarial Training [34]. Additionally, production systems can leverage detectors [17, 43] to identify adversarial inputs, an approach increasingly important in light of emerging regulatory frameworks such as ISO 27090 and the EU AI Act.

N. Bunzel (✉) · M. Steinebach
Fraunhofer SIT, Darmstadt, Germany

TU-Darmstadt, Darmstadt, Germany

ATHENE, Darmstadt, Germany
e-mail: niklas.bunzel@sit.fraunhofer.de; martin.steinebach@sit.fraunhofer.de

R. A. Frick · L. Graner · N. Göller
Fraunhofer SIT, Darmstadt, Germany

ATHENE, Darmstadt, Germany
e-mail: raphael.frick@sit.fraunhofer.de; lukas.graner@sit.fraunhofer.de; nicolas.goeller@sit.fraunhofer.de
https://www.sit.fraunhofer.de/
https://www.tu-darmstadt.de
https://www.athene-center.de/

E. Nowroozi et al. (eds.), *Adversarial Example Detection and Mitigation Using Machine Learning*, https://doi.org/10.1007/978-3-031-99447-0_14

The cybersecurity domain routinely faces automated attacks that exploit known vulnerabilities using pentesting tools. It is anticipated that similar automated techniques will be applied to neural networks, prompting a focus on detecting transferred adversarial black-box attacks executed with default parameters. In this work, we present a fast, explainable multi-class detector based on image statistics that is capable of identifying various adversarial attacks, including Attack on Attention (AoA) and Fast Gradient Sign Method (FGSM), while maintaining a low false positive rate [7].

Furthermore, evasion attacks specifically targeting deepfake detection systems pose a significant threat. Unlike adversarial noise that is distributed across the entire image, adversarial patches [6] are confined to specific regions, potentially making them more suitable for real-world applications. These patches, which may be physically printed and integrated into everyday objects such as clothing, water bottles, or plant pots [27, 46, 50, 57], are designed to deceive autonomous systems; though they are not completely imperceptible to human observers. While current research predominantly emphasizes noise perturbation attacks on video deepfake detection [37, 45, 51], our focus on patch attacks offers novel insights into counteracting these threats. The objective of this work is to develop a countermeasure that detects and mitigates adversarial patch attacks, thereby ensuring reliable assessments by deepfake detectors. Our proposed method involves an initial patch detection and, if necessary, patch mitigation followed by an evaluation by a deepfake detection system. We validate our approach using the FaceForensics++ dataset and discuss directions for future research aimed at enhancing patch detection methods.

14.2 Related Work

14.2.1 Attacks

Adversarial Noise Attacks Adversarial noise attacks add perturbations to pixels across the whole image. In [19] the authors proposed a one-step gradient optimization approach to produce such a noise-based attack. For transfer-based attacks [39] the attacker generates adversarial examples on a white-box surrogate model, which then transfer to the target model. The AoA [12] attack demonstrated how adversarial perturbations can be optimized to target attention mechanisms specifically, significantly optimizing transferability.

Adversarial Patch Attacks An adversarial patch attack is the manipulation of a specific part of an image by the addition of a localized adversarial perturbation. In image classification, the most salient object typically determines the output. Adversarial patches exploit this by generating a patch more salient than real-world objects. While easily detectable by humans, such patches can mislead machine learning models [6] with high confidence. RobustDPatch [30] extends DPatch [32] providing means for generating an adversarial patch that is resilient to transformations such as scaling and cropping, increasing its effectiveness in real-world settings.

14.2.2 Detectors

Noise Detection Adversarial noise detection leverages statistical and neural network-based techniques to distinguish adversarial examples from benign inputs. In this study we focus on detectors leveraging statistical features, as neural network-based detectors are inherently vulnerable to adversarial attacks. Hendrycks and Gimpel [21] proposed two model-independent methods, whereby the first exploits differences in variance across principal components, where adversarial examples exhibit higher variance in later components. This enables threshold-based detection of attacks, assuming the attacker lacks knowledge of the defense. The second method analyzes the softmax output by computing the Kullback-Leibler divergence from a uniform distribution. Benign images typically exhibit higher divergence due to confident classification, whereas adversarial examples often yield lower confidence scores, facilitating detection [21]. Xu et al. [56] addressed adversarial perturbations by exploiting the redundancy in high-dimensional feature spaces. Their approach detects adversarial inputs by comparing classifier predictions on the original image and its feature-reduced versions. Kherchouche et al. [26] introduced a two-step framework integrating detection and denoising. They found that adversarial examples exhibit distinct histograms compared to clean images. To model these differences, they employed generalized Gaussian (GGD) and asymmetric generalized Gaussian (AGGD) distributions.

Adversarial Noise Detection for Deepfakes Detection An adversarial attack detector specifically designed for the deepfake detection domain is proposed in [42], which utilizes an explainable AI (xAI) technique to extract features and a ResNet50 model to classify if the image contains an adversarial perturbation.

Patch Detection The detection of adversarial patches has been extensively studied, leading to various defense techniques [8, 24, 25, 31, 54]. Liu et al. proposed Segment and Complete (SAC) [33], which segments adversarial patches at the pixel level and removes them using shape completion. Chen et al. [13] introduced a universal patch defense that leverages feature attribution methods [1, 44, 47]. By embedding salient features into benign images, changes in predictions indicate the presence of an adversarial patch. JPEG compression has also been explored as a feature-squeezing technique [56], exploiting its ability to alter high-frequency components, which are characteristic of adversarial perturbations. However, patches trained with additional constraints may evade this approach. Bunzel et al. [9] proposed a method based on edge-detection, taking high entropy and edge density of adversarial patches into account. Potential patches are identified via thresholding and mitigated through inpainting. A significant change in the model's output after inpainting indicates the presence of an adversarial patch.

14.3 Adversarial Noise Detection: Image Classification

14.3.1 Approach

AoA Detector AoA adversarial perturbations are more similar to randomly drawn edges than random noise. However, these edges are not randomly chosen, but are based on the AoA-algorithm using the SGLRP-heatmap [22]. In less detailed images, these edges are conspicuous, while in detailed images they become less noticeable. This observation motivated the development of a detector (EAD) that distinguishes adversarial from legitimate images by analyzing edge patterns. EAD comprises three main components: edge detection, edge smoothing, and image similarity comparison with optional brightness reduction. The process begins by converting the input to grayscale, followed by sharpening to enhance edges. The grayscale image is then split into two paths, $g1$ and $g2$. On $g1$, edges are detected using the Canny algorithm [11]. Meanwhile, $g2$ undergoes brightness reduction (optional) and smoothing, using an average filter, to suppress artificially introduced edges before edge detection. The resulting edge images from $g1$ and $g2$ are compared using the SSIM algorithm [52], yielding an index $I \in [0, 1]$ that is thresholded to classify the image. The threshold, as well as the parameters (notably those for the Canny algorithm), were determined experimentally.

FGSM Detector The detection of FGSM attacks follows the detector approach of [26]. Here, we decided to rely on the shape controlling parameter β of the generalized Gaussian distribution to detect FGSM attacks. As the shape is the main difference between benign and FGSM images, as shown by Kherchouche et al. [26]. Simplicity and explainability benefit greatly from reduction to one key feature.

Multi-Class Detector The multi-class detector stacks the detectors for each individual attack. First the FGSM detector is calculated and then the AoA detector. Only if the FGSM detector returns a negative result, the AoA detector is calculated. Other techniques for concatenating the single detector results like min or max are also possible for normalized scores. We decided to stack the binary results, this is the most easily extendable option and allows for faster execution time on adversarial examples.

14.3.2 Evaluation

Detecting AoA For the evaluation of our AoA detection scheme, we used the ImageNet [15] validation set and DAmageNet[1] [12] datasets. To generate dev and test sets for our detectors, we performed a 70/30 random split of both data sets. For fixed parameters of the sharpening and smoothing operations as well as the

[1] AoA attack executed on the ImageNet validation set.

Fig. 14.1 Comparison of edge detection of legitimate images with different levels of detail. For all images, the same Canny parameters $t1 = 19$ and $t2 = 67$ were used. For image (**a**), this translates to near-perfect edge detection (**b**), while for image (**c**), it produces much too fine-granular edge detections (**d**)

brightness reduction, the following parameters were determined experimentally: For image sharpening, the sharpness enhancer inside the Python module Pillow [14] with a factor of 1.3 was used. A 3x3 average filter was used for smoothing. As aforementioned, the brightness reduction is performed on a grayscale image, where each pixel can take a value between 0 and 255. The brightness reduction was performed by decreasing all pixels by a constant value of 5. This corresponds to a reduction of the total brightness by $\sim$2%. When using the Canny algorithm with the parameters, $t1 = 19$ and $t2 = 67$, a decision threshold of $T = 0.3026$ resulted in the best performance. The EAD was able to achieve a detection rate of 69.60% on each of 35,000 previously unseen images from ImageNet and DAmageNet. EAD records the highest AUC with a value of 0.9406, as displayed in Fig. 14.2. An open issue is the varying levels of detail in naturally occurring images. The frequency distribution of SSIM indices obtained from images in ImageNet and DAmageNet clearly shows that the indices from ImageNet are much more widely distributed. This phenomenon seems to be related to the different levels of detail of natural images, which cannot all be handled equally well with fixed Canny parameters. Figure 14.1 shows an example of this behavior showcasing four images with different levels of detail. These are two legitimate images with very different levels of detail. With fixed Canny parameters, the edge detection performs worse in comparison to the usage of dynamic parameters leading to an increment of false positives.

Detecting FGSM To evaluate the FGSM detection we used the FGSM implementation from the adversarial robustness toolbox (ART) [38] with it's default parameters. We used 15,000 images of ImageNet and 9000 of their corresponding FGSM counter parts to select the threshold. The proposed approach achieves an AUC of 0.96 as displayed in Fig. 14.2 and when using a threshold value of 2.89, a true positive rate of 74% and a false positive rate of 5% is obtained.

Multi-Class Detector For the multi-class detection we stack the individual detectors, by classifying the input samples using the FGSM detector first and then using the AoA detector. The FGSM detector classifies 15% of the AoA attacks as FSGM, whereas the AoA detector classifies 19% of FGSM attacks as AoA. With this

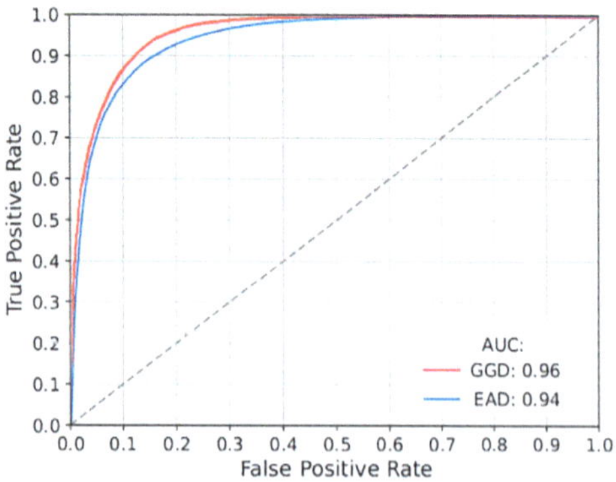

Fig. 14.2 ROC curve and AUC of EAD against AoA. Canny parameters $t1 = 19$ and $t2 = 67$ were chosen for the detector. AUC is 0.9416

Table 14.1 Confusion matrix of the multi-class detector

	Benign	FGSM	AoA
Benign	**90.75%**	20.53%	23.98%
FGSM	5.02%	**74.45%**	15.29%
AoA	4.23%	5.02%	**60.73%**

Note: The best performance per method is highlighted in bold

Table 14.2 Performance of our multi-class detector vs. feature squeezing as a percentage of the samples in a binary classification task

		Benign	Adversarial
Feature Squeezing	Benign	**95.13%**	85.18%
	Adversarial	**4.87%**	14.82%
Multi-class Detection	Benign	90.75%	**22.69%**
	Adversarial	9.25%	**77.31%**

Note: The best performance per method is highlighted in bold

stacking order, 77.3% of the adversarial images were classified as adversarial and 90.75% of the benign images were correctly identifed as benign (Table 14.1). The 9.25% false positives are classified as 5% FGSM and 4% AoA. In the multi-class setting we detect 74% of the FGSM attacks correctly and 60% of the AoA attacks. 15% of the AoA attacks were classified as FGSM attacks and 5% of FGSM as AoA. We compared our approach to Feature Squeezing [56], we detect far more adversarial images as adversarial and are almost as good at recognising benign images as can be seen in Table 14.2.

14.4 Adversarial Patch Detection: Deepfake Detection

14.4.1 Patch Detection and Mitigation

A growing challenge in deepfake detection is the deliberate application of adversarial patches, crafted to undermine detection algorithms, making it harder for detectors to distinguish deepfakes from authentic media. To counter this, we first identify candidate regions likely affected by these patches and then mitigate their influence by removing them from the frame. This combined approach enhances the robustness of deepfake classifiers against adversarial attacks. Existing adversarial patch detection techniques often use deep learning models to estimate targeted regions. Our method, however, utilizes signal processing features, avoiding deep learning to prevent the detection method itself from being targeted by adversarial attacks. Signal-based methods can also be more resource-efficient and faster in inference compared to deep learning approaches.

14.4.2 Feature Selection

To identify signal-based features for detection, we first examine the characteristics introduced by adversarial patches in media. These patches are predominantly compact with high color intensity and brightness variations, causing sharp contours on its edge in the target video frame. For digital images, patches often have a different compression history compared to authentic regions. When applying lossy compression like JPEG, the distribution of DCT-coefficients in both authentic and modified regions is altered based on the compression history, such as the number of applied compressions and the strength of compression. Compressing with lower quality settings results in stronger compression, which removes more high frequencies, acting as a block-based low-pass filter. As such, when an image compressed at low quality is subtracted from the original high frequencies are isolated.

Based on the identified properties, we propose a strategy to detect patches using signal-based segmentation based on an Error-Level Analysis (ELA) that showcases its suitability to detect digital and physical adversarial patches [10].

The ELA algorithm [28], a multimedia forensics technique, can identify inserted image parts post-capture, detecting digital and physical adversarial patches captured by cameras in systems like driver assistance. It exploits resulting characteristics of the JPEG compression process, where pixel data is transformed into the frequency domain using DCT and then quantized, typically resulting in high-frequency coefficients being set to zero and removed from the final image.

ELA recompresses the input image, computes the difference from the original, revealing removed high frequencies. As adversarial patches contrast with smoother areas like human skin or backgrounds due to significant high-frequency components, this can be taken advantage of. Figure 14.3b illustrates the ELA result for an example image, which contains a deepfaked face and an adversarial patch. As can be seen, the patch area is the only highly textured region.

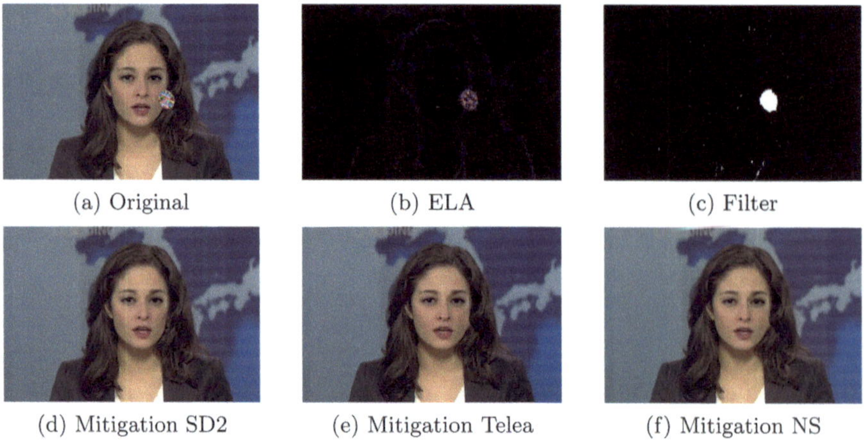

Fig. 14.3 Example of a deepfaked face image with an adversarial patch (**a**), the intermediate ELA result of the pipeline, magnified by factor of 10 (**b**) and the filtered mask of the ELA result (**c**), representing detected regions, which presumably contain patches. The mitigation results based on (**a**) and (**c**) are shown in (**d**), (**e**), (**f**) using Stable Diffusion 2, Telea and NS inpainting techniques, respectively

14.4.3 Proposed Patch Detection

We present a pipeline for detecting adversarial patches in deepfake detection systems using ELA, building on our previous work [9]. Our test results show the pipeline's effectiveness against patch attacks in object detection and face recognition models across various datasets. This method can be applied to various AI-models without adjustments and does not require a trained AI model for detection, preventing it from being targeted by additional adversarial patches.

The following list summarizes the proposed steps for identifying and mitigating any adversarial patches present in images or video frames:

1. **ELA Feature Extraction:** We apply ELA with a JPEG quality factor of 95 to identify digitally inserted regions or areas with high frequencies. The high quality factor is used as it is highly unlikely that digital patches will be inserted with high compression, since the compression artifacts may affect the ability to fool a classifier. It also isolates high-frequency noise typical of complex adversarial patches. The resulting heatmap highlights patches, while the environment shows minimal or no noise, as visible in Fig. 14.3b.
2. **Filtering:** For segmentation of regions of interest in the ELA output, it is filtered by binarization. In experiments, an appropriate threshold of 3 (regarding a value range between 0 and 255, where 0 represents no high frequency measured) was estimated. To reduce outliers, the morphological operations erosion and then dilation (both with a uniform kernel of size 11 pixels) are applied. By this, patches smaller than a few pixels cannot be expected and can be ignored. An example is shown in Fig. 14.3c. For simplicity, the combination of this processing step with the previous one are denoted as "patch detection".

3. **Patch Mitigation:** We remove and refill the identified candidate regions using inpainting as proposed in [9]. The inpainting methods include Stable Diffusion v2 (SD2) for inpainting, as well as non-deep-learning-based inpainting techniques like the fluid dynamics inspired Navier-Stokes technique [3] and Telea [49], based on a fast marching method. Figure 14.3 features an exemplary image to which our method for detection and mitigation was applied, using the three different inpainting methods, as shown in (d), (e) and (f), respectively.

14.4.4 Evaluation

We assessed our mitigation method using 500 keyframes from 200 deepfaked videos from the FaceForensics++ dataset. Introducing adversarial patches reduced the deepfake detection rate of the frames from 99.80 to 0%, demonstrating the effectiveness of the patches. We further investigated the effects of both lossy compression and the "centrality" of the patch (measured by a value between 0 and 1, where 0 represents a patch placed at the border of an image and 1 at the exact center). After applying our mitigation, detection rates improved on average to 93.85% for uncompressed videos and 87.43% with H.264 compression. The detailed results are summarized in Table 14.3.

While SD2 inpainting might look better to the human eye in comparison to the other two inpainting methods, it did not work as well with H.264 compression as the other two. A reason for this may be that Stable Diffusion 2 synthesizes features amplified through the compression and which appear realistic to the human eye, but are considered as synthesized and thus non-authentic by the deepfake detector. Overall, the Navier-Stokes based inpainting provided the best performance in terms of mitigation under different settings in our experiments. Mitigation effectiveness increased as patches moved away from the face center. Additionally, our method

Table 14.3 Deepfake detection rates of original unpatched Deepfake images and without mitigation strategy (left column) and of patched Deepfake images with different mitigation inpainting techniques and compressions applied over varying patch centrality values

Original deepfake	Patched deepfake				
			Patch centrality		
	Mitigation	Compression	1.0	0.7	0.3
	None	None	0.00%	0.00%	4.96%
99.80%	None	H.264	0.00%	0.00%	3.19%
	SD2	None	88.45%	93.59%	99.21%
	SD2	H.264	75.77%	77.45%	89.31%
	NS	None	90.84%	93.32%	98.81%
	NS	H.264	92.06%	93.97%	91.68%
	Telea	None	88.77%	92.86%	98.81%
	Telea	H.264	86.60%	89.73%	90.30%

maintained low false detection rates on authentic videos, with deepfake detection rates remaining around 1.4%, ensuring classification accuracy for genuine content was unaffected.

14.5 Conclusion

We present a model-independent and explainable detectors that identify adversarial noise perturbations and adversarial patch attacks. For the adversarial noise perturbation we showcased a multi-class detection approach that leverages image statistics to identify the specific attacks. The multi-class detector was evaluated against off-the-shelf transferred black box attacks, achieving a 90% true negative rate along with true positive rates of 74 and 60% for FGSM and AoA attacks, respectively. The detection of FGSM attacks is based on the distinct shape of the generalized Gaussian distribution of benign versus adversarial images, whereas the detection of AoA attacks relies on identifying artificial edges introduced into natural images. The multi-class approach is easily extendible and has a low FPR. Challenges remain in detecting and differentiating between various attack types and in further enhancing the true positive rate. Furthermore, adversarial patches have been shown to control the output of neural networks, rendering deep learning–based deepfake detection methods particularly vulnerable. To address patch attacks on deepfake classifiers, we introduce a two-step mitigation strategy: first, potential adversarial patches are detected using traditional ELA to identify regions of high entropy within facial areas; next, these regions are "removed" via inpainting, thereby reconstructing the image content and mitigating the adversarial influence before passing the image to the deepfake detection model. Evaluations on a subset of the FaceForensics++ dataset indicate that this method preserves classifier efficacy on authentic data while mitigating adversarial effects. The binarization threshold of the ELA output and the parameters for morphological operations are highly dependent on video quality and facial region characteristics, suggesting that dynamic adaptation of these hyperparameters may be necessary.

Acknowledgments This research work has been partly funded by the German Federal Ministry of Education and Research and the Hessian Ministry of Higher Education, Research, Science and the Arts within their joint support of the National Research Center for Applied Cybersecurity ATHENE as well as by the Federal Office for Information Security (BSI) under the research project "SecMedID".

References

1. Ancona, M., Ceolini, E., Öztireli, C., Gross, M.: Towards better understanding of gradient-based attribution methods for deep neural networks. In: International Conference on Learning Representations (2018). https://openreview.net/forum?id=Sy21R9JAW

2. Andriushchenko, M., Croce, F., Flammarion, N., Hein, M.: Square attack: a query-efficient black-box adversarial attack via random search. In: Vedaldi, A., Bischof, H., Brox, T., Frahm, J. (eds.) Computer Vision - ECCV 2020 - 16th European Conference, Glasgow, UK, August 23–28, 2020, Proceedings, Part XXIII. Lecture Notes in Computer Science, vol. 12368, pp. 484–501. Springer, Berlin (2020). https://doi.org/10.1007/978-3-030-58592-1_29
3. Bertalmio, M., Bertozzi, A.L., Sapiro, G.: Navier-stokes, fluid dynamics, and image and video inpainting. In: Proceedings of the 2001 IEEE Computer Society Conference on Computer Vision and Pattern Recognition. CVPR 2001. vol. 1. IEEE, Piscataway (2001)
4. Bojarski, M., del Testa, D.W., Dworakowski, D., Firner, B., Flepp, B., Goyal, P., Jackel, L.D., Monfort, M., Muller, U., Zhang, J., Zhang, X., Zhao, J., Zieba, K.: End to end learning for self-driving cars (2016). ArXiv abs/1604.07316
5. Bourzac, K.: Bringing big neural networks to self-driving cars, smartphones, and drones. IEEE Spectrum, 13–29 (2016)
6. Brown, T.B., Mané, D., Roy, A., Abadi, M., Gilmer, J.: Adversarial patch (2017). CoRR abs/1712.09665
7. Bunzel, N., Böringer, D.: Multi-class detection for off the shelf transfer-based black box attacks. In: Proceedings of the 2023 Secure and Trustworthy Deep Learning Systems Workshop, SecTL 2023, Melbourne, July 10–14, 2023. pp. 5:1–5:6. ACM, New York (2023). https://doi.org/10.1145/3591197.3591305. https://doi.org/10.1145/3591197.3591305
8. Bunzel, N., Hamborg, J.: Adversarial patch detection: leveraging depth contrast for enhanced threat visibility. In: 54th Annual IEEE/IFIP International Conference on Dependable Systems and Networks, DSN 2024 - Workshops, Brisbane, Australia, June 24–27, 2024, pp. 39–45. IEEE, Piscataway (2024). https://doi.org/10.1109/DSN-W60302.2024.00020. https://doi.org/10.1109/DSN-W60302.2024.00020
9. Bunzel, N., Siwakoti, A., Klause, G.: Adversarial patch detection and mitigation by detecting high entropy regions. In: 53nd Annual IEEE/IFIP International Conference on Dependable Systems and Networks, DSN Workshops 2023, Porto, Portugal, June 27–30, 2023. pp. 124–128. IEEE, Los Alamitos (2023)
10. Bunzel, N., Frick, R.A., Klause, G., Schwarte, A., Honermann, J.: Signals are all you need: detecting and mitigating digital and real-world adversarial patches using signal-based features. In: Proceedings of the 2nd ACM Workshop on Secure and Trustworthy Deep Learning Systems, SecTL'24, pp. 24–34. Association for Computing Machinery, New York (2024). https://doi.org/10.1145/3665451.3665530
11. Canny, J.: A computational approach to edge detection. IEEE Trans. Pattern Anal. Mach. Intell. **PAMI-8**(6), 679–698 (1986). https://doi.org/10.1109/TPAMI.1986.4767851
12. Chen, S., He, Z., Sun, C., Huang, X.: Universal adversarial attack on attention and the resulting dataset damagenet. IEEE Trans. Pattern Anal. Mach. Intell. (2020)
13. Chen, Z., Dash, P., Pattabiraman, K.: Jujutsu: A two-stage defense against adversarial patch attacks on deep neural networks. In: Proceedings of the 2023 ACM Asia Conference on Computer and Communications Security, ASIA CCS'23, pp. 689–703. Association for Computing Machinery, New York (2023)
14. Clark, A.: Pillow (pil fork) documentation (2015)
15. Deng, J., Dong, W., Socher, R., Li, L.J., Li, K., Fei-Fei, L.: Imagenet: a large-scale hierarchical image database. In: 2009 IEEE Conference on Computer Vision and Pattern Recognition, pp. 248–255. IEEE, Piscataway (2009)
16. Dziugaite, G., Ghahramani, Z., Roy, D.M.: A study of the effect of jpg compression on adversarial images (2016). ArXiv abs/1608.00853
17. Feinman, R., Curtin, R.R., Shintre, S., Gardner, A.B.: Detecting adversarial samples from artifacts (2017). ArXiv abs/1703.00410
18. Giusti, A., Guzzi, J., Cireşan, D.C., He, F.L., Rodríguez, J.P., Fontana, F., Faessler, M., Forster, C., Schmidhuber, J., Caro, G.D., Scaramuzza, D., Gambardella, L.M.: A machine learning approach to visual perception of forest trails for mobile robots. IEEE Robot. Autom. Lett. **1**(2), 661–667 (2016). https://doi.org/10.1109/LRA.2015.2509024
19. Goodfellow, I.J., Shlens, J., Szegedy, C.: Explaining and harnessing adversarial examples. In: Bengio, Y., LeCun, Y. (eds.) 3rd International Conference on Learning Representations, ICLR

2015, San Diego, CA, USA, May 7–9, 2015, Conference Track Proceedings (2015). https://doi.org/10.48550/ARXIV.1412.6572
20. Graese, A., Rozsa, A., Boult, T.E.: Assessing threat of adversarial examples on deep neural networks. In: 15th IEEE International Conference on Machine Learning and Applications, ICMLA 2016, Anaheim, CA, USA, December 18–20, 2016. pp. 69–74. IEEE Computer Society, Washington (2016). https://doi.org/10.1109/ICMLA.2016.0020
21. Hendrycks, D., Gimpel, K.: Early methods for detecting adversarial images. In: 5th International Conference on Learning Representations, ICLR 2017, Toulon, France, April 24–26, 2017, Workshop Track Proceedings. OpenReview.net (2017)
22. Iwana, B.K., Kuroki, R., Uchida, S.: Explaining convolutional neural networks using softmax gradient layer-wise relevance propagation. In: 2019 IEEE/CVF International Conference on Computer Vision Workshops, ICCV Workshops 2019, Seoul, Korea (South), October 27–28, 2019. pp. 4176–4185. IEEE, Piscataway (2019). https://doi.org/10.1109/ICCVW.2019.00513
23. Janglova, D.: Neural networks in mobile robot motion. Int. J. Adv. Robot. Syst. **1**, 15–22 (2004). https://doi.org/10.5772/5615
24. Ji, N., Feng, Y., Xie, H., Xiang, X., Liu, N.: Adversarial yolo: defense human detection patch attacks via detecting adversarial patches (2021). https://doi.org/10.48550/ARXIV.2103.08860
25. Jing, L., Wang, R., Ren, W., Dong, X., Zou, C.: Pad: patch-agnostic defense against adversarial patch attacks. In: Proceedings of the IEEE/CVF Conference on Computer Vision and Pattern Recognition, pp. 24472–24481 (2024)
26. Kherchouche, A., Fezza, S.A., Hamidouche, W.: Detect and defense against adversarial examples in deep learning using natural scene statistics and adaptive denoising. Neural Comput. Appl. **34**(24), 21567–21582 (2022). https://doi.org/10.1007/s00521-021-06330-x
27. Komkov, S., Petiushko, A.: AdvHat: real-world adversarial attack on ArcFace face ID system. In: 2020 25th International Conference on Pattern Recognition (ICPR). IEEE, Los Alamitos, CA, USA (Jan 2021). https://doi.org/10.1109/icpr48806.2021.9412236
28. Krawetz, N., Solutions, H.F.: A picture's worth. Technical Report, Hacker Factor Solutions (presented at Black Hat Briefings, 2007) **6**(2), 2 (2007)
29. Kurakin, A., Goodfellow, I.J., Bengio, S.: Adversarial examples in the physical world. In: 5th International Conference on Learning Representations, ICLR 2017, Toulon, France, April 24–26, 2017, Workshop Track Proceedings. OpenReview.net (2017)
30. Lee, M., Kolter, Z.: On physical adversarial patches for object detection (2019). https://arxiv.org/abs/1906.11897
31. Liang, B., Li, J., Huang, J.: We can always catch you: detecting adversarial patched objects with or without signature (2021). https://doi.org/10.48550/ARXIV.2106.05261
32. Liu, X., Yang, H., Liu, Z., Song, L., Chen, Y., Li, H.: DPATCH: an adversarial patch attack on object detectors. In: Espinoza, H., hÉigeartaigh, S.Ó., Huang, X., Hernández-Orallo, J., Castillo-Effen, M. (eds.) Workshop on Artificial Intelligence Safety 2019 Co-located with the Thirty-Third AAAI Conference on Artificial Intelligence 2019 (AAAI-19), Honolulu, Hawaii, January 27, 2019. CEUR Workshop Proceedings, vol. 2301. CEUR-WS.org (2019). https://ceur-ws.org/Vol-2301/paper_5.pdf
33. Liu, J., Levine, A., Lau, C.P., Chellappa, R., Feizi, S.: Segment and complete: defending object detectors against adversarial patch attacks with robust patch detection. In: Proceedings of the IEEE/CVF Conference on Computer Vision and Pattern Recognition, pp. 14973–14982 (2022)
34. Madry, A., Makelov, A., Schmidt, L., Tsipras, D., Vladu, A.: Towards deep learning models resistant to adversarial attacks. In: 6th International Conference on Learning Representations, ICLR 2018, Vancouver, BC, Canada, April 30–May 3, 2018, Conference Track Proceedings. OpenReview.net (2018)
35. Mnih, V., Kavukcuoglu, K., Silver, D., Rusu, A.A., Veness, J., Bellemare, M.G., Graves, A., Riedmiller, M., Fidjeland, A.K., Ostrovski, G., Petersen, S., Beattie, C., Sadik, A., Antonoglou, I., King, H., Kumaran, D., Wierstra, D., Legg, S., Hassabis, D.: Human-level control through deep reinforcement learning. Nature **518**(7540), 529–533 (2015). https://doi.org/10.1038/nature14236

36. Moosavi-Dezfooli, S., Fawzi, A., Fawzi, O., Frossard, P.: Universal adversarial perturbations. In: 2017 IEEE Conference on Computer Vision and Pattern Recognition, CVPR 2017, Honolulu, HI, USA, July 21–26, 2017, pp. 86–94. IEEE Computer Society, Washington (2017). https://doi.org/10.1109/CVPR.2017.17
37. Neekhara, P., Dolhansky, B., Bitton, J., Canton-Ferrer, C.: Adversarial threats to deepfake detection: a practical perspective. In: IEEE Conference on Computer Vision and Pattern Recognition Workshops, CVPR Workshops 2021, virtual, June 19–25, 2021. pp. 923–932. Computer Vision Foundation/IEEE, Piscataway (2021). https://doi.org/10.1109/CVPRW53098.2021.00103
38. Nicolae, M.I., Sinn, M., Tran, M.N., Buesser, B., Rawat, A., Wistuba, M., Zantedeschi, V., Baracaldo, N., Chen, B., Ludwig, H., Molloy, I., Edwards, B.: Adversarial robustness toolbox v1.2.0 (2018). CoRR 1807.01069
39. Papernot, N., McDaniel, P.D., Goodfellow, I.J., Jha, S., Celik, Z.B., Swami, A.: Practical black-box attacks against machine learning. In: Karri, R., Sinanoglu, O., Sadeghi, A., Yi, X. (eds.) Proceedings of the 2017 ACM on Asia Conference on Computer and Communications Security, AsiaCCS 2017, Abu Dhabi, United Arab Emirates, April 2–6, 2017, pp. 506–519. ACM, New York (2017). https://doi.org/10.1145/3052973.3053009
40. Papernot, N., Mcdaniel, P., Jha, S., Fredrikson, M., Celik, Z.B., Swami, A.: The limitations of deep learning in adversarial settings. 2016 IEEE European Symposium on Security and Privacy (EuroS&P), pp. 372–387 (2016)
41. Pascanu, R., Stokes, J.W., Sanossian, H., Marinescu, M., Thomas, A.: Malware classification with recurrent networks. In: 2015 IEEE International Conference on Acoustics, Speech and Signal Processing (ICASSP), pp. 1916–1920 (2015). https://doi.org/10.1109/ICASSP.2015.7178304
42. Pinhasov, B., Lapid, R., Ohayon, R., Sipper, M., Aperstein, Y.: XAI-based detection of adversarial attacks on deepfake detectors. Transact. Mach. Learn. Res. (2024)
43. Roth, K., Kilcher, Y., Hofmann, T.: The odds are odd: a statistical test for detecting adversarial examples. In: International Conference on Machine Learning (ICML) (2019)
44. Selvaraju, R.R., Cogswell, M., Das, A., Vedantam, R., Parikh, D., Batra, D.: Grad-CAM: visual explanations from deep networks via gradient-based localization. Int. J. Comput. Vision **128**(2), 336–359 (2019). https://doi.org/10.1007/s11263-019-01228-7
45. Shahriyar, S.A., Wright, M.: Evaluating robustness of sequence-based deepfake detector models by adversarial perturbation. In: Proceedings of the 1st Workshop on Security Implications of Deepfakes and Cheapfakes, WDC'22, pp. 13–18. Association for Computing Machinery, New York (2022). https://doi.org/10.1145/3494109.3527194
46. Sharif, M., Bhagavatula, S., Bauer, L., Reiter, M.K.: Accessorize to a crime: real and stealthy attacks on state-of-the-art face recognition. In: Weippl, E.R., Katzenbeisser, S., Kruegel, C., Myers, A.C., Halevi, S. (eds.) Proceedings of the 2016 ACM SIGSAC Conference on Computer and Communications Security, Vienna, Austria, October 24–28, 2016, pp. 1528–1540. ACM, New York (2016). https://doi.org/10.1145/2976749.2978392
47. Smilkov, D., Thorat, N., Kim, B., Viégas, F., Wattenberg, M.: Smoothgrad: removing noise by adding noise (2017). https://arxiv.org/abs/1706.03825
48. Szegedy, C., Zaremba, W., Sutskever, I., Bruna, J., Erhan, D., Goodfellow, I.J., Fergus, R.: Intriguing properties of neural networks. In: Bengio, Y., LeCun, Y. (eds.) 2nd International Conference on Learning Representations, ICLR 2014, Banff, AB, Canada, April 14–16, 2014, Conference Track Proceedings (2014)
49. Telea, A.: An image inpainting technique based on the fast marching method. J. Graph. Tools **9**(1), 23–34 (2004)
50. Thys, S., Ranst, W.V., Goedemé, T.: Fooling automated surveillance cameras: adversarial patches to attack person detection. In: IEEE Conference on Computer Vision and Pattern Recognition Workshops, CVPR Workshops 2019, Long Beach, CA, USA, June 16–20, 2019, pp. 49–55. Computer Vision Foundation/IEEE, Piscataway (2019). https://doi.org/10.1109/CVPRW.2019.00012
51. Vo, N.H., Phan, K.D., Tran, A., Dang-Nguyen, D.: Adversarial attacks on deepfake detectors: a practical analysis. In: Jónsson, B.T., Gurrin, C., Tran, M., Dang-Nguyen, D., Hu, A.M., Binh,

H.T.T., Huet, B. (eds.) MultiMedia Modeling - 28th International Conference, MMM 2022, Phu Quoc, Vietnam, June 6–10, 2022, Proceedings, Part II. Lecture Notes in Computer Science, vol. 13142, pp. 318–330. Springer, Berlin (2022). https://doi.org/10.1007/978-3-030-98355-0_27
52. Wang, Z., Bovik, A., Sheikh, H., Simoncelli, E.: Image quality assessment: from error visibility to structural similarity. IEEE Trans. Image Process. **13**(4), 600–612 (2004). https://doi.org/10.1109/TIP.2003.819861
53. Wang, X., Wang, S., Chen, P.Y., Wang, Y., Kulis, B., Lin, X., Chin, S.: Protecting neural networks with hierarchical random switching: towards better robustness-accuracy trade-off for stochastic defenses. In: Proceedings of the Twenty-Eighth International Joint Conference on Artificial Intelligence, IJCAI-19, pp. 6013–6019. International Joint Conferences on Artificial Intelligence Organization, Montreal (2019). https://doi.org/10.24963/ijcai.2019/833
54. Xiang, C., Mittal, P.: Detectorguard: provably securing object detectors against localized patch hiding attacks. In: Proceedings of the 2021 ACM SIGSAC Conference on Computer and Communications Security (CCS '21), pp. 3177–3196, Association for Computing Machinery, New York (2021). https://doi.org/10.1145/3460120.3484757
55. Xiao, C., Zhong, P., Zheng, C.: Enhancing adversarial defense by k-winners-take-all. In: 8th International Conference on Learning Representations, ICLR 2020, Addis Ababa, Ethiopia, April 26–30, 2020. OpenReview.net (2020)
56. Xu, W., Evans, D., Qi, Y.: Feature squeezing: detecting adversarial examples in deep neural networks. In: 25th Annual Network and Distributed System Security Symposium, NDSS 2018, San Diego, California, USA, February 18–21, 2018. The Internet Society, Reston (2018)
57. Xu, K., Zhang, G., Liu, S., Fan, Q., Sun, M., Chen, H., Chen, P., Wang, Y., Lin, X.: Adversarial t-shirt! Evading person detectors in a physical world. In: Vedaldi, A., Bischof, H., Brox, T., Frahm, J. (eds.) Computer Vision - ECCV 2020 - 16th European Conference, Glasgow, UK, August 23–28, 2020, Proceedings, Part V. Lecture Notes in Computer Science, vol. 12350, pp. 665–681. Springer, Berlin (2020). https://doi.org/10.1007/978-3-030-58558-7_39
58. Yuan, Z., Lu, Y., Wang, Z., Xue, Y.: Droid-sec: deep learning in android malware detection. SIGCOMM Comput. Commun. Rev. **44**(4), 371–372 (2014). https://doi.org/10.1145/2740070.2631434

Chapter 15
Probabilistic Robustness in Deep Learning: A Concise Yet Comprehensive Guide

Xingyu Zhao

15.1 Introduction

Deep learning (DL) has demonstrated transformative potential across various industries, including transportation, healthcare, and energy. Its ability to process complex data and make accurate predictions makes it a promising candidate for integration into safety-critical applications [36, 38, 42], such as autonomous driving, medical diagnosis, and energy grid management. However, the *robustness* of DL models in these high-stakes environments remains a significant concern, posing a risk to their safe deployment [5, 24, 53].

Robustness in DL generally refers to a model's ability to maintain consistent predictions despite small input perturbations [8, 14, 23]. In other words, a robust model should produce the same output even when its input is slightly altered. Adversarial examples (AEs) [20, 41] are inputs that have been perturbed to cause a model to make incorrect predictions. These perturbations are often imperceptible to humans but can lead to significant errors in the DL model's output. Indeed, robustness has become a central focus in DL research, encompassing a wide range of topics like the study of adversarial attacks in malicious contexts [8, 20, 29, 57], methods for detecting AEs by testing [18, 21], and techniques for verifying model robustness with formal guarantees [24, 40]. Additionally, various approaches have been developed to improve robustness, such as distillation [34], randomized smoothing [10], gradient masking [1], and adversarial training (AT) [20, 29]. As multi-modal foundation models gain popularity, recent research has increasingly shifted toward understanding their robustness [17, 51, 52].

X. Zhao (✉)
WMG, University of Warwick, Coventry, UK
e-mail: xingyu.zhao@warwick.ac.uk

E. Nowroozi et al. (eds.), *Adversarial Example Detection and Mitigation Using Machine Learning*, https://doi.org/10.1007/978-3-031-99447-0_15

Notably, the majority of robustness research focuses on *adversarial robustness* (AR), which addresses *worst-case* and *deterministic* scenarios. The terms "worst-case" and "deterministic" refers to the fact that AR metrics and their corresponding evaluation problems are typically formulated around the following three (distinct yet interrelated) types of questions [52]:

- a binary question, "Does an AE exist within the perturbation norm-ball?"
- an optimisation question on maximising the safe perturbation distance, "What is the largest perturbation radius within which no AE exists?"
- an optimisation question on maximising the prediction loss, "Which AE within the given norm-ball yields the highest prediction loss?"

In contrast, a more recent and *practical* (cf. Sect. 15.2) view considers *probabilistic robustness* (PR) [3, 11, 35, 39, 43–47, 49, 52]—it answers:

- a statistical inference question, "What is the *probability* of discovering an AE when the input is stochastically perturbed according to a specified distribution within a given norm-ball?"

A "frequentist" interpretation of this probability is the *limiting relative frequency* of perturbations for which the DL model's output remains unchanged, within an infinite sequence of independently generated perturbations [52]. In other words, it represents the *proportion* of non-AEs in the infinite set of all possible input perturbations. Thus, more intuitively, the statistical inference question of PR becomes "What is the proportion of AEs in the given perturbation norm-ball?"

In this paper, our main contribution is to deliver a concise yet comprehensive overview of all critical aspects of PR research. This includes its formal definition and associated metrics, existing evaluation methods, improvement techniques, and its significance for system safety assurance and reliability modelling. Specifically, we begin by summarising and contrasting the key distinctions between PR and AR in Sect. 15.2. Following this, we provide an in-depth analysis of existing evaluation methods for PR in Sect. 15.3. In Sect. 15.4, we revisit existing AT techniques and propose a reformulated "min-max" optimisation framework as the theoretical foundation for AT specifically designed to enhance PR. Furthermore, from a systems engineering perspective, we explore potential pathways for integrating PR verification evidence into AI system-level assurance to enhance safety and reliability in Sect. 15.5. Finally, we identify open questions and key challenges in the field, outlining promising directions for future research in Sect. 15.6.

15.2 Definition of Probabilistic Robustness

Although definitions of robustness vary across different DL tasks and model types, it generally refers to a DL model's ability to maintain consistent predictions despite small input perturbations. Typically it is defined as all inputs in a region η have the same prediction, where η is a small norm ball (in a L_p-norm distance) of radius γ

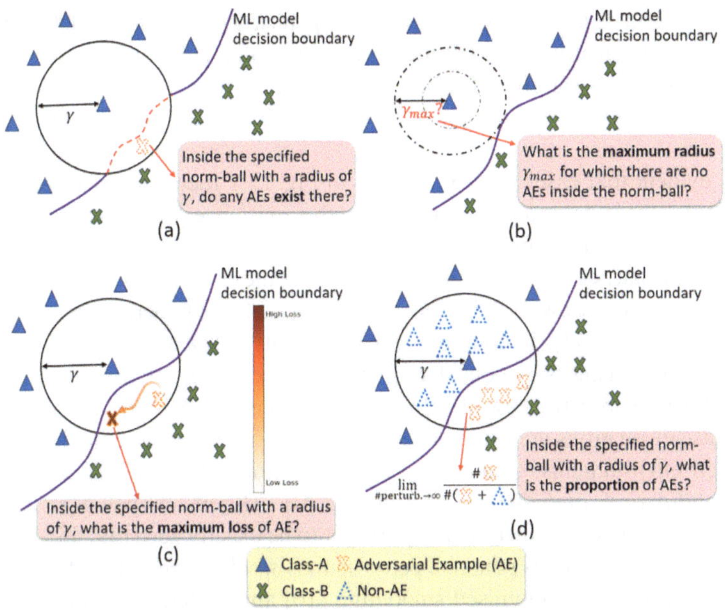

Fig. 15.1 Four types of problem formulations of robustness evaluation problems [52]

around an input x. A perturbed input (e.g., by adding noise on x) x' within η is an AE if its prediction label differs from x.

Robustness are typically measured on metrics defined around AEs [23, 57]. As discussed earlier in Sect. 15.1, four types of robustness metrics, as summarised in [14, 52] and Fig. 15.1, address: *(1)* a binary question, e.g., formulated as a satisfiability problem that usually can be solved by SAT/SMT solvers [24, 26]; *(2)* a maximum safe perturbation radius question, e.g., formulated as reachability analysis problems [40, 48]; *(3)* a maximum prediction loss question, e.g., formulated as adversarial attacks leveraging gradient information [8, 13, 20, 29]; and *(4)* a statistical inference question estimating the "proportion" (a *limiting relative frequency* interpretation of the probability of detecting AEs when generating random perturbations) of AEs within a given perturbation norm-ball.

Definition 15.1 (Probabilistic Robustness) For a DL model f_θ that takes input x, the local PR of an input x in a norm ball of radius γ is:

$$PR(x, \gamma) = \mathbb{E}_{\substack{x' \sim Pr(\cdot|x) \\ x' : \|x - x'\| \leq \gamma}} [I_{\{f_\theta(x') = f_\theta(x)\}}(x')] \tag{15.1}$$

where $I_S(x)$ is an indicator function—it is equal to 1 when S is true and 0 otherwise; $Pr(\cdot \mid x)$ is the local distribution of how perturbations x' are generated from x, which is precisely the "input model" used by, e.g., [46, 47, 52].

While PR in Eq. (15.1) represents the *expected probability* of detecting AEs, alternative tail-sensitive local PR measures can be defined, e.g., ρ-essential supremum (ρ-ess sup) for a given PR threshold and Conditional Value-at-Risk (CVaR) for tail averages [39], and Entropic Value-at-Risk (EVaR) for exponential tail bounds [50].

It is evident that cases (a), (b), and (c) in Fig. 15.1 focus on *extreme scenarios*, where robustness is evaluated based on the existence of a *deterministic* AE, reflecting a *worst-case* perspective. This approach, originally termed "adversarial" robustness, is particularly relevant in malicious contexts and *security* studies. While a comprehensive review of AR is out of the scope of this paper, we refer readers to surveys [9, 23, 32, 57]. In contrast, case (d) adopts a *probabilistic* view, assessing *overall* local robustness by considering the presence of AEs statistically. Compared to AR, this probabilistic view of PR is more practical for real-world applications, as worst-case evaluations are often unnecessary, especially for large models operating in *benign* environments. Quantifying the proportion/risk of AEs is more relevant for *reliability and safety*, where ensuring the AE risk remains below an acceptable threshold is sufficient, rather than requiring it to be exactly zero [39, 45, 46, 49, 52].

Although some studies generate perturbations *stochastically*, this does not imply a focus on PR, e.g., randomised smoothing [10] aim to certify a safe perturbation radius (i.e., case (b) in Fig. 15.1) by replacing standard model inference with the averaging of predictions over a set of randomly perturbed inputs.

Furthermore, the PR definition of Eq. 15.1 is a type of *local* robustness focusing on model stability around an given *individual input*, as opposed to *global* robustness across the entire input space. A closely related concept to local PR but defined at the global level is the *probabilistic Lipschitzness* [30, 31]:

Definition 15.2 (Probabilistic Lipschitzness) Given a distribution D over the entire input space of model f, its probabilistic Lipschitzness is formal stated as:

$$Pr_{x,x' \sim D \otimes D}(||f(x) - f(x')|| \leq k * ||x - x'|| \mid ||x - x'|| \leq \gamma) \geq 1 - \epsilon, \tag{15.2}$$

which is defined globally by focusing on pairs of γ-close inputs such that their cumulative probability on the Lipschitz property is at least $1 - \epsilon$.

Similarly, [45] explicitly defines global robustness probabilistically, referred to as *total statistical robustness* (or simply termed "reliability" in [14, 54]):

Definition 15.3 (Total Statistical Robustness) Reusing the notations from aforementioned definitions, the total statistical robustness of a model f is:

$$\mathbb{E}_{X \sim D}[\mathbb{E}_{X' \sim Pr(\cdot|x)}[I_{\{f(x')=f(x)\}}(x')]] \tag{15.3}$$

which is the expectation of local PR over the distribution of model's input space.

Again, the "inner-expectation" can be replaced by other risk measures like ρ-ess sup and CVaR in [39], and EVaR in [50].

15.3 Evaluation of Probabilistic Robustness

For AR, evaluation is typically conducted by measuring adversarial accuracy on a test set, where strong adversarial attacks are applied, such as those in [12, 16]. While no benchmark for PR currently exists, we provide a summary of existing works that evaluate[1] PR-related metrics in Definitions 15.1, 15.2, and 15.3 in Table 15.1. 15 papers (including journal extension of 2 conference papers) are organised chronologically (since 2019) and explicitly categorised based on the specific PR metric studied, how input perturbations generated, the core techniques or theoretical foundations underlying their estimators, and the DL tasks considered.

Webb et al. [46] introduces the first *black-box* estimator based on adaptive multi-level splitting, a Monte Carlo technique for estimating the probability of rare events. Similarly [43] proposes a more efficient black-box estimator by using a variant of adaptive multi-level splitting and a sequential Monte Carlo sampling method known as "last particle". Another black-box estimator is presented in [3], which combines Chernoff bounds with simple Monte Carlo sampling.

In contrast, PROVEN [47] is the first *white-box* analytical approach, leveraging the fact that a deep learning model's output can be lower/upper-bounded by two linear functions, and then applying Hoeffding's inequality to derive PR guarantees. Similarly, [11] presents a white-box method that derives upper bounds on tail probabilities using the Cramér-Chernoff concentration inequality to analyse uncertainty propagation through the DL model layers. More recently, [44] proposes a highly efficient estimator for local PR by integrating fast gradient computations with Hamiltonian Monte Carlo sampling.

While the aforementioned studies primarily focus on perturbations modeled as random L_p noise, [49] and [35] extend PR analysis to *semantic perturbations* (referred to as functional perturbations in [49]), e.g., rotation and translation. In [49], simple Monte Carlo sampling and an adaptive version of Hoeffding's inequality are employed to establish confidence bounds on local PR. In [35], a black-box approach is adopted, utilising the Cramér-Chernoff inequality alongside simple Monte Carlo sampling.

At the global-level, [30, 31] is the first work look at the property of Definition 15.2 which is checked probabilistically by an algorithm based on abstract interpretation and importance sampling techniques. The related but more straightforward metric of Definition 15.3, was first introduced in [45, 54], where the latter refers to it as "reliability". In [45], the estimator from [46] is used at the local level, with global estimation performed via simple Monte Carlo sampling (as its primary focus is on improving global robustness (cf. Sect. 15.4) through data augmentation rather than estimation). In contrast, [54] explicitly connects this

[1] We use the term "evaluation" broadly to encompass related concepts such as certification, verification, and assessment, without distinguishing their subtle differences. Here, "evaluation" refers to any study that evaluates a model's robustness based on PR metrics without modifying the DL model itself.

Table 15.1 Estimators on PR-related metrics: local PR (LPR, Definition 15.1), global Probabilistic Lipschitzness (GPL, Definition 15.2) and total statistical robustness (TSR, Definition 15.3)

Paper	Metric	Perturb.	Estimator	DL Task
ICLR'19 [46]	LPR	L_p noise	Adaptive multi-level splitting, Metropolis–Hasting algo.	Img. classific.
ICML'19 [47]	LPR	L_p noise	Layer-wise bound propagation, Hoeffding's inequality	Img. classific.
ICSE'19 [30, 31]	GPL	L_p noise	Abstract interpretation, importance sampling	Img. classific.
UAI'21 [45]	TSR	L_p noise	Simple Monte Carlo and [46]	Img. classific.
AISafety'21 [14, 54]	TSR	L_p noise	Input space partition, operational profile	Img. classific.
Elsevier-NN [11]	LPR	L_p noise	Cramer-Chernoff inequality, layer-wise uncertainty propaga.	Reg.&classific.
ICSE'21 [3]	LPR	L_p noise	Chernoff bounds, simple Monte Carlo	Img. classific.
NeurIPS'21 [43]	LPR	L_p noise	Adaptive multi-level splitting, "last particle" simulation	Img. classific.
ECML'22 [49]	LPR	Semantic	Adaptive Hoeffding's inequality, simple Monte Carlo	Img. classific.
AAAI'22 [35]	LPR	Semantic	Cramer-Chernoff inequality, simple Monte Carlo	Img. classific.
ICML'22 [39]	ρ-ess sup CVaR	L_p noise	Stochastic gradient decent, simple Monte Carlo	Img. classific.
AIStats'23 [44]	LPR	L_p noise	Gradient-informed Hamiltonian Monte Carlo	Img. classific.
ICCV'23 [22]	LPR	L_p noise	Adaptive multi-level splitting, Metropolis–Hasting algo.	XAI
IJCAI'24 [50]	EVaR	L_p noise	Adaptive stochastic search, simple Monte Carlo	Img. classific.
ECCV'24 [52]	LPR	text char.	Adaptive Hoeffding's inequality, sequential statistical testing	Text2Image

metric to the *operational profile* (OP) concept from software reliability engineering [33], approximating the OP from data after partitioning the input space.

Most existing research on PR has focused on image classification tasks. The first and, to date, only extension of PR to explanations from eXplainable AI (XAI) methods is [22]. In [22], the PR of Definition 15.1 is redefined in terms of wrong explanations after random perturbations of images, it applies the same techniques as [46]. More recently, the first study on PR in text-to-image generative AI was introduced in [52]. This work employs the adaptive Hoeffding's inequality and sequential statistical testing with early stopping rules to enhance efficiency in identifying AEs by comparing two sets of generated images—one from the original input text and the other from perturbed text characters.

15.4 Enhancement of Probabilistic Robustness

Despite the relatively greater focus on evaluating PR, to the best of our knowledge, limited works explicitly aimed at improving the PR of DL models are [39, 45, 50]. [45] demonstrates—both empirically and theoretically—that simply adding a randomly perturbed inputs significantly enhances PR and leads to better generalisation (i.e., the robustness improvement in training set can be better generalised to the test set). Both [39, 50] design loss functions using common risk measures—ρ-essential supremum, Conditional Value-at-Risk, and Entropic Value-at-Risk—then train models to minimize them on the training dataset.

However, these three approaches provide *quite limited improvement in AR*. We believe that, akin to the well-established AT framework for AR, a dedicated AT framework for PR is necessary, which remains absent in the literature. Developing such an AT for PR scheme would *enable significant joint improvement of both PR and AR*. AT [2, 20, 29] is a widely used empirical approach for improving robustness, typically formulated as a "min-max" optimisation. The "inner max" often empirically solved by adversarial attacks, e.g., PGD [29] has become a critical benchmark and is widely regarded as the standard method for AT in practice [2]. The objective of AT for AR is to identify the AEs that maximises model loss and incorporate it into AT, cf. top-middle illustration in Fig. 15.2. However, to enhance local PR, the goal shifts to identifying the *widest peak*—the region with the highest proportion of AEs in the loss landscape—rather than the *tallest peak* (cf. top-right illustration in Fig. 15.2). Incorporating the AE from this region into AT can potentially reduce loss for AEs in its vicinity, thereby lowering the overall

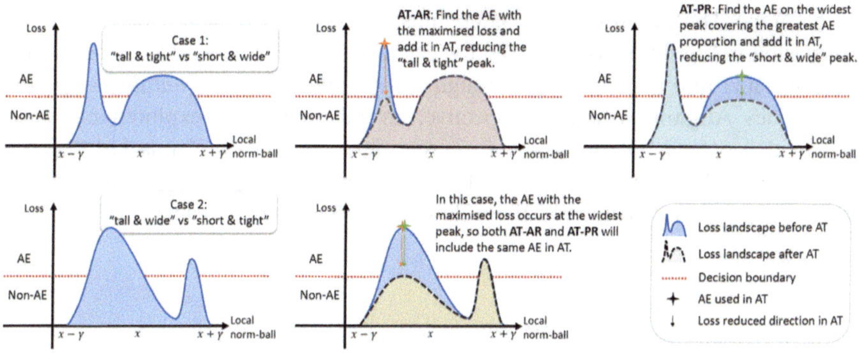

Fig. 15.2 Two cases of local loss landscape and AEs identified by AT-AR and AT-PR

AE proportion (cf. PR interpretation in Definition 15.1). In some cases, AT-WCR and AT-PR may select the same AE when the widest and tallest peaks coincide, as depicted in the second row of Fig. 15.2. The two scenarios illustrated in Fig. 15.2 representatively cover all possible cases, though additional peaks may exist in practice.

Given the difference between the goals of improving AR and PR, we formulate a new min-max problem termed as AT-PR:

Definition 15.4 (AT for PR) For a DL model f_θ, trained on a dataset $\mathcal{S}$ consisting of pairs (x, y), The objective of AT-PR can be formulated as:

$$\min_{\theta} \mathbb{E}_{(x,y)\sim\mathcal{S}} \left[\max_{\|\delta\|\leq\gamma,\, PR(x+\delta,k)=0} k \right], \tag{15.4}$$

where k and δ are the variables in the new inner max optimization, $PR(\cdot, \cdot)$ is the metric defined in Eq. (15.1).

The intuition behind the new "inner max" is to find an optimal perturbation δ that produces an AE $x' = x + \delta$ that maximises the radius k of a smaller norm-ball centred on this AE x', such that the PR within this k-sized norm-ball is 0, i.e., all inputs within this new and widest norm-ball are AEs. The high non-linearity and non-convexity of DL models makes it intractable to analytically solve both the established min-max for AR and our new formulation in Eq. (15.4). Novel and *efficient numerical algorithms* to solve the new framed min-max problem of AT-PR will be an important research direction.

15.5 System-Level Integration of Probabilistic Robustness

Safety[2] is a *system property*, not a component property—focusing solely on "safety at DL model levels" without considering the full system and application context is misleading [7]. A safe system accounts for redundancy, fail-safes, human oversight, and regulatory constraints, none of which are captured by evaluations at DL model levels alone. Thus, discussions of safety must extend beyond the DL model itself. As illustrated in Fig. 15.3, AR is linked to system-level security, while PR metrics can be translated into system safety through *system analysis* and *reliability modelling* techniques. Aligning with this perspective, several studies have explored integrating PR evidence from DL models into system-level reliability claims [14, 15, 53]. However, establishing a rigorous methodology for such integration remains an open research question. Specifically, the following questions need to answered.

[2] In system engineering, safety and reliability, though distinct, do not require separate statistical reasoning in safety-critical applications. This is because any failure in such systems can lead to unsafe outcomes. Consequently, as in [55], reliability specifically refers to the probabilities/rates of failures directly impacting system safety, thereby unifying the two concepts in this context.

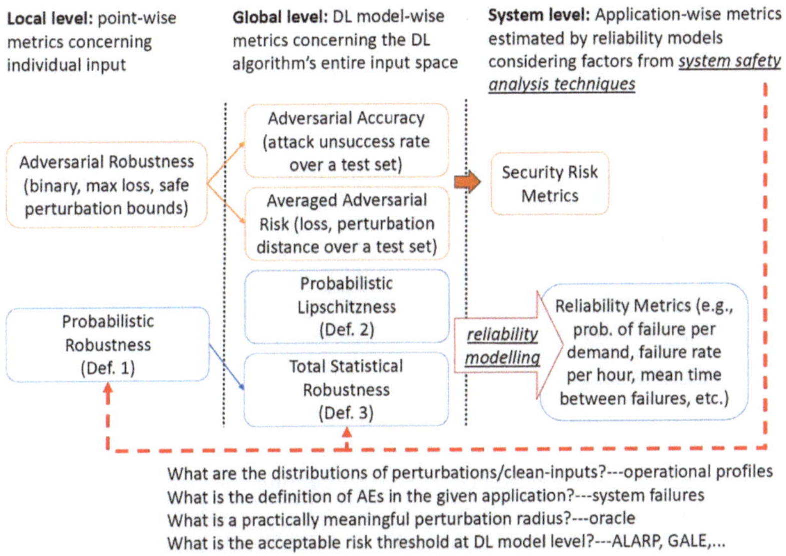

Fig. 15.3 Metrics studied ranging from local norm-ball, DL model input domain to system safety space. *Safety analysis* and *reliability modelling* techniques, considering factors like system architecture, redundancy design, operational environment/profiles, discrete/continuous nature, failure modes/criticality, should be applied at the system level. Then map them to application-specific estimator settings at the local and global levels of DL models, and propagate estimation results with informed confidence

What Is the Most Suitable Reliability Metric for This Specific Application? And What Are the Acceptable Risk Thresholds for the Given Metric? Reliability in safety-critical systems is fundamentally a quantitative claim, concerned with the likelihood/rate of failures that have safety consequences. Systems can generally be categorized into continuous-time systems, which operate continuously to control a process (e.g., flight control software), and on-demand systems, which respond to discrete activation signals (e.g., the emergency shutdown system of a nuclear plant). Reliability metrics differ based on system type: failure rate (failures per unit time) is typically used for continuous-time systems, while the *probability of failure per demand* (*pfd*) is relevant for on-demand systems [28, 53]. Other reliability measures may also be applicable depending on the application context, such as the probability of mission success for spacecraft launch systems, the probability of fatalities per mile and disengagement rate as studied for autonomous vehicles in [25, 55, 56]. Safety standards such as IEC 61508 and DO-178C define domain-specific reliability metrics and establish acceptable risk thresholds based on system criticality (e.g., Safety Integrity Levels). More broadly, risk acceptability is guided by regulatory principles such as ALARP and GALE [14, 27].

How Can System-Level Reliability Metrics be Translated into Corresponding Metrics/Definitions at the DL Model Global and Local Levels? Once an appropriate system-level reliability metric is defined, reliability modelling must account for the system architecture—how the DL component interacts with traditional software and hardware, and whether "defence-in-depth" mechanisms, as suggested by Bloomfield and Rushby [7], are in place. The goal is to express *system reliability as a function of local/global PR-related metrics* while considering system architecture and failure dependencies. Key hyperparameters in PR-related metrics must be aligned with system-level considerations: perturbation/clean-input distributions should reflect the system's operational profile, AE definitions should correspond to system-level failures, and perturbation radius should be derived from system specifications and testing oracles. While initial attempts have been made in [14, 15, 53, 54], these considerations must be modeled on a case-by-case basis.

How Can PR Estimates and Their Uncertainties be Propagated from Local and Global Levels to Support System-Level Reliability Claims with Informed Confidence? Intuitively, plugging PR estimates into a reliability modelling function yields system reliability claims. However, accounting for estimation uncertainties is crucial. Fortunately, most estimators in Table 15.1 provide confidence measures. The greater challenge lies in systematically propagating these uncertainties through the reliability model [4]. This aligns with the state-of-the-art safety assurance methodology "Assurance 2.0" [6], which aims for *indefeasible* confidence through *probabilistic assessment* and *residual risk evaluation*.

15.6 Challenges and Open Problems

Benchmarking PR Estimators The absence of a standardised framework for evaluating PR hinders accurate assessment and comparison of their effectiveness. Establishing rigorous benchmarking protocols is essential, requiring well-defined reference metrics, models, and datasets, along with computationally feasible gold-standard baselines. A key technical challenge lies in ensuring usability, reproducibility, and scalability across diverse models and datasets.

Extending PR to Frontier AI Tasks PR research has predominantly focused on image classification tasks, as highlighted in Table 15.1, with limited exploration into other domains. Only one study has investigated text-to-image models [52], and another has addressed XAI [22]. Expanding PR research to multi-modal and generative AI tasks is essential for achieving broader applicability and impact. However, unlike classification, which typically involve continuous image inputs and discrete categorical outputs, these tasks often deal with diverse modalities, e.g., images, text, and audio, as both inputs and outputs. A key open problem is the development of task-specific PR formulations that are computationally feasible while effectively capturing perturbations relevant to each modality. Furthermore, a recent work [19] introduce PR for conformal prediction techniques.

Improving PR by Dedicated AT Unlike AR, which has well-established AT techniques, no dedicated AT scheme exists for PR. In this paper Sect. 15.4, we have reformulated the min-max optimisation specifically for PR, establishing a new theoretical foundation for PR-focused AT. Designing AT schemes with efficient numerical algorithms to solve this new optimisation problem is challenging and forms an important future work.

Theoretical Foundation for PR-Focused AT A critical area for future research is the theoretical analysis for PR optimised models. Understanding how PR generalises across diverse data distributions and model architectures remains an open challenge. This necessitates advancements in learning theories, particularly in areas such as Rademacher complexity and PAC-Bayesian bounds [37], to provide rigorous theoretical foundations.

End-to-End Case Studies of System-Level Integration As per Sect. 15.5, PR holds significant potential for integration into system-level safety claims, but its practical applicability requires end-to-end case study demonstrations. This endeavour demands a highly interdisciplinary skill set, combining expertise in AI/DL, reliability modelling, safety assurance, and application domain knowledge.

15.7 Conclusion

This paper provides a concise yet comprehensive overview of PR in DL, covering its definitions, evaluation methods, and enhancement strategies. We introduce a reformulated AT framework tailored for PR and explore its integration into system-level safety and reliability assurance. Despite progress, key challenges remain, including benchmarking PR evaluation methods, extending PR to emerging AI tasks, and developing rigorous system-level integration methodologies. Addressing these challenges is crucial for ensuring the trustworthiness of AI in safety-critical applications.

Acknowledgments This work is supported by the UK EPSRC New Investigator Award through Harnessing Synthetic Data Fidelity for Assured Perception of Autonomous Vehicles [EP/Z536568/1] and the NVIDIA Academic Grant Program.

References

1. Athalye, A., Carlini, N., Wagner, D.: Obfuscated gradients give a false sense of security: circumventing defenses to adversarial examples. In: International Conference on Malignant Lymphoma (ICML'18) (2018)
2. Bai, T., Luo, J., Zhao, J., Wen, B., Wang, Q.: Recent advances in adversarial training for adversarial robustness. In: Proceedings of the Thirtieth International Joint Conference on Artificial Intelligence (IJCAI'21), pp. 4312–4321 (2021)

3. Baluta, T., Chua, Z.L., Meel, K.S., Saxena, P.: Scalable quantitative verification for deep neural networks. In: International Conference on Software Engineering (ICSE'21), pp. 312–323 (2021)
4. Bensalem, S., Cheng, C.H., Huang, W., Huang, X., Wu, C., Zhao, X.: What, indeed, is an achievable provable guarantee for learning-enabled safety-critical systems. In: International Conference on Bridging the Gap Between AI and Reality, pp. 55–76 (2023)
5. Bloomfield, R., Khlaaf, H., Conmy, P.R., Fletcher, G.: Disruptive innovations and disruptive assurance: assuring machine learning and autonomy. Computer **52**(9), 82–89 (2019)
6. Bloomfield, R., Rushby, J.: Confidence in assurance 2.0 cases. In: The Practice of Formal Methods: Essays in Honour of Cliff Jones, Part I, pp. 1–23 (2024)
7. Bloomfield, R., Rushby, J.: Where AI assurance might go wrong: initial lessons from engineering of critical systems (2025). arXiv:2502.03467
8. Carlini, N., Wagner, D.: Towards evaluating the robustness of neural networks. In: IEEE Symposium on Security and Privacy (SP), pp. 39–57 (2017)
9. Chakraborty, A., Alam, M., Dey, V., Chattopadhyay, A., Mukhopadhyay, D.: A survey on adversarial attacks and defences. CAAI Trans. Intell. Technol. **6**(1), 25–45 (2021)
10. Cohen, J., Rosenfeld, E., Kolter, Z.: Certified adversarial robustness via randomized smoothing. In: International Conference on Malignant Lymphoma (ICML'19), pp. 1310–1320 (2019)
11. Couellan, N.: Probabilistic robustness estimates for feed-forward neural networks. Neural Netw. **142**, 138–147 (2021)
12. Croce, F., Andriushchenko, M., Sehwag, V., Debenedetti, E., Flammarion, N., Chiang, M., Mittal, P., Hein, M.: Robustbench: a standardized adversarial robustness benchmark (2020). arXiv:2010.09670
13. Croce, F., Hein, M.: Reliable evaluation of adversarial robustness with an ensemble of diverse parameter-free attacks. In: International Conference on Malignant Lymphoma (ICML'20), pp. 2206–2216 (2020)
14. Dong, Y., Huang, W., Bharti, V., Cox, V., Banks, A., Wang, S., Zhao, X., Schewe, S., Huang, X.: Reliability assessment and safety arguments for machine learning components in system assurance. ACM TECS **22**(3), 1–48 (2023)
15. Dong, Y., Zhao, X., Wang, S., Huang, X.: Reachability verification based reliability assessment for deep reinforcement learning controlled robotics and autonomous systems. IEEE Robot. Autom. Lett. **9**(4), 3299–3306 (2024)
16. Dong, Y., Fu, Q.A., Yang, X., Pang, T., Su, H., Xiao, Z., Zhu, J.: Benchmarking adversarial robustness on image classification. In: IEEE/CVF Computer Vision and Pattern Recognition Conference (CVPR'20), pp. 321–331 (2020)
17. Du, C., Li, Y., Qiu, Z., Xu, C.: Stable diffusion is unstable. In: Neural Information Processing Systems (NeurIPS'24), vol. 36 (2024)
18. Du, X., Xie, X., Li, Y., Ma, L., Liu, Y., Zhao, J.: Deepstellar: model-based quantitative analysis of stateful deep learning systems. In: ACM Joint Meeting on European Software Engineering Conference and Symposium on the Foundations of Software Engineering (ACM ESEC/FSE'19) (2019)
19. Ghosh, S., Shi, Y., Belkhouja, T., Yan, Y., Doppa, J., Jones, B.: Probabilistically robust conformal prediction. In: UAI '23: Conference on Uncertainty in Artificial Intelligence (UAI'23), pp. 681–690. PMLR (2023)
20. Goodfellow, I.: Explaining and harnessing adversarial examples. In: International Conference on Learning Representations (ICLR'15) (2015)
21. Huang, W., Zhao, X., Banks, A., Cox, V., Huang, X.: Hierarchical distribution-aware testing of deep learning. ACM Trans. Softw. Eng. Methodol. **33**(2), 1–35 (2023)
22. Huang, W., Zhao, X., Jin, G., Huang, X.: Safari: versatile and efficient evaluations for robustness of interpretability. In: IEEE/CVF International Conference on Computer Vision (ICCV'23), pp. 1988–1998 (2023)
23. Huang, X., Kroening, D., Ruan, W., et al.: A survey of safety and trustworthiness of deep neural networks: verification, testing, adversarial attack and defence, and interpretability. Comput. Sci. Rev. **37**, 100270 (2020)

24. Huang, X., Kwiatkowska, M., Wang, S., Wu, M.: Safety verification of deep neural networks. In: International Conference on Computer-Aided Verification (CAV'17), pp. 3–29 (2017)
25. Kalra, N., Paddock, S.M.: Driving to safety: how many miles of driving would it take to demonstrate autonomous vehicle reliability? Transp. Res. A Policy Practice **94**, 182–193 (2016)
26. Katz, G., Barrett, C., Dill, D.L., Julian, K., Kochenderfer, M.J.: Reluplex: an efficient smt solver for verifying deep neural networks. In: International Conference on Computer-Aided Verification (CAV'17) (2017)
27. Littlewood, B., Salako, K., Strigini, L., Zhao, X.: On reliability assessment when a software-based system is replaced by a thought-to-be-better one. Reliabil. Eng. Syst. Safety **197**, 106752 (2020)
28. Littlewood, B., Strigini, L.: Validation of ultra-high dependability for software-based systems. Commun. ACM **36**(11), 69–80 (1993)
29. Madry, A., Makelov, A., Schmidt, L., Tsipras, D., Vladu, A.: Towards deep learning models resistant to adversarial attacks. In: International Conference on Learning Representations (ICLR'18) (2018)
30. Mangal, R., Nori, A.V., Orso, A.: Robustness of neural networks: a probabilistic and practical approach. In: ICSE-NIER'19: Proceedings of the 41st International Conference on Software Engineering: New Ideas and Emerging Results (ICSE-NIER'19), pp. 93–96. IEEE, Piscataway (2019)
31. Mangal, R., Sarangmath, K., Nori, A.V., Orso, A.: Probabilistic lipschitz analysis of neural networks. In: The International Static Analysis Symposium (SAS), pp. 274–309 (2020)
32. Meng, M.H., Bai, G., Teo, S.G., Hou, Z., Xiao, Y., Lin, Y., Dong, J.S.: Adversarial robustness of deep neural networks: a survey from a formal verification perspective. IEEE Trans. Depend. Secur. Comput. (2022). https://doi.org/10.1109/TDSC.2022.3179131
33. Musa, J.D.: Operational profiles in software-reliability engineering. IEEE Softw. **10**(2), 14–32 (1993)
34. Papernot, N., McDaniel, P., Wu, X., Jha, S., Swami, A.: Distillation as a defense to adversarial perturbations against deep neural networks. In: 2016 IEEE Symposium on Security and Privacy (SP'16). IEEE, Piscataway (2016)
35. Pautov, M., Tursynbek, N., Munkhoeva, M., Muravev, N., Petiushko, A., Oseledets, I.: CC-Cert: a probabilistic approach to certify general robustness of neural networks. In: AAAI Conference on Artificial Intelligence (AAAI'22), vol. 36, pp. 7975–7983 (2022)
36. Rabe, M., Milz, S., Mader, P.: Development methodologies for safety critical machine learning applications in the automotive domain: a survey. In: 2021 IEEE/CVF Conference on Computer Vision and Pattern Recognition Workshops (CVPR) (2021)
37. Raman, V., Tewari, A., Subedi, U.: Probabilistically robust PAC learning. In: NeurIPS ML Safety Workshop (2022)
38. Rech, P.: Artificial neural networks for space and safety-critical applications: reliability issues and potential solutions. IEEE Trans. Nucl. Sci. **71**(4), 377–404 (2024)
39. Robey, A., Chamon, L., Pappas, G.J., Hassani, H.: Probabilistically robust learning: balancing average and worst-case performance. In: International Conference on Machine Learning (ICML'22) (2022)
40. Ruan, W., Huang, X., Kwiatkowska, M.: Reachability analysis of deep neural networks with provable guarantees. In: Proceedings of the 27th International Joint Conference on Artificial Intelligence (IJCAI'18), pp. 2651–2659 (2018)
41. Szegedy, C., Zaremba, W., Sutskever, I., Bruna, J., Erhan, D., Goodfellow, I., Fergus, R.: Intriguing properties of neural networks. In: International Conference on Learning Representations (ICLR'14) (2014)
42. Tambon, F., Laberge, G., An, L., Nikanjam, A., Mindom, P.S.N., Pequignot, Y., Khomh, F., Antoniol, G., Merlo, E., Laviolette, F.: How to certify machine learning based safety-critical systems? A systematic literature review. Autom. Softw. Eng. **29**(2), 38 (2022)

43. Tit, K., Furon, T., Rousset, M.: Efficient statistical assessment of neural network corruption robustness. In: 35th Conference on Neural Information Processing Systems (NeurIPS'21), vol. 34, pp. 9253–9263 (2021)
44. Karim, T.I.T., Furon, T., Rousset, M.: Gradient-informed neural network statistical robustness estimation. In: Proceedings of The 26th International Conference on Artificial Intelligence and Statistics (AISTATS'23), vol. 206, pp. 323–334. PMLR (2023)
45. Wang, B., Webb, S., Rainforth, T.: Statistically robust neural network classification. In: Uncertainty in Artificial Intelligence, pp. 1735–1745. PMLR (2021)
46. Webb, S., Rainforth, T., Teh, Y.W., Kumar, M.P.: A statistical approach to assessing neural network robustness. In: 7th International Conference on Learning Representations (LCLR'19) (2019)
47. Weng, L., Chen, P.Y., Nguyen, L., Squillante, M., Boopathy, A., Oseledets, I., Daniel, L.: Proven: verifying robustness of neural networks with a probabilistic approach. In: International Conference on Machine Learning (ICML'19), pp. 6727–6736 (2019)
48. Zhang, H., Weng, T.W., Chen, P.Y., Hsieh, C.J., Daniel, L.: Efficient neural network robustness certification with general activation functions. In: 32nd Conference on Neural Information Processing Systems (NeurIPS'18) (2018)
49. Zhang, T., Ruan, W., Fieldsend, J.E.: ProA: a probabilistic robustness assessment against functional perturbations. In: European Conference on Machine Learning (ECML'23), pp. 154–170 (2023)
50. Zhang, T., Zhang, Y., Mu, R., Liu, J., Fieldsend, J., Ruan, W.: PRASS: probabilistic risk-averse robust learning with stochastic search. In: International Joint Conference on Artificial Intelligence (IJCAI'24) (2024)
51. Zhang, Y., Chen, Z., Cheng, C.H., Ruan, W., Huang, X., Zhao, D., Flynn, D., Khastgir, S., Zhao, X.: Trustworthy text-to-image diffusion models: a timely and focused survey. arXiv:2409.18214 (2024)
52. Zhang, Y., Tang, Y., Ruan, W., Huang, X., Khastgir, S., Jennings, P., Zhao, X.: Protip: probabilistic robustness verification on text-to-image diffusion models against stochastic perturbation. In: European Conference on Computer Vision (ECCV'24) (2024)
53. Zhao, X., Banks, A., et al.: A safety framework for critical systems utilising deep neural networks. In: Computer Safety, Reliability, and Security: 39th International Conference, SAFECOMP 2020 (SafeComp'20) (2020)
54. Zhao, X., Huang, W., Banks, A., Cox, V., Flynn, D., Schewe, S., Huang, X.: Assessing the reliability of deep learning classifiers through robustness evaluation and operational profiles. In: Proceedings of the Workshop on Artificial Intelligence Safety 2021 co-located with IJCAI 2021, vol. 2916 (2021). https://ceur-ws.org/Vol-2916/paper_16.pdf
55. Zhao, X., Robu, V., Flynn, D., Salako, K., Strigini, L.: Assessing the safety and reliability of autonomous vehicles from road testing. In: 2019 IEEE 30th International Symposium on Software Reliability Engineering (ISSRE'19), pp. 13–23 (2019)
56. Zhao, X., Salako, K., Strigini, L., Robu, V., Flynn, D.: Assessing safety-critical systems from operational testing: a study on autonomous vehicles. Inf. Softw. Technol. **128**, 106393 (2020)
57. Zühlke, M.M., Kudenko, D.: Adversarial robustness of neural networks from the perspective of lipschitz calculus: a survey. ACM Comput. Surv. **57**(6), 1–41 (2025)

Part IV
Federated Learning Under Attack and Defense

Chapter 16
Enhancing Federated Learning Security: Cluster-Based Strategies to Counter GAN-Poisoned Attacks

Ehsan Nowroozi, Yoosef Habibi, Abdul Basit Mughal, Mohammad Saedi, and Mohsen Jafari

16.1 Introduction

Federated learning (FL) is a decentralized machine learning approach to train distributed models without centralized data collection or sharing, as demonstrated by Zhang et al. [27]. This differs from the traditional centralized ML, in which all data must be sent to a central location for training. FL provides better privacy to clients because their sensitive information is not uploaded to a server, but is kept decentralized on clients' devices [25]. This design also reduces communication costs compared to standard centralized training methods. According to Hao et al. [10], FL facilitates ML collaboration in ways that allow privacy preservation; this is mostly in health and personal domains where the confidentiality of data is very high. Given that FL is decentralized, it also introduces new security challenges that must be addressed to guarantee better protection of user privacy and data.

E. Nowroozi
Centre for Sustainable Cyber Security (CS2), University of Greenwich, London, UK
e-mail: ehsan.nowroozi65@ieee.org

Y. Habibi (✉)
University of Padova, Padova, Italy

A. B. Mughal
Ravensbourne University London, London, UK
e-mail: a.mughal@students.rave.ac.uk

M. Saedi
Department of Computer Science City St George's, University of London, London, UK
e-mail: Mohammad.saedi@city.ac.uk

M. Jafari
School of Computing and Engineering, University of West London, London, UK
e-mail: 32140699@student.uwl.ac.uk

E. Nowroozi et al. (eds.), *Adversarial Example Detection and Mitigation Using Machine Learning*, https://doi.org/10.1007/978-3-031-99447-0_16

The decentralized architecture is not adequately defended because it contains new potential attack vectors. This chapter focuses on the security of FL systems. Specifically, a data poisoning threat scenario, whereby adversaries willfully share corrupted updates to compromise the global model, is sufficiently dangerous. Recent advances in generative adversarial networks (GANs) capable of synthesizing highly realistic fake images and data have led to the need to study how such techniques could be a data-poisoning attack against FL [14]. Despite advances in defenses for FL modesl, still gaps exist to completely mitigate the risks from GAN-motivated poisoning that aim to either degrade models or induce misclassifications through synthesized poisoned samples.

In this chapter, we focus on contributions toward the advancement of security for federated learning, specifically against data poisoning attacks in FL: model degradation attacks that gradually weaken the overall model performance and targeted misclassification attacks that induce errors on classes or samples [3]. Also, we study a clustering-based poisoned update detection method at the model-aggregation stage in FL. The idea is simply to flag anomalous updates with a clustering algorithm before they can influence the global model to filter out contaminated gradients contributed by adversarial parties [17].

16.1.1 Contribution

- The value of the contributions will be demonstrated through experimental results regarding the performance of the proposed clustering defense.
- The results will show that the detection method has a significantly limiting effect on accuracy degradation and misclassifications emerging from two GAN data poisoning attacks introduced.
- This will help create a viable setting for using this defense in practice.

16.1.2 Organization

The chapter is organized as follows:

- The Sect. 16.1 of the research in developing new attack models and defensive techniques and describes how the rest of the report is organized.
- The Sect. 16.2 provides context with a literature review of prior work.
- The Sect. 16.3, describes the methodology.
- The Sect. 16.4 presents the results.
- Finally, Sect. 16.5 presents conclusions and recommendations for future work.

16.2 Related Studies

In this section, we briefly review state-of-the-art (SOTA) studies, focusing on GAN-Based Attacks, Anomaly Detection Approaches, and Adversarial Training

Strategies. Additionally, we discuss other attack and defense methodologies such as Backdoor Attacks, Model Manipulation Attack, and Robust Algorithm Methods.

16.2.1 GAN-Based Attacks

The explosively increased modeling power of GANs has further empowered new poisoning threats using synthetic fake data. Chakraborty et al. [5] first experimentally demonstrated the possibility of poisoning GAN-generated images to attack centralized image classifiers. More recently, Wang et al. [26] wrote the first GAN attack against FL for image classification, which trained a GAN model on accurate data to generate poisoned images labeled as target classes. The poisoned samples can then be fed in FL training, which degrades the accuracy of the global model because GAN modeling is in continuous evolution, where its power to synthesize very realistic fake data is increasingly updated. This creates the risk of poisoning an ML system with a contaminated training dataset, leading to an engineer trying to protect it with adequate defenses, an ever-shifting generative adversarial threat.

16.2.2 Anomaly Detection Approaches

Some studies have applied anomaly detection by profiling the normal behaviors of users to flag poisoning events; for example, techniques for detecting changes in data distributions or divergent updates. However, because of non-IID (Independent and Identically Distributed) and heterogeneous data across devices in FL, it is significantly more difficult to establish standard patterns [6]. Therefore, the results will be false detections when legitimate users have different local data or training dynamics [16]. These statistical anomaly detection strategies are difficult to apply because of the decentralized and dynamic nature of the FL.

16.2.3 Adversarial Training Strategies

Promising research involves incorporating adversarial training into federated optimization, which can enhance the model's robustness by simulating attack scenarios at the local training stage [12]. Such defenses involve extra difficulties concerning the additional computation costs and complex coordination needed across many participating devices that are very different in federated networks [24]. Finally, these approaches focus on anticipated threat models that might be circumvented over time by adaptive solid attackers.

16.2.4 Other Methodologies

In attack methods, Backdoor Attacks inject poisoned training samples associated with benign inputs with the target labels. Chai et al. [4] demonstrated backdoors against FL by poisoning a subset of the client updates, where even with a single malicious user, the global model can be compromised by misclassifying the backdoor images. In the Model Manipulation Attack, instead of poisoning the training data, adversaries can directly poison the collaborative model training by manipulating their local updates before global sharing. Shejwalkar and Houmansadr [21] explored poisoning attacks on specific words or classes through update corruption against BERT fine-tuning. Early defense methods, including Krum and multi-Krum, used robust aggregation for malicious update identification instead of simple averaging [8]. Krum selects the update closest to the majority, and multi-Krum generalizes this to multiple adversaries. However, these methods incur extensive communication and computational costs per round because every client update must be communicated and computed with all other client updates. A research study conducted by Nowroozi et al. [20], highlighted vulnerabilities in federated learning systems caused by data poisoning attacks in computer networks, further emphasizing the need for robust security measures to mitigate these threats.

16.3 Methodology

In this section, we outline two primary categories of data-poisoning attacks against FL systems: Model Degradation Attacks and Targeted Misclassification Attacks. Model Degradation Attacks gradually weaken the overall performance of the FL model by injecting poisoned data, leading to cumulative accuracy degradation over successive training rounds. Targeted Misclassification Attacks, however, strategically induce specific errors by misclassifying target classes through adversarial samples. The proposed defense methodology introduces a clustering-based detection method that employs unsupervised learning techniques to identify and filter anomalous updates from compromised clients. Finally, this approach is compared with state-of-the-art defenses, highlighting its adaptability and efficiency in maintaining the integrity and reliability of FL models.

16.3.1 Attack Methodology

In this chapter, we consider two principal categories for data poisoning attacks against FL learning systems: model degradation attacks and targeted misclassification attacks. Each of the aforementioned attacks leverages the decentralization inherent in FL to inject malicious updates into the training process, ultimately undermining the integrity and performance of the global model.

Model Degradation Attack The Model Degradation Attack weakens the FL model over time. Adversaries are those in charge of one or many client devices involved in the FL process. Adversaries generate poisoned data samples through GANs or other similar technologies to craft inputs that might appear benign that degrade the model's performance [7].

- **Poisoned Data Generation:** GANs provide synthetic samples of data resembling legitimate data.
- **Poisoned data injection:** The poisoned samples are injected into the local training datasets of compromised clients.
- **Malicious update transmission:** The poisoned clients transmit the poisoned model updates to the central server which contain gradients influenced by poisoned data and subtly malfunction the global model [30].
- **Aggregation and propagation:** Full updates from all participating clients, including malicious ones, are mixed and sent to the central server. Then, an aggregate global model is propagated back to all clients.
- **Cumulative degradation:** Successive rounds of training result in accumulation of poisoned updates and reduced accuracy, manifesting as increased error rates over legitimate data. Since this process happens gradually, its detection becomes more challenging (Zhang et al. 2020).

Targeted Misclassification Attack In this approach, adversaries synthesize adversarial samples that seem similar to legitimate data but are classified with the wrong target classes. Then, These samples are fed into the compromised clients' local training data. Thus, model updates contain misclassification-inducing gradients which will be sent to the central server. The server aggregates all updates, including manipulated ones. The updates with the misclassification-inducing gradients push global model predictions toward elicited misclassification. Zhang et al. [28]. As a result, global models misclassify inputs as specified target classes. The success rate of this attacks lies on the adversarial examples' effectiveness in influencing the model without being detected. These attacks exploit the decentralized nature of federated learning, highlighting the need for robust defenses to maintain model integrity.

16.3.2 Defense Methodology

In this section, we discuss a defending mechanism against data poisoning attacks using a clustering-based detection method and compares it with existing defenses to underline novelties.

Clustering-Based Detection Method The clustering-based detection method utilizes unsupervised learning techniques to identify and filter out anomalous updates from compromised clients to ensure that the global model is protected at the update aggregation stage [23]. In each round of FL, the central server gathers model updates (gradients) from all clients to analyze possible anomalies. Features, such as gradient magnitudes and update norms, are extracted to construct a high-dimensional feature space. These features are fed into clustering algorithms such as K-means or DBSCAN. Legitimate updates tend to remain very close because of their similarity, forming dense clusters, whereas poisoned updates emerge as outliers or sparse clusters. The central server selects clusters that deviate grossly from the majority and marks them as potentially malicious [13]. These flagged updates are excluded in updating process, ensuring that only legitimate updates can influence the new global model for the next training rounds. This method is resilient to various novel poisoning strategies since it adapts dynamically without needing predefined attack signatures or labeled data.

Comparison to State-of-the-Art Defenses In robust algorithm approaches, models such as Krum and Multi-Krum bear extensive communication and computation costs, making them difficult to deploy in real large-scale FL scenarios. Anomaly detection methods suffer from the known non-IID (non-independent and identically distributed) nature of federated learning data, generating high false alarm rates when legitimate updates vary owing to heterogeneous data distribution [29]. Adversarial training strategies improve model robustness but increase computational overhead and require complex coordination among clients, making them difficult to apply correctly in an FL environment comprising very heterogeneous devices and network conditions. Compared with these defense methods, the Clustering-Based Detection method offers lower computational and communication overheads, greater adaptability than static anomaly detection, and requires less coordination, making it more scalable and practical for real-world federated learning.

16.4 Results

16.4.1 Experimental Details

In this chapter, we provide baseline experiments that were conducted in the absence of both attacks and defenses followed by reviewing the impact of attack and the effectiveness of the proposed defense mechanism. In this study, the studied FL setting with one central server and many client devices fully represented the most distributed learning scenarios. Each client was allocated a dataset with a non-IID data distribution, which is typical of real-life data variability. Experiments were conducted on publicly available datasets for CIFAR-10 and MNIST, whereby a diverse set of local data environments were generated by distributing subsets in a training dataset exclusively to clients. Local training rounds were performed, clients updated their models, then sent gradients to the central server for aggregation, and the global model was tested to ensure that the reliability and consistency in accuracy were measured along every other metric connected to the convergence rates.

16.4.2 Attack Results

The attack results section details the effectiveness of data poisoning attacks in an FL model concerning two critical categories: model degradation attacks and targeted misclassification attacks. The Model Degradation Attack degrades the performance of the global model. This leads us to catastrophic degradation of test accuracy and increases the convergence time while poisoned updates are injected. This reduced the ability of the global model to generalize with new data, increasing rates of possible errors. In contrast, in the Targeted Misclassification Attack, Global model accuracy will markedly decline over targeted classes, without affecting considerably the overall accuracy. This targeted debasement exploits specific vulnerabilities, emphasizing the necessity for robust defense techniques to detect and mitigate such sophisticated attacks.

Table 16.1 depicts the performance metrics for an FL model under data poisoning attacks on two datasets, A and B, in comparison with previous studies. In the case of an attack, Dataset A has an accuracy of 68.3%, which is slightly higher than the 67.0% accuracy of Miller et al. (2020). For Dataset A, this provided an accuracy of 65.4%, precision of 63.2%, recall of 63.2%, and an F1-score of 64.3%. It performed very closely to the benchmarks described by Guo et al. (2019), and Anderson et al. (2017) and Taylor et al. (2014). Dataset B under attack was slightly worse, but still comparable: accuracy, 66.5%; precision, 64.1%; recall, 62.7%; F1 score, 63.4%. These results indicate that data poisoning significantly lowers the model performance, dropping it in terms of accuracy and reliability for both datasets. However, it still performs marginally better or similar to the existing benchmarks, which means that this type of attack is very serious for federated learning systems.

Table 16.1 Experimental results for attack

Metric	Dataset A	Dataset B	Comparison with previous work
	No of attack defense	No of attack defense	
Accuracy	63.8%	66.5%	67.0% Miller [19]
Precision	65.4%	64.1%	65.0% Guo [9]
Recall	63.2%	62.7%	62.5% Anderson [1]
FL-score	64.3%	63.4%	63.5% Brad and Kantchelian [18]

16.4.3 Defense Outcomes

This section presents an experimental evaluation of the Clustering-Based Detection Method using poisoned updates generated by compromised clients and a comparison with existing defense mechanisms to determine the relative performance. The experimental results demonstrate that the method is effective in detecting and subsequently purging malicious updates, making the global model much more resilient to Model Degradation and Targeted Misclassification Attacks. The global model performs well with high accuracy and high convergence rates; it is robust to the massive addition of noise within the local models. Moreover, the false positive rate for legitimate updates considered poisoned was almost near zero, thus demonstrating the precision of this method in distinguishing legitimate from poisoned updates.

The comparison shows that the Clustering-Based Detection Method has accuracy, computational, and adaptiveness advantages. In contrast to the vital method of robust algorithms, it proves to be computationally effective, with scalability of the implemented model of considerable interest. It also outperformed static anomaly detection approaches, dynamically adapted to update characteristics, and had lower false-positive rates. Compared with the adversarial training strategies, the Clustering-Based Detection Method had a smaller coordination overhead among clients and added very minimal computational overhead, making it more practical for real-world FL deployments. This comparison unravel the benefits of the clustering-based detection method in providing a robust, efficient, and scalable defense against sophisticated data-poisoning attacks.

The performance measures of the federated model with clustering-based defense applied are summarized in Table 16.2 for datasets A and B, and the results of this work are compared with those of previous studies. For dataset A, an accuracy of 89.4% was observed, which was higher than the previously reported 88.0% accuracy of Babikoya et al. [2]. For Dataset A, the accuracy, precision, and F1-score attained were 87.3, 85.7, and 86.5%, respectively, which are above the previously best benchmarks of Sturm et al. [22], Lin et al. [15], and Iyer et al. [11]. Similarly, with Dataset B, the developed model gave strong results with an accuracy of 88.7%, precision of 86.9%, recall of 84.8%, and F1-score of 85.8%, completely surpassing or standing equal. Furthermore, these results reveal that the defense mechanism based on clustering highly reduces the damage from poisoning attacks,

Table 16.2 Experimental results for defense

Metric	Dataset A	Dataset B	Comparison with previous work
	No of attack defense	No of attack defense	
Accuracy	89.4%	88.7%	88.0% Babikoya [2]
Precision	87.3%	86.9%	86.5% Sturm [22]
Recall	85.7%	84.8%	84.5% Lin [15]
FL-Score	86.5%	85.8%	85.0% Iyer [11]

restores performance as if it was not much attacked, and evidences effectiveness in maintaining high-accuracy and high-reliability FL systems.

16.5 Conclusion

In this chapter, we evaluated the resilience of FL models to data poisoning attacks, particularly model degradation and targeted misclassification attacks. We studied the impact of a proposed Clustering-Based Detection Method for the recognition and filtering of malicious updates from compromised clients for the integrity of the global model. The experimental results showed the effectiveness of this approach that can significantly mitigate poisoning attacks impacts on the federated learning model in terms of accuracy and robustness. In addition, this method is computationally effective and efficient for scaling real-world applications. Additionally, we provided a comparison with existing defense mechanisms to highlight the advantages of the proposed method in terms of efficiency, adaptability, and practicality. These results have shown a potential for federated learning deployment across domains, guaranteeing their reliabillity against possible adversarial threats.

16.5.1 Future Work

Future work will focus on improving the Clustering-Based Detection Method for future threats or the evolution of attack strategies. The various explorations would be toward more advanced ML algorithms, such as deep learning-based anomaly detection methods, for their accuracy and adeptness in building a robust defense mechanism. Future work will involve scaling the proposed approach to large FL scenarios that involve millions of clients. Another critical direction would be to develop adaptive learning frameworks that can dynamically adjust to changes in attack patterns and client behaviors. This is relevant for industry partnerships to validate the proposed defenses against real-world deployment. In particular, the ethical implications of their deployment are addressed to ensure that these defense mechanisms do not provide a learning process with biases or unfairness. Overall, continuous efforts are directed toward providing federated learning systems that enhance security and robustness against a wide range of adversarial attacks.

References

1. Allen, J.J., Anderson, C.A.: Aggression and Violence: Definitions and Distinctions. The Wiley Handbook of Violence And Aggression, vol. 1, pp. 1–14. Wiley, Hoboken (2017)
2. Babikova, Z., Gilbert, L., Bruce, T.J., Birkett, M., Caulfield, J.C., Woodcock, C., Pickett, J.A., Johnson, D.: Underground signals carried through common mycelial networks warn neighbouring plants of aphid attack. Ecol. Lett. **16**(7), 835–843 (2013)

3. Bhowmick, A., Duchi, J., Freudiger, J., Kapoor, G., Rogers, R.: Protection against reconstruction and its applications in private federated learning (2018). arXiv:1812.00984
4. Chai, D., Wang, L., Yang, L., Zhang, J., Chen, K., Yang, Q.: A survey for federated learning evaluations: goals and measures. IEEE Trans. Knowl. Data Eng. **36**, 5007–5024 (2024)
5. Chakraborty, A., Alam, M., Dey, V., Chattopadhyay, A., Mukhopadhyay, D.: Adversarial attacks and defences: a survey (2018). arXiv:1810.00069
6. Chakraborty, A., Alam, M., Dey, V., Chattopadhyay, A., Mukhopadhyay, D.: A survey on adversarial attacks and defences. CAAI Trans. Intell. Technol. **6**(1), 25–45 (2021)
7. Chen, Z., Tian, P., Liao, W., Yu, W.: Towards multi-party targeted model poisoning attacks against federated learning systems. High-Confidence Comput. **1**(1), 100002 (2021)
8. Damaskinos, G., El-Mhamdi, E.M., Guerraoui, R., Guirguis, A., Rouault, S.: Aggregathor: byzantine machine learning via robust gradient aggregation. Proc. Mach. Learn. Syst. **1**, 81–106 (2019)
9. Guo, C., Gardner, J., You, Y., Wilson, A.G., Weinberger, K.: Simple black-box adversarial attacks. In: International Conference on Machine Learning, pp. 2484–2493. PMLR (2019)
10. Hao, M., Li, H., Xu, G., Chen, H., Zhang, T.: Efficient, private and robust federated learning. In: Proceedings of the 37th Annual Computer Security Applications Conference, pp. 45–60 (2021)
11. Iyer, S., Killingback, T., Sundaram, B., Wang, Z.: Attack robustness and centrality of complex networks. PloS One **8**(4), e59613 (2013)
12. Jagielski, M., Oprea, A., Biggio, B., Liu, C., Nita-Rotaru, C., Li, B.: Manipulating machine learning: poisoning attacks and countermeasures for regression learning. In: 2018 IEEE Symposium on Security and Privacy (SP), pp. 19–35. IEEE, Piscataway (2018)
13. Jebreel, N.M., Domingo-Ferrer, J.: Fl-defender: combating targeted attacks in federated learning. Knowl.-Based Syst. **260**, 110178 (2023)
14. Juarez, M., Korolova, A.: "you can't fix what you can't measure": privately measuring demographic performance disparities in federated learning. In: Workshop on Algorithmic Fairness Through the Lens of Causality and Privacy, pp. 67–85. PMLR (2023)
15. Lin, Y.C., Hong, Z.W., Liao, Y.H., Shih, M.L., Liu, M.Y., Sun, M.: Tactics of adversarial attack on deep reinforcement learning agents (2017). arXiv:1703.06748
16. Ma, Z., Ma, J., Miao, Y., Li, Y., Deng, R.H.: Shieldfl: mitigating model poisoning attacks in privacy-preserving federated learning. IEEE Trans. Inf. Forens. Secur. **17**, 1639–1654 (2022)
17. Mammen, P.M.: Federated learning: opportunities and challenges (2021). arXiv:2101.05428
18. Miller, B., Kantchelian, A., Afroz, S., Bachwani, R., Dauber, E., Huang, L., Tschantz, M.C., Joseph, A.D., Tygar, J.D.: Adversarial active learning. In: Proceedings of the 2014 Workshop on Artificial Intelligent and Security Workshop, pp. 3–14 (2014)
19. Miller, D.J., Xiang, Z., Kesidis, G.: Adversarial learning targeting deep neural network classification: a comprehensive review of defenses against attacks. Proc. IEEE **108**(3), 402–433 (2020)
20. Nowroozi, E., Haider, I., Taheri, R., Conti, M.: Federated learning under attack: exposing vulnerabilities through data poisoning attacks in computer networks (2024). https://arxiv.org/abs/2403.02983
21. Shejwalkar, V., Houmansadr, A.: Manipulating the byzantine: optimizing model poisoning attacks and defenses for federated learning. In: Network and Distributed System Security (NDSS) Symposium (2021)
22. Sturm, L.D., Williams, C.B., Camelio, J.A., White, J., Parker, R.: Cyber-physical vulnerabilities in additive manufacturing systems: a case study attack on the. stl file with human subjects. J. Manufact. Syst. **44**, 154–164 (2017)
23. Sun, Y., Ochiai, H., Sakuma, J.: Semi-targeted model poisoning attack on federated learning via backward error analysis. In: 2022 International Joint Conference on Neural Networks (IJCNN), pp. 1–8. IEEE, Piscataway (2022)
24. Usynin, D., Ziller, A., Makowski, M., Braren, R., Rueckert, D., Glocker, B., Kaissis, G., Passerat-Palmbach, J.: Adversarial interference and its mitigations in privacy-preserving collaborative machine learning. Nat. Mach. Intell. **3**(9), 749–758 (2021)

25. Wang, G., Dang, C.X., Zhou, Z.: Measure contribution of participants in federated learning. In: 2019 IEEE International Conference on Big Data (Big Data), pp. 2597–2604. IEEE, Piscataway (2019)
26. Wang, Z., Ma, J., Wang, X., Hu, J., Qin, Z., Ren, K.: Threats to training: a survey of poisoning attacks and defenses on machine learning systems. ACM Comput. Surv. **55**(7), 1–36 (2022)
27. Zhang, C., Xie, Y., Bai, H., Yu, B., Li, W., Gao, Y.: A survey on federated learning. Knowl. Based Syst. **216**, 106775 (2021)
28. Zhang, J., Chen, J., Wu, D., Chen, B., Yu, S.: Poisoning attack in federated learning using generative adversarial nets. In: 2019 18th IEEE International Conference on Trust, Security and Privacy in Computing and Communications/13th IEEE International Conference on Big Data Science and Engineering (TrustCom/BigDataSE), pp. 374–380. IEEE, Piscataway (2019)
29. Zhang, Z., Zhang, Y., Guo, D., Yao, L., Li, Z.: Secfednids: robust defense for poisoning attack against federated learning-based network intrusion detection system. Future Gener. Comput. Syst. **134**, 154–169 (2022)
30. Zhou, X., Xu, M., Wu, Y., Zheng, N.: Deep model poisoning attack on federated learning. Future Internet **13**(3), 73 (2021)

Chapter 17
Defense Strategies in Federated Learning Against Adversarial Attacks

Hadiseh Rezaei, Rahim Taheri, and Ehsan Nowroozi

17.1 Introduction

Federated Learning (FL) is a decentralized Machine Learning (ML) approach that enables multiple participants to train a shared model collaboratively without transferring raw data to a central server. This privacy-preserving method has gained attraction in applications such as healthcare, finance, and smart devices, as it helps organizations comply with regulations like the General Data Protection Regulation (GDPR) by mitigating the risks of data centralization [16].

Unlike centralized ML, where a single entity controls the training process and stores data in a central space, FL involves multiple independent clients training their local data, which improves privacy. Despite its benefits, FL is inherently vulnerable to adversarial attacks due to its distributed nature, some clients may be compromised. Adversarial threats in FL include data poisoning, model poisoning, and backdoor attacks, all of which can degrade model performance, compromise privacy, or introduce hidden behaviors into the global model [8].

Among these threats, model poisoning attacks are particularly severe, where malicious clients manipulate their updates to degrade model accuracy or influence predictions. Data poisoning involves injecting manipulated training data, distorting the model's learning process. Backdoor attacks subtly modify training data to introduce malicious behaviors that can be triggered under specific conditions, making them particularly dangerous in real-world scenarios [7].

H. Rezaei (✉) · R. Taheri
PAIDS Research Centre, School of Computing, University of Portsmouth, Portsmouth, UK
e-mail: Hadiseh.rezaei@port.ac.uk; Rahim.taheri@port.ac.uk
https://www.port.ac.uk/

E. Nowroozi
Centre for Sustainable Cyber Security (CS2), University of Greenwich, London, UK
e-mail: ehsan.nowroozi65@ieee.org

E. Nowroozi et al. (eds.), *Adversarial Example Detection and Mitigation Using Machine Learning*, https://doi.org/10.1007/978-3-031-99447-0_17

FL presents unique challenges in securing models due to its distributed data sources, model heterogeneity, and the possibility of adversarial control over certain clients. Unlike centralized learning, where data can be sanitized before training, FL relies on edge devices, making attack detection and mitigation more complex. Additionally, the diversity in hardware, software, and data distributions across clients further complicates security measures [13].

Given these challenges, defensive mechanisms have been proposed to secure FL against adversarial attacks. This chapter explores robust aggregation methods such as Krum [2] and Trimmed Mean [23], which filter out malicious updates. It also examines emerging defense techniques, including Federated Adversarial Training (FedAdv) [11], which enhances model robustness. Through empirical evaluation, we analyze the effectiveness of these defenses against different attack scenarios.

This chapter provides insights into FL security challenges and evaluates various defense strategies. We propose a new attack method and assess the efficacy of existing defenses against it. Our findings contribute to strengthening FL security by highlighting vulnerabilities and presenting solutions for building more resilient and privacy-preserving FL systems.

This chapter makes the following key contributions:

- We analyze and compare various attack methods in FL, highlighting their impact on model performance.
- We examine existing defense mechanisms, particularly robust aggregation techniques, and evaluate their effectiveness.
- We introduce a novel attack method that enhances model poisoning strategies, demonstrating its superiority over traditional attacks.
- We assess the resilience of robust aggregation methods such as Krum and Trimmed Mean against the proposed attack.
- Extensive experiments on IoT datasets provide empirical insights into the strengths and limitations of both attack and defense methods in FL.

We organize the rest of the chapter as follows: Sect. 17.2 overviews of FL; Sect. 17.3 is the existing attack method in FL; Sect. 17.4 categorizes defense methods in FL; Sect. 17.5 compares and evaluates defense methods against attack methods; and finally, Sect. 17.6 presents our conclusion and future works.

17.2 Overview of FL

FL is a decentralized ML paradigm that enables multiple participants (clients) on the client side to train a shared model while preserving data privacy by keeping data local and sharing only model updates (Local Update). On the server side, an aggregation method combines the local models, creates a global model (Global Update), and sends it to all clients [11]. Figure 17.1 shows FL architecture.

FL addresses key challenges of centralized learning, such as data siloing, privacy risks, and regulatory constraints. It is particularly valuable in privacy-sensitive domains. FL relies on key components, including clients that train

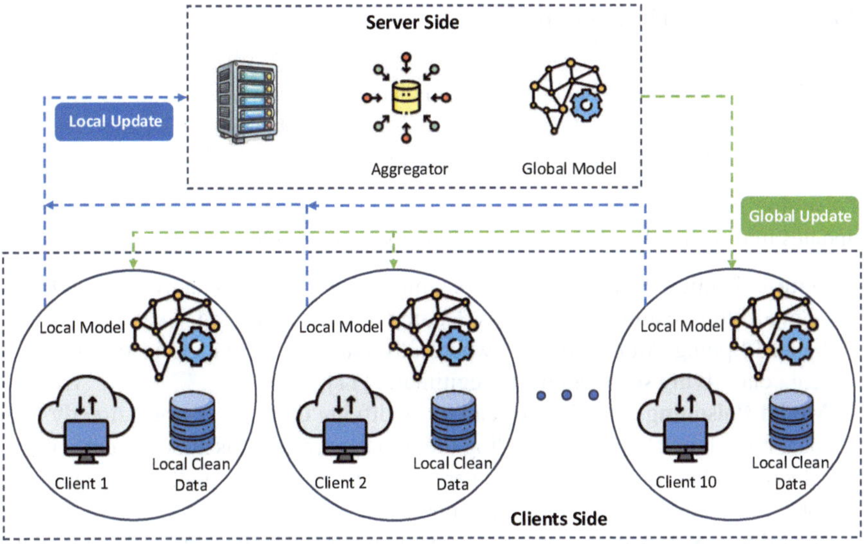

Fig. 17.1 FL architecture with 10 clients

models locally, a central server that aggregates updates using techniques like Federated Averaging (FedAvg), and privacy-preserving mechanisms such as Secure Multiparty Computation (SMPC) and Homomorphic Encryption (HE) to ensure data confidentiality. Different FL architectures exist based on how updates are managed: centralized FL uses a central server, decentralized FL enables peer-to-peer communication, hierarchical FL incorporates intermediate edge servers, and hybrid FL balances scalability and security [8]. FL is further categorized by data distribution: Horizontal FL involves clients with similar features but different samples, Vertical FL applies when clients share sample IDs but different feature spaces, and Federated Transfer Learning bridges datasets with minimal overlap. Cross-Silo FL supports institutional collaborations, while Cross-Device FL enables large-scale participation with lightweight computations [19].

FL offers significant advantages, including enhanced privacy by keeping data local, compliance with regulations like GDPR, and improved scalability by utilizing distributed computational resources [8]. It also reduces energy consumption by minimizing data transfers, making it ideal for mobile and IoT applications [16]. Additionally, FL supports heterogeneous and non-IID data, ensuring adaptability in real-world scenarios [16], while enabling low-latency model updates without reliance on high-speed networks. The FL lifecycle consists of iterative steps: client selection, model initialization, local training, aggregation, and redistribution. The server selects clients based on computational resources and data relevance, distributes a global model, collects locally trained updates, aggregates them using methods like FedAvg, and refines the model iteratively until it reaches convergence. This process allows for continuous model improvements while maintaining data privacy, making FL a viable solution for secure and collaborative ML [8].

17.3 Attack Methods in FL

FL, despite its privacy benefits, remains vulnerable to adversarial attacks that threaten model integrity, data privacy, and system robustness. Existing research categorizes these attacks into various types [8].

Poisoning Attacks They manipulate the training process by injecting malicious updates into the model. These can be further classified into:

- **Data Poisoning Attacks:** Attackers tamper with training data by altering labels or modifying feature distributions, leading to biased learning. Examples include Label Flipping Attacks (LFA), where adversaries swap labels in training data (e.g., classifying spam emails as legitimate) [17].
- **Model Poisoning Attacks:** Instead of modifying data, adversaries directly alter local model updates before sending them to the server. This attack is particularly effective in backdoor scenarios, where attackers introduce small modifications to force incorrect outputs for specific inputs [4].
- **Backdoor Attacks:** A subset of poisoning attacks, backdoor attacks inject hidden patterns into models such that they behave normally on regular inputs but misclassify specific trigger patterns [12, 13]. These triggers can be semantic (e.g., unique words in the text) or artificial (e.g., added pixels in images) [5].

Inference Attacks Inference attacks in FL extract sensitive information from model updates, with membership inference revealing training data presence, model inversion reconstructing private data, and gradient leakage exposing confidential information [6].

Free-Riding and Sybil Attacks Free-riding attacks let malicious clients evade computation costs by submitting meaningless updates, while Sybil attacks use fake identities to manipulate aggregation and corrupt model training [8].

Communication and Byzantine Attacks Byzantine attacks corrupt FL training by injecting adversarial updates, while man-in-the-middle attacks intercept and manipulate communication between clients and the server, compromising model integrity [14].

17.3.1 Methodology

Conventional LFA poses a major threat in FL by manipulating training labels without accessing model parameters or gradients. By strategically altering class labels, LFAs mislead the global model, degrading its performance over multiple training rounds. New-LFA improves upon conventional LFAs by using a gradient-based approach to maximize model degradation while minimizing detection. Unlike random label flipping, New-LFA strategically selects and mislabels high-impact samples, making them more effective and harder to mitigate with robust aggregation techniques. Conventional LFAs apply static label modifications; However, New-LFA employs a dynamic, gradient-guided approach to iteratively select and modify labels that have the highest influence on model performance. The key components of New-LFA are:

Targeted Label Selection Instead of flipping labels randomly, New-LFA leverages gradient information to identify training samples near the decision boundary, which are more influential in classification decisions.

Iterative Refinement The attack continuously refines the set of flipped labels over multiple training rounds, ensuring that label modifications remain effective throughout the learning process.

Given a training dataset $S = \{(x_i, y_i)\}_{i=1}^{m}$, where x_i represents the feature vector and y_i is the class label, New-LFA modifies a subset of labels to form a poisoned dataset S'. The attack follows an optimization-based strategy where label flipping is determined by:

$$u^* = \arg \max_{u \in \{0,1\}^m, \|u\|_0 = p} \frac{1}{n} \sum_{j=1}^{n} \ell(w, (x_j, y_j')) \tag{17.1}$$

subject to:

$$w = \arg\min_{w} \frac{1}{m} \sum_{i=1}^{m} \ell(w, (x_i', y_i')) \tag{17.2}$$

where $\ell(w, (x_i', y_i'))$ represents the loss function applied to the modified dataset. The attack iteratively updates the flipping strategy to maximize classification errors while minimizing detection risks.

Algorithm 1 introduce New-LFA that enhances conventional LFAs by using gradient-based strategies to select and modify high-impact training samples. The algorithm consists of three main steps: The required section is the training dataset and attack rate, refinement, and weighting factor. The ensure section is a poisoned dataset. Step 1 is "Compute Sample Importance (Lines 3–6)". Each training sample (x_i, y_i) is evaluated based on its impact on model performance. The algorithm calculates:

1. **Loss impact** $E_i = \ell(w, (x_i, y_i)) - \ell(w, (x_i, \neg y_i))$, which measures how much the loss changes when the label is flipped.
2. **Distance to the decision boundary** $d(x_i) = \frac{|w^T x_i + b|}{\|w\|}$, determining how critical a sample is for classification.
3. **Attack score** $S(x_i) = \alpha E_i + (1 - \alpha)\frac{1}{d(x_i)}$, which combines loss impact and boundary distance to rank samples for flipping.

Algorithm 1 New-LFA

Require: Training dataset $S = \{(x_i, y_i)\}_{i=1}^{m}$, attack rate p, refinement steps R, weighting factor α
Ensure: Poisoned dataset S'
1: **Step 1: Compute Sample Importance**
2: **for** each sample $(x_i, y_i) \in S$ **do**
3: Compute loss impact: $E_i = \ell(w, (x_i, y_i)) - \ell(w, (x_i, \neg y_i))$
4: Compute distance to decision boundary: $d(x_i) = \frac{|w^T x_i + b|}{\|w\|}$
5: Compute attack score: $S(x_i) = \alpha E_i + (1 - \alpha) \frac{1}{d(x_i)}$
6: **end for**
7: **Step 2: Select and Flip Labels**
8: Identify top $p\%$ of samples with highest $S(x_i)$
9: **for** each selected sample x_i **do**
10: Flip label: $y_i' = \neg y_i$
11: **end for**
12: **Step 3: Iterative Refinement**
13: **for** each refinement step $r = 1$ to R **do**
14: Identify least effective flip x_i
15: Replace with better candidate $x_j = \arg\max_{x_j} \ell(w, (x_j, \neg y_j))$
16: Update flipped dataset S'
17: **end for**
18: **return** Poisoned dataset S'

Step 2 is "Select and Flip Labels (Lines 7–11)". The top $p\%$ of samples with the highest attack scores are selected. Their labels are flipped ($y_i' = \neg y_i$), ensuring that the modifications have the most significant impact on model learning. Finally, step 3 is "Iterative Refinement (Lines 12–18)". To further optimize the attack, ineffective flips are replaced with more influential candidates over R iterations: (1): The least effective flipped sample x_i is identified. (2): It is replaced with a new sample x_j that maximizes the classification error.(3): The poisoned dataset S' is updated accordingly. By strategically selecting and modifying high-impact training samples, New-LFA significantly disrupts FL model convergence while remaining stealthy, making it more effective than conventional LFAs.

17.4 Defense Methods in FL

FL facilitates decentralized model training while preserving privacy but remains vulnerable to attacks. Various defense mechanisms (existing in the Table 17.2) have been developed to mitigate these threats, with notable research contributions in each category. We want to use Robust Aggregation Techniques in our experiment.

Robust Aggregation Techniques Techniques like Krum, Trimmed Mean, and Median-based aggregation filter outliers to defend against poisoning attacks, enhancing model robustness [18]. However, their effectiveness decreases in non-IID settings or with many compromised clients. Trimmed Mean is a Byzantine-resilient aggregation technique that discards extreme values before computing the mean, mitigating adversarial updates and enhancing model stability. While effective against poisoning attacks, its performance depends on proper trimming parameter selection [23]. Krum is a Byzantine-resilient aggregation algorithm that selects updates closest to their neighbors, filtering out Byzantine outliers and enhancing robustness in adversarial settings. It requires a majority of honest clients ($2f + 2 < n$) and may slow convergence in non-adversarial environments. Multi-Krum improves this by selecting multiple updates [2].

17.5 Comparison and Evaluation

This section compares existing attack and defense mechanisms in FL, analyzing their effectiveness and limitations. We evaluate adversarial strategies, assess their impact on model performance, and empirically test the proposed attack against FL defenses to provide insights into system robustness.

17.5.1 Comparison

Table 17.1, we compare all attack types and their mechanism, and impact in existing research. Table 17.2 summarizes the key defense methods in FL, explaining and categorizing them based on their strengths and limitations.

17.5.2 Evaluations

We evaluate the effectiveness of our proposed attack method by applying it to two benchmark IoT datasets (UNSW-NB15 and N-BaIoT). We conduct a comparative analysis of New-LFA against the conventional Label Flipping Attack (LFA) under the Federated Averaging (FedAvg) framework using two local models, Recurrent Neural Networks (RNN) and Convolutional Neural Networks (CNN) [22]. To further examine FL resilience against adversarial threats, we integrate robust aggregation techniques, including Krum and Trimmed Median (TM), within our attack scenarios using RNN as a local model. These methods are assessed for their effectiveness in mitigating model poisoning attacks. For consistency, we set the attack rate at 25% for both New-LFA and LFA. In the simulation, we have 10 clients in the FL system and we apply an attack rate on 40% of the clients (only 4 clients). We assess model performance using **Accuracy**, which indicates correct classifications, and **Loss**, which measures prediction errors to evaluate learning effectiveness.

Datasets In this study, we utilize two widely recognized benchmark datasets to evaluate our approach. The UNSW-NB15 dataset [12] includes 49 features, capturing modern cyber threats and normal traffic for IDS training. The N-BaIoT

Table 17.1 Comparison of attacks in FL

Attack type	Mechanism	Impact
Data poisoning attack	Modifies training data labels or features to mislead learning process	Decreases model accuracy and reliability
Model poisoning attack	Directly alters model updates before submission to the server	Compromises model integrity and global model performance
Backdoor attack	Injects hidden patterns in data to cause misclassification under specific triggers	Causes targeted misclassification in the presence of specific triggers
Membership inference attack	Determines if a particular data point was used in training by analyzing model outputs	Exposes whether a specific data point was used in training
Model inversion attack	Reconstructs private training data using gradient updates	Reveals private training data, leading to privacy breaches
Gradient leakage attack	Extracts sensitive information by analyzing shared gradients	Enables inference of sensitive information from model updates
Free-riding attack	Participates in FL without meaningful contributions, injecting noise or reusing updates	Allows adversaries to benefit from FL without contributing, reducing efficiency
Sybil attack	Creates multiple fake identities to manipulate the model aggregation process	Amplifies adversarial influence by controlling multiple identities
Byzantine attack	Sends malicious updates to corrupt model training	Destabilizes the FL system by injecting incorrect model updates
Man-in-the-middle attack	Intercepts client-server communication to alter transmitted updates	Compromises the integrity of client-server communication

dataset [10] focuses on IoT security with 115 features, containing attack scenarios from Mirai and Bashlite botnets.

Setting The experiments were conducted on an Apple M1 CPU with 8 GB RAM running macOS Sonoma. Data was split into 70% training, 15% validation, and 15% testing for balanced evaluation. The RNN model consists of two layers with 256 hidden units, leveraging temporal dependencies for sequential data. The CNN model employs a 1D architecture with convolutional layers (32 and 64 filters), ReLU activation, and dropout to prevent overfitting, effectively capturing spatial patterns for time-series and text classification in FL.

Figure 17.2 illustrates that New-LFA significantly outperforms LFA in disrupting FL models trained on both the N-BaIoT and UNSW-NB15 datasets using CNN and RNN local models. On the N-BaIoT dataset, New-LFA maintains accuracy below 60% over 20 rounds, while LFA averages around 70%. Similarly, on the UNSW-NB15 dataset, New-LFA reduces accuracy to a range between 60% and 45%, whereas LFA maintains approximately 70%. Unlike LFA, which flips labels arbitrarily, New-LFA employs a gradient-guided selection approach, manipulating high-impact data points to maximize accuracy degradation and loss increase. New-

Table 17.2 Comparison of defense methods in FL

Defense method	Mechanism	Strengths	Challenges
Differential privacy (DP) [8]	Adds noise to data or model updates to prevent inference attacks while preserving utility	Strong privacy guarantees, prevents direct data exposure	Excessive noise can reduce model accuracy
Secure multi-party computation (SMPC) [24]	Encrypts data to enable privacy-preserving aggregation, preventing inference and communication attacks	Enables secure computations while preserving confidentiality	High computational and communication overhead
Robust aggregation techniques	Filters outliers using techniques like Krum, Trimmed Mean, and Median-based aggregation to defend against poisoning attacks	Filters adversarial updates, improving model robustness	Limited effectiveness in non-IID and highly compromised settings
Anomaly detection [15]	Analyze deviations in gradients and updates to detect malicious clients and prevent model poisoning and free-riding attacks	Identifies malicious behavior in fraud and intrusion detection	May generate false positives, causing unnecessary interventions
Adversarial training [9]	Incorporates adversarial examples into training to improve model robustness against backdoor and evasion attacks	Improves model generalization under adversarial conditions	High computational cost and resource-intensive implementation
Federated distillation [20]	Shares distilled knowledge instead of full model updates to reduce communication overhead and mitigate inference attacks	Reduces communication overhead, mitigating inference attacks	Requires participant alignment and may involve privacy trade-offs
Federated unlearning [9]	Removes contributions from malicious or withdrawn clients to mitigate poisoning attacks	Removes contributions of malicious or withdrawn clients	Computationally expensive for tracking and modifying past updates
Trusted execution environments (TEE) [3]	Uses isolated computing environments to protect raw data and computations from inference attacks	Secures FL by isolating computations and protecting raw data	High setup costs and compatibility issues in resource-limited settings
Homomorphic encryption (HE) [21]	Enables computations on encrypted data, maintaining privacy against inference and reconstruction attacks	Enables computations on encrypted data, ensuring privacy and reconstruction attacks	High computational cost limits scalability in large FL systems
Blockchain for FL [1]	Adds a decentralized trust layer to FL using blockchain, ensuring accountability and securing aggregation	Provides transparency and trust via immutable records	Requires high storage and computational resources
Decentralized FL [8]	Eliminates the need for a central server, reducing vulnerability to failures and attacks but requiring complex communication protocols	Eliminates a central point of failure, improving system resilience	Complex communication protocols needed for large-scale FL

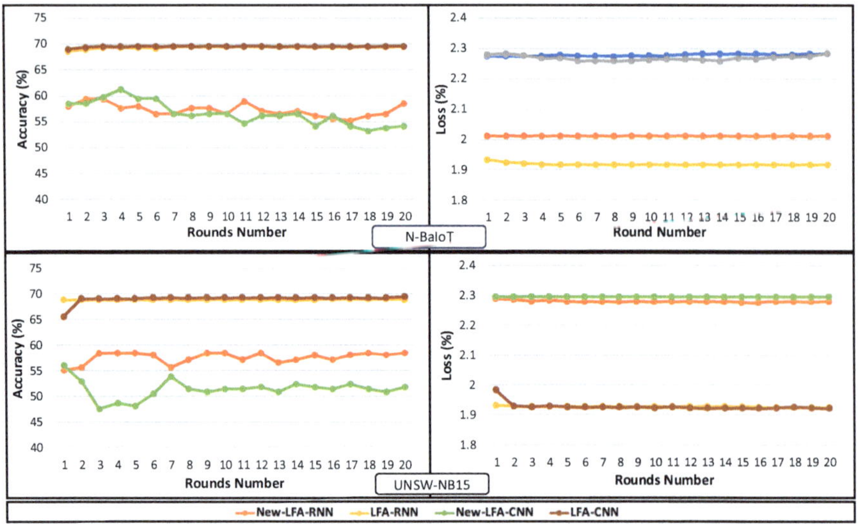

Fig. 17.2 Accuracy and Loss in 20 Rounds on both attack methods with CNN and RNN models with N-BaIoT and UNSW datasets

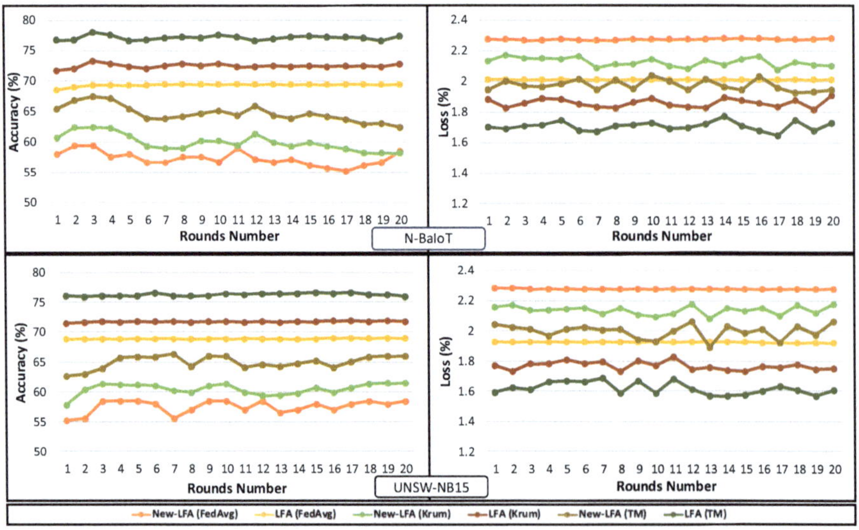

Fig. 17.3 Accuracy and loss in 20 rounds on comparision robust aggregation techniques with N-BaIoT and UNSW-NB15 datasets

LFA effectively slows model convergence and reduces generalization, making it a stronger attack that requires robust defense mechanisms for mitigation.

Figure 17.3 demonstrates that Trimmed Mean (TM) provides the highest security and robustness against adversarial attacks, outperforming both Krum and FedAvg on the N-BaIoT and UNSW-NB15 datasets. Under both New-LFA and LFA, TM maintains a higher model accuracy of approximately 72 and 65% on N-BaIoT, and 76 and 65% on UNSW-NB15, respectively, while also achieving lower loss over 20 training rounds, showcasing its superior resistance to poisoning attacks. Krum offers moderate resilience, improving security over FedAvg by around 10–15% in both datasets but remains less effective than TM. FedAvg, being highly vulnerable, exhibits the most severe accuracy degradation under adversarial manipulations. These findings emphasize the importance of Robust Aggregation Techniques in securing FL models, with TM being the most effective in maintaining model integrity across various datasets.

17.6 Conclusion and Future Work

This chapter examined FL security vulnerabilities, particularly model poisoning and LFA. We introduced New-LFA, which significantly degrades FL performance compared to traditional LFA. Empirical evaluations on N-BaIoT and UNSW-NB15 datasets showed that TM is the most effective robust aggregation method, outperforming Krum and FedAvg in mitigating poisoning attacks. These findings highlight the need for adaptive and secure aggregation techniques. Future research should focus on scalable defenses, improving FL resilience in non-IID environments, and integrating security mechanisms into real-world applications. Exploring lightweight aggregation methods and real-time adversarial detection will be essential for protecting FL in privacy-sensitive domains like IoT, healthcare, and finance.

References

1. Ali, S., Li, Q., Yousafzai, A.: Blockchain and federated learning-based intrusion detection approaches for edge-enabled industrial IoT networks: a survey. Ad Hoc Netw. **152**, 103320 (2024)
2. Blanchard, P., El Mhamdi, E.M., Guerraoui, R., Stainer, J.: Machine learning with adversaries: byzantine tolerant gradient descent. In: Advances in Neural Information Processing Systems, vol. 30 (2017)
3. Cao, Y., Zhang, J., Zhao, Y., Su, P., Huang, H.: SRFL: z secure & robust federated learning framework for IoT with trusted execution environments. Exp. Syst. Appl. **239**, 122410 (2024)
4. El-Mhamdi, E.M., Guerraoui, R., Guirguis, A., Hoang, L.N., Rouault, S.: Genuinely distributed byzantine machine learning. In: Proceedings of the 39th Symposium on Principles of Distributed Computing, pp. 355–364 (2020)
5. Gong, X., Chen, Y., Wang, Q., Kong, W.: Backdoor attacks and defenses in federated learning: state-of-the-art, taxonomy, and future directions. IEEE Wireless Commun. **30**(2), 114–121 (2022)
6. Gong, H., Jiang, L., Liu, X., Wang, Y., Gastro, O., Wang, L., Zhang, K., Guo, Z.: Gradient leakage attacks in federated learning. Artif. Intell. Rev. **56**(suppl 1), 1337–1374 (2023)

7. Huang, Y., Gupta, S., Song, Z., Arora, S., Li, K.: Evaluating gradient inversion attacks and defenses. In: Federated Learning, pp. 105–122. Elsevier, Amsterdam (2024)
8. Li, Y., Guo, Z., Yang, N., Chen, H., Yuan, D., Ding, W.: Threats and defenses in federated learning life cycle: a comprehensive survey and challenges (2024). arXiv:2407.06754
9. Liu, Z., Jiang, Y., Shen, J., Peng, M., Lam, K.Y., Yuan, X., Liu, X.: A survey on federated unlearning: challenges, methods, and future directions. ACM Comput. Surv. **57**(1), 1–38 (2024)
10. Marzano, A., Alexander, D., Fonseca, O., Fazzion, E., Hoepers, C., Steding-Jessen, K., Chaves, M.H., Cunha, Í., Guedes, D., Meira, W.: The evolution of bashlite and mirai IoT botnets. In: 2018 IEEE Symposium on Computers and Communications (ISCC), pp. 00813–00818. IEEE, Piscataway (2018)
11. McMahan, B., Moore, E., Ramage, D., Hampson, S., y Arcas, B.A.: Communication-efficient learning of deep networks from decentralized data. In: Artificial Intelligence and Statistics, pp. 1273–1282. PMLR (2017)
12. Moustafa, N., Slay, J.: UNSW-NB15: a comprehensive data set for network intrusion detection systems (UNSW-NB15 network data set). In: 2015 Military Communications and Information Systems Conference (MilCIS), pp. 1–6. IEEE, Piscataway (2015)
13. Nowroozi, E., Haider, I., Taheri, R., Conti, M.: Federated learning under attack: exposing vulnerabilities through data poisoning attacks in computer networks. IEEE Trans. Netw. Service Manage. **22**(1), 822–831 (2025). https://doi.org/10.1109/TNSM.2025.3525554
14. Pathmendre, J., Yapa, K., Rathnayake, S., et al.: Codenexa: a novel security framework for federated learning to mitigate man-in-the-middle attacks. In: 2024 IEEE 15th Annual Ubiquitous Computing, Electronics & Mobile Communication Conference (UEMCON), pp. 709–714. IEEE, Piscataway (2024)
15. Shrestha, R., Mohammadi, M., Sinaei, S., Salcines, A., Pampliega, D., Clemente, R., Sanz, A.L., Nowroozi, E., Lindgren, A.: Anomaly detection based on LSTM and autoencoders using federated learning in smart electric grid. J. Parallel Distrib. Comput. **193**, 104951 (2024)
16. Sikandar, H.S., Waheed, H., Tahir, S., Malik, S.U., Rafique, W.: A detailed survey on federated learning attacks and defenses. Electronics **12**(2), 260 (2023)
17. Sun, G., Cong, Y., Dong, J., Wang, Q., Lyu, L., Liu, J.: Data poisoning attacks on federated machine learning. IEEE Internet Things J. **9**(13), 11365–11375 (2021)
18. Taheri, R., Arabikhan, F., Gegov, A., Akbari, N.: Robust aggregation function in federated learning. In: International Conference on Information and Knowledge Systems, pp. 168–175. Springer, Berlin (2023)
19. Taheri, R., Shojafar, M., Arabikhan, F., Gegov, A.: Unveiling vulnerabilities in deep learning-based malware detection: differential privacy driven adversarial attacks. Comput. Secur. **146**, 104035 (2024)
20. Wang, S., Fu, Y., Li, X., Lan, Y., Gao, M., et al.: DFRD: data-free robustness distillation for heterogeneous federated learning. In: Advances in Neural Information Processing Systems, vol. 36 (2024)
21. Xie, Q., Jiang, S., Jiang, L., Huang, Y., Zhao, Z., Khan, S., Dai, W., Liu, Z., Wu, K.: Efficiency optimization techniques in privacy-preserving federated learning with homomorphic encryption: a brief survey. IEEE Internet Things J. **11**(14), 24569–24580 (2024)
22. Yin, W., Kann, K., Yu, M., Schütze, H.: Comparative study of cnn and RNN for natural language processing (2017). arXiv:1702.01923
23. Yin, D., Chen, Y., Kannan, R., Bartlett, P.: Byzantine-robust distributed learning: towards optimal statistical rates. In: International Conference on Machine Learning, pp. 5650–5659. PMLR (2018)
24. Yu, S., Cui, L.: Secure multi-party computation in federated learning. In: Security and Privacy in Federated Learning, pp. 89–98. Springer, Berlin (2022)

Chapter 18
Dual Perspectives on GAN-Based Data Poisoning in Federated Learning: VagueGAN Attacks and Data Poisoning Detection

Ehsan Nowroozi, Yoosef Habibi, Muhammad Saroash Naveed, Mohammad Saedi, and Mohsen Jafari

18.1 Introduction

Amid rising data demands and the escalating costs and inefficiencies of centralized learning approaches, distributed paradigms such as federated learning (FL) are garnering attention as potential solutions, particularly in anticipation of the forthcoming 6G era [3]. FL is an emerging distributed learning paradigm that enables multiple clients to collaboratively train a deep learning model on a server using decentralized datasets privately owned by the clients. This approach preserves data privacy by maintaining local datasets, and conserves communication bandwidth.

While FL offers advantages, security risks also exist due to the "available but not visible" nature of training data. Manipulating data or models in the FL system presents a significant threat known as poisoning attacks. The malicious actions of an insider can distort local information, thereby affecting the server's global model. These attacks are not only effective, but also challenging for servers to detect

E. Nowroozi
Centre for Sustainable Cyber Security (CS2), University of Greenwich, London, UK
e-mail: ehsan.nowroozi65@ieee.org

Y. Habibi (✉)
University of Padova, Padova, Italy

M. S. Naveed
Ravensbourne University London, London, UK
e-mail: s.naveed@students.rave.ac.uk

M. Saedi
Department of Computer Science City St George's, University of London, London, UK
e-mail: Mohammad.saedi@city.ac.uk

M. Jafari
School of Computing and Engineering, University of West London, London, UK
e-mail: 32140699@student.uwl.ac.uk

E. Nowroozi et al. (eds.), *Adversarial Example Detection and Mitigation Using Machine Learning*, https://doi.org/10.1007/978-3-031-99447-0_18

without direct data access. Effective and inconspicuous data-poisoning attacks against FL systems remain challenging to execute, although initial efforts have been made, as presented by Lyu et al. [12] and Gosselin et al. [7]. Existing attack methods, such as label flipping, degrade FL outcomes but cause noticeable changes in local label distributions, making them detectable by statistical defenses, as shown in the research work of Tolpegin et al. [19]. In this study, VagueGAN utilizes generative adversarial networks (GANs), which were investigated by Goodfellow et al. [6]. VagueGAN generates ambiguous poisoned data that effectively undermines global models while maintaining statistical characteristics.

This study examines a novel algorithm for mitigating these threats by utilizing Shapley Additive Explanation (SHAP) values, a concept investigated by Lundberg et al. [11]. SHAP values function as a metric for evaluating feature importance. The algorithm assesses the modifications induced by both benign and malicious clients using SHAP values, thereby facilitating the classification of the majority of clients with high precision through linear classifiers. Notably, the methodology demonstrated an accuracy rate of 90–100% when implemented with Support Vector Machines (SVMs) [2].

18.1.1 Contributions

In this study, a VagueGAN-based model poisoning attack technique was employed to generate vague poisoned data to target the model effectively. A series of experiments was conducted using CIFAR-10, MNIST, and Fashion-MNIST to demonstrate that the VagueGAN attack effectively degrades the FL performance while remaining less detectable. A data poisoning detection technique introduced to identify malicious clients through SHAP values processed as detection features presents an intuitive approach despite the inherent complexity of ANNs.

18.2 Related Studies

In the following, we briefly discuss studies related to attacks in FL utilizing GAN-based models. Further, we review data poisoning attack and defense mechanisms introduced in SOTA.

18.2.1 GAN-Based Attacks in Federated Learning

In a study by Hitaj et al. [8] employs a GAN model to extract features from local data using GAN-generated pseudo data, initiated by the global model. To expand on this, another attack strategy proposed by Zhang et al. [22] introduces PoisonGAN, which enlarges a local dataset with authentic pseudo samples employing a discriminator established by the global model. However, these attack techniques significantly modify the local label distributions, rendering them easily identifiable by the corresponding defense mechanisms.

18.2.2 Data Poisoning Attack and Detection in Federated Learning

Studies conducted in [8] and Shi et al. [17] have researched extensively on data poisoning attacks in traditional deep learning systems. Schwarzschild et al. [15] posited that the investigation of such attacks on FL systems remains in its early stages. Current research primarily concentrates on employing label-flipping methods to generate contaminated local data by modifying specific labels within the training dataset. For instance, Tolpegin et al. [19] proposed that label flipping can be utilized to reorganize the label-sample relationships in a local dataset and produce corrupted data. For defense approaches, contemporary research primarily emphasizes the analysis of statistical characteristics across models submitted by various clients to detect anomalies. The proposed method by Tolpegin et al. [19] and Upreti et al. [20] employs dimensionality reduction techniques such as principal component analysis (PCA) and uniform manifold approximation and projection (UMAP) to reduce model dimensions into two-dimensional space for abnormality detection. Additionally, Li et al. [10] introduced Local Malicious Factor (LoMar), which relies on kernel density estimation and k-nearest neighbor techniques, presenting an alternative approach to defense.

Several critical arguments have been identified in the literature. Fang et al. [5] contend that defenses such as Krum and Bulyan are ineffective against specific attacks where adversaries send gradient updates opposing model adjustments from honest clients, thus posing a significant threat that necessitates sophisticated solutions for detection and prevention. Moreover, Nowroozi et al. [14] studied mitigation strategies for label-flipping attacks in malicious URL detection systems, offering valuable insights into enhancing data poisoning defenses.

This study proposes a dual-approach strategy providing a GAN-based attack and data-poisoning defense mechanism that strengthens the security of FL systems but also enhances resilience by maintaining model robustness against evolving threats. Consequently, this study developed a comprehensive GAN-based attack and data poisoning defense solution, which is essential for safeguarding the reliability and trustworthiness of FL environments.

18.3 Methodology

This investigation presents an attack and defense framework FL systems. The offensive component employs VagueGAN, a purpose-built generative adversarial network designed to compromise FL integrity by introducing deceptive data that mimics authentic samples, thus circumventing detection mechanisms. Conversely, the defensive strategy implements a novel detection approach that combines Shapley Additive Explanations (SHAP) and Support Vector Machines (SVM) to effectively identify and counteract the influence of contaminated data. This dual-pronged approach enhances the robustness and dependability of FL systems. Together, these components constitute a comprehensive paradigm for analyzing and mitigating the challenges posed by data poisoning attacks within FL contexts.

18.3.1 Attack Model

This section delineates the objectives and assumptions related to data-poisoning attacks on FL. In the context of adversarial attacks on FL, malicious entities aim to remain undetected by the server while simultaneously compromising the efficacy of the global model. These adversaries may either infiltrate existing clients or present themselves as legitimate participants to gain unauthorized access to training protocols and local datasets. It is important to note that attackers are unable to access the server's datasets and models or those of other uncompromised clients. The server is assumed to maintain integrity and security throughout the process. To achieve their objectives, adversaries implement data poisoning strategies. The resulting poisoned dataset is then utilized to train a local model, which is subsequently uploaded to the server. If the server fails to identify the compromised model, it may inadvertently incorporate it into the global model during the aggregation phase, potentially compromising overall system integrity. A generative adversarial network approach was employed for the vagueGAN model.

VagueGAN employs a novel strategy to achieve an optimal balance between attack effectiveness and covertness by leveraging the capabilities of Generative Adversarial Networks to produce seemingly authentic vague data containing carefully calibrated amounts of poisonous noise. To execute a surreptitious attack, the VagueGAN must generate genuine poisoned data. Building on the Conditional GAN (CGAN) model, we developed a method to create poisoned data from authentic local data [13]. To enhance the practicality of VagueGAN and ensure superior results, even with limited local datasets, we implemented advanced techniques such as batch normalization and LeakyReLU, along with significantly deeper architectures in both generator and discriminator components [9]. However, for VagueGAN to execute a successful attack, it must generate noisy poisoned data, as data noise can substantially degrade the quality of the training data and, consequently, the trained model. This approach results in the production of data that is highly susceptible to noise. The sample distribution generated by a GAN approximates the real sample distribution, balancing seemingly legitimate and noisy outcomes, as illustrated in Fig. 18.1 using image data [18]. In contrast, the sample distribution generated by VagueGAN shows only minimal resemblance to the actual sample distribution. PoisonGAN, which functions more as a data augmentation technique than a genuine

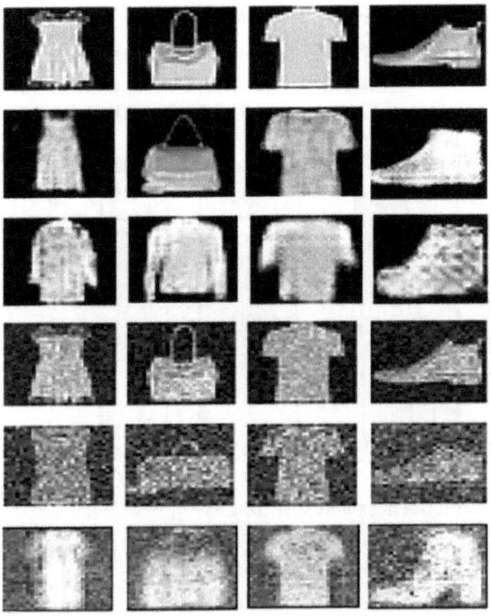

Fig. 18.1 GAN noise samples [18]

attack method, produces samples of relatively high quality when its discriminator is initialized using the global model.

VagueGAN, on the other hand, offers potential computational advantages over traditional GANs by focusing on generating low-quality and vague data. It is important to differentiate between real data with superimposed noise and the hazy data produced by VagueGAN. Adding noise to real data creates poisoned data with attack quality that is difficult to control. While increasing noise levels may improve attack effectiveness, it also increases the likelihood of noise detection, compromising attack stealth. This is because manually added noise (e.g., Gaussian noise) lacks statistical correlation with the original data. In contrast, VagueGAN generates noise guided by the original data's statistical properties, resulting in more nuanced noise created by the generative adversarial network.

An analysis of VagueGAN's experimental outcomes, as illustrated in Fig. 18.2, was conducted to investigate the inverse relationship between the attack potency A_t and imperceptibility S_t. This investigation revealed a fundamental correlation between this trade-off and the fidelity of the data generated by VagueGAN. By leveraging GAN technology, VagueGAN can produce adulterated data with diverse qualities. As the synthetic data approach the verisimilitude of authentic samples, represented in Fig. 18.2's upper-right region, a significant decline in attack potency A_t is observed, coupled with enhanced imperceptibility S_t. In contrast, when the adulterated data become increasingly indistinct and chaotic, shifting towards the lower-left sector of Fig. 18.2, attack potency A_t increases while imperceptibility S_t diminishes. This dynamic exemplifies VagueGAN's ability to orchestrate attacks

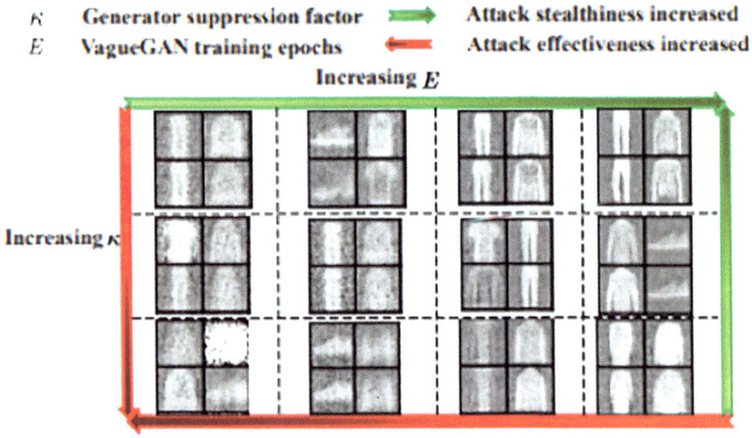

Fig. 18.2 Interchange between attack effectiveness and stealthiness

that are either more clandestine but less impactful or more potent but more discernible, contingent on its proficiency in replicating legitimate data attributes. By precisely calibrating the training iterations, E, that controls the degree of model fitting to the training data and constraining the suppression factor, κ, that controls the generator's ability to produce high-quality samples, VagueGAN compromises between attack potency A_t and imperceptibility S_t.

18.3.2 Defence Model

The model for the data poisoning detection algorithm is explained in this section. We begin by outlining the experimental setup, the threat model, and the system model. After that, in order to comprehend the impact of poisoning attacks and extract new features from the outcomes, we carry out experiments pertaining to adversary capabilities and SHAP values in various FL settings. Finally, we developed a detection algorithm model to study attacks that contaminate data in horizontal FL settings. Distributed learning with identical feature spaces and model structures is made possible by the architecture in a horizontal FL setting. In label-flipping attacks, the attacker aims to change the local labels such that the global model predicts the images of one class as belonging to another. This adjustment results in a decline in one class's evaluation metrics, while the performance of the other classes remains the same.

The MNIST [4] and Fashion MNIST [21] datasets, which are publicly accessible, were employed in our experimental investigations. These datasets are commonly utilized in studies examining poisoning attacks within FL environments. CNNs were developed for both datasets. Following the research of Tolpegin et al. [19], we set the initial momentum to $\lambda = 0.5$ and the learning rate to $\eta = 0.01$. The batch size is initially configured as 10 for training and 1000 for testing. During the aggregation phase, the server selects $m_A = 5$ clients per round. Our implementation of the CNN models with the specified hyperparameters on the MNIST and Fashion MNIST datasets yielded comparable results. To calculate SHAP values, we utilized 100 test images as background data.

This study evaluates the impact of poisoning attacks on FL systems through two distinct experiments, focusing on the performance across varying numbers of rounds. The initial experiment assesses the effects of altering the number of poisoned clients over multiple rounds. We conduct 200 rounds of the base algorithm, employing different quantities of malicious clients. To gauge the server model's performance, we measure the accuracy and recall of the aggregated model on the test set at ten-round intervals. Two distinct attacks are utilized in this experiment: $\alpha_{5,4}$ 5,4, which alters the MNIST dataset's label from 5 to 4, and another that changes the Fashion MNIST dataset's label from sandals to coats. Our findings indicate that an increase in malicious clients correlates with a higher frequency of significant accuracy fluctuations. Poisoning attacks may cause substantial performance degradation at specific points during training, their long-term effects on the model may be limited after several attack-free rounds. However, these attacks can still lead to considerable temporary performance losses, as evidenced by the MNIST model, where the recall for digit 5 drops to 0% in certain instances. Previous experiments show that a linear model can detect the majority of poisoning attacks. Consequently, as a large-margin classifier, we suggest using an SVM.

18.4 Results

18.4.1 Attack Results

This study evaluates the effectiveness of VagueGAN's poisoning capabilities using three real-world datasets: CIFAR-10, Fashion-MNIST, and MNIST. For MNIST and Fashion-MNIST, the architecture incorporated batch normalization and a CNN with two convolutional layers. The CIFAR-10 model utilizes a more complex CNN structure consisting of six convolutional layers, batch normalization, and two fully connected layers. The training included an initial learning rate of 0.01, scheduler step size of 50, momentum of 0.5, and decaying learning rate. The FL setup comprises 60 clients and a central server employing the FedAvg algorithm to randomly select 10 clients per round for model aggregation. The FL training process spans 200 rounds. Client datasets, which could be either IIDs or non-IIDs, were partitioned into 60 subsets. To benchmark the performance of VagueGAN, we compared it with two prevalent data poisoning techniques. The first method involved a label-flipping attack that targeted specific labels in the training set for alteration. The second approach combines label flipping with PoisonGAN, which produces

Table 18.1 Comparison of experimental results for different data poisoning attacks

Attack method	Dataset	No attack accuracy	Attack accuracy	Reference	Year
Label flipping	MNIST	98.75%	94.50%	[16]	2021
Label flipping	CIFAR-10	86.20%	78.40%	[16]	2021
PoisonGAN	MNIST	98.75%	93.80%	[16]	2021
PoisonGAN	CIFAR-10	86.20%	76.70%	[16]	2021
Noise superimposition	MNIST	98.75%	91.30%	[16]	2021
Noise superimposition	CIFAR-10	86.20%	74.60%	[16]	2021
Backdoor attack	MNIST	98.75%	89.00%	[19]	2020
Backdoor attack	CIFAR-10	86.20%	70.50%	[19]	2020
VagueGAN	MNIST	98.75%	90.20%	This work	2024
VagueGAN	CIFAR-10	86.20%	72.80%	This work	2024

valid pseudo-samples to enhance the local dataset before label manipulation. In addition, we examined a Noise Superimposition Attack, where Gaussian noise was overlaid onto genuine data to create contaminated samples. Table 18.1 presents the comparative results of our study with those of recent studies.

Comparison of Attack Performance and Efficacy of Poisoning Attacks The effectiveness of data poisoning is quantified by the global model accuracy (a_{pt}) under attack. Three attack methods were evaluated with varying malicious client percentages (α) from 5% to 30%, averaged over 10 iterations. VagueGAN exhibits performance attack potency. Non-IID settings and a higher proportion of malicious clients resulted in reduced global model accuracy. CIFAR-10 exhibited greater vulnerability to data poisoning than Fashion-MNIST and MNIST. A noise superimposition attack with $\alpha = 20\%$ on Fashion-MNIST under IID conditions, yielding accuracies of 88.12%, 87.97%, 87.16%, and 86.66% for no noise, light noise, high noise, and VagueGAN, respectively. Heavy noise achieves comparable effectiveness to VagueGAN, but lacks stealth; light noise proves insufficient for an effective attack.

Poisoning Attack In poisoning attacks, the statistical distances between legitimate and compromised models are measured using Principal Component Analysis (PCA) to assess attack concealment. VagueGAN surpasses benchmark techniques in stealth, making detection more difficult even with PCA-based defenses. Accuracy fluctuation analysis related to attack concealment showed significant variations in benchmark methods, whereas VagueGAN exhibited minor fluctuations, similar to non-attack scenarios. The study also evaluated the stealth of noise-superimposition attacks and found poor concealment owing to divergent noise distributions affecting authentic data. Accuracy fluctuations under high-noise conditions evidence the superior stealth of the VagueGAN.

18.4.2 Defence Results

To evaluate the efficacy of the detection model in identifying malicious models, we established a dataset comprising 300 malicious and 300 pristine clients, utilizing parameters from three converged models. These clients generated the feature vectors and labels for the training set. Malicious clients were compromised using known attacks from previous experiments, and SHAP-based features were extracted for training purposes. Unidentified and novel images were incorporated into the testing set to assess the robustness of the model. The training set contained 600 feature vectors, and the testing set included 200. The detection models were trained by using a sample size of 200 individuals. In calculating performance per threshold, we utilized scikit-learn to compute approximately 130 thresholds ($n_{\text{thresh}} \approx 130$) for the model performance evaluation. The MNIST model with an RBF kernel has thresholds ranging from $\epsilon \in [-0.04, 1.8]$, whereas the linear kernel thresholds span $\epsilon \in [-2.28, 6.4]$. For the Fashion MNIST models, the RBF and linear kernel thresholds were $\epsilon \in [-1.13, 2.19]$ and $\epsilon \in [-2.46, 7.44]$, respectively. Table 18.2 presents a comparative analysis of this study and previous research on the effectiveness of defense techniques.

This study assessed three detection models using I.I.D databases to ascertain average recall, precision, and accuracy. The results in Table 18.2 indicate that the MNIST model achieved an average accuracy of 97.3% average accuracy with the RBF kernel and 99.1% with the linear kernel. The MNIST model using a linear kernel demonstrated the highest accuracy of 100%. The fashion MNIST models exhibited comparatively lower accuracy, averaging 94.2% with a linear kernel and 93.8% with an RBF kernel. The lowest accuracy of 93.5% was observed in the Fashion MNIST model employing a linear kernel. Table 18.3 illustrates that recall values ranged from approximately 94% to 100%, while the precision ranged from 92% to 100%.

The remarkable efficacy of malicious model detection can be attributed to the utilization of SHAP-based features and emphasis on relative modifications during feature engineering. Malicious models exhibit substantial divergence from converged global models while demonstrating similarities to other malicious instances.

Table 18.2 Comparison of experimental results

Study	Dataset	Kernel	Accuracy	Recall	Precision
VagueGAN	MNIST	Linear	99.1%	98.3%	100%
	MNIST	RBF	97.3%	98.3%	96.5%
	Fashion MNIST	Linear	93.8%	94.0%	93.7%
	Fashion MNIST	RBF	94.2%	95.7%	92.9%
[19]	MNIST	–	97.0%	–	–
	Fashion MNIST	–	93.5%	–	–
[1]	MNIST	–	96.8%	97.5%	96.0%
	Fashion MNIST	–	92.7%	93.0%	92.0%

Table 18.3 Average accuracy, recall, precision considering three models with $\epsilon = 0$

Kernel	Model	Accuracy	Recall	Precision
RBF	MNIST	97.3%	98.3%	96.5%
	Fashion MNIST	94.2%	95.7%	92.9%
Linear	MNIST	99.1%	98.3%	100%
	Fashion MNIST	93.8%	94.0%	93.7%

This characteristic facilitates more effective detection through comparisons between local models and the global model rather than inter-local model evaluations. The detection framework employs features derived from discrepancies between the global model (F_{r-1}) and the received local model (f_r) to enhance similarity quantification. Furthermore, the selection of local model parameters that closely align with the global model parameters helps mitigate the aggregation. The transformation of the data poisoning detection problem into a linear formulation, facilitated by the SHAP values, elucidates the high accuracy achieved and corroborates the suitability of the proposed detection methodology.

18.4.3 Experimental Limitations

Although this study's experiments demonstrate favorable accuracy in controlled environments, they exhibit several limitations. The experiments utilized IID datasets; non-IID datasets, which more accurately represent real-world distributions, have not yet been examined. Additionally, the study solely considered horizontal FL settings; further investigation is required to assess the algorithm in vertical FL scenarios. The research indicates that label-flipping attacks result in substantial short-term consequences but no long-term effects, suggesting the need for additional research and mitigation of long-term impacts. The limited applicability of the detection algorithm to other poisoning attack types can be attributed to its specific design for label-flipping attacks. Furthermore, the FL algorithm's susceptibility to catastrophic forgetting, wherein performance on previous tasks deteriorates after training on new tasks, underscores the significant challenge of maintaining consistent performance across tasks.

18.5 Conclusion

In this chapter, we explored the potential of VagueGAN, a GAN model designed for data-poisoning attacks in FL systems. VagueGAN generates ambiguous data that appear authentic while containing an appropriate amount of malicious noise, thereby achieving a balance between attack efficacies. Furthermore, we investigated the effects of label-flipping attacks, revealing that despite their short-term impact, they can significantly degrade the FL model's performance after a single training iteration. Finally, we provided a defense mechanism highlighting that combining Shapley additive explanation (SHAP) serving as feature spaces for detection algorithms with models such as SVM could be a counterfeit mechanism against such attacks.

References

1. Awan, S., Luo, B., Li, F.: Contra: Defending against poisoning attacks in federated learning. In: Computer Security–ESORICS 2021: 26th European Symposium on Research in Computer Security, Darmstadt, Germany, October 4–8, 2021, Proceedings, Part I 26, pp. 455–475. Springer, Cham (2021)
2. Boser, B.E., Guyon, I.M., Vapnik, V.N.: A training algorithm for optimal margin classifiers. In: Proceedings of the Fifth Annual Workshop on Computational Learning Theory, pp. 144–152 (1992)
3. Chen, Y., Zhu, P., He, G., Yan, X., Baligh, H., Wu, J.: From connected people, connected things, to connected intelligence. In: 2020 2nd 6G Wireless Summit (6G SUMMIT), pp. 1–7. IEEE, Piscataway (2020)
4. Deng, L.: The MNIST database of handwritten digit images for machine learning research [best of the web]. IEEE Signal Process. Mag. **29**(6), 141–142 (2012)
5. Fang, M., Cao, X., Jia, J., Gong, N.: Local model poisoning attacks to {Byzantine-Robust} federated learning. In: 29th USENIX Security Symposium (USENIX Security 20), pp. 1605–1622 (2020)
6. Goodfellow, I., Pouget-Abadie, J., Mirza, M., Xu, B., Warde-Farley, D., Ozair, S., Courville, A., Bengio, Y.: Generative adversarial networks. Commun. ACM **63**(11), 139–144 (2020)
7. Gosselin, R., Vieu, L., Loukil, F., Benoit, A.: Privacy and security in federated learning: a survey. Appl. Sci. **12**(19), 9901 (2022)
8. Hitaj, B., Ateniese, G., Perez-Cruz, F.: Deep models under the GAN: information leakage from collaborative deep learning. In: Proceedings of the 2017 ACM SIGSAC Conference on Computer and Communications Security, pp. 603–618 (2017)
9. Ioffe, S., Szegedy, C.: Batch normalization: accelerating deep network training by reducing internal covariate shift. In: International Conference on Machine Learning, pp. 448–456. PMLR (2015)
10. Li, X., Qu, Z., Zhao, S., Tang, B., Lu, Z., Liu, Y.: LoMar: a local defense against poisoning attack on federated learning. IEEE Trans. Depend. Secur. Comput. **20**(1), 437–450 (2021)
11. Lundberg, S.M., Lee, S.I.: A unified approach to interpreting model predictions. Adv. Neural Inf. Process. Syst. **30** (2017)
12. Lyu, L., Yu, H., Yang, Q.: Threats to federated learning: a survey. Preprint. arXiv:2003.02133 (2020)
13. Mirza, M., Osindero, S.: Conditional generative adversarial nets. Preprint. arXiv:1411.1784 (2014)
14. Nowroozi, E., Jadalla, N., Ghelichkhani, S., Jolfaei, A.: Mitigating label flipping attacks in malicious URL detectors using ensemble trees (2024). https://arxiv.org/abs/2403.02995
15. Schwarzschild, A., Goldblum, M., Gupta, A., Dickerson, J.P., Goldstein, T.: Just how toxic is data poisoning? A unified benchmark for backdoor and data poisoning attacks. In: International Conference on Machine Learning, pp. 9389–9398. PMLR (2021)
16. Shejwalkar, V., Houmansadr, A.: Manipulating the Byzantine: optimizing model poisoning attacks and defenses for federated learning. In: NDSS (2021)
17. Shi, Y., Erpek, T., Sagduyu, Y.E., Li, J.H.: Spectrum data poisoning with adversarial deep learning. In: MILCOM 2018–2018 IEEE Military Communications Conference (MILCOM), pp. 407–412. IEEE, Piscataway (2018)

18. Sun, W., Gao, B., Xiong, K., Lu, Y., Wang, Y.: VagueGAN: A GAN-based data poisoning attack against federated learning systems. In: 2023 20th Annual IEEE International Conference on Sensing, Communication, and Networking (SECON), pp. 321–329. IEEE, Piscataway (2023)
19. Tolpegin, V., Truex, S., Gursoy, M.E., Liu, L.: Data poisoning attacks against federated learning systems. In: Computer Security–ESORICs 2020: 25th European Symposium on Research in Computer Security, ESORICs 2020, Guildford, UK, September 14–18, 2020, Proceedings, Part I 25, pp. 480–501. Springer, Cham (2020)
20. Upreti, D., Kim, H., Yang, E., Seo, C.: Defending against label-flipping attacks in federated learning systems with UMAP (2022). https://doi.org/10.21203/rs.3.rs-1984301/v1
21. Xiao, H., Rasul, K., Vollgraf, R.: Fashion-MNIST: a novel image dataset for benchmarking machine learning algorithms. Preprint. arXiv:1708.07747 (2017)
22. Zhang, J., Chen, B., Cheng, X., Binh, H.T.T., Yu, S.: Poisongan: Generative poisoning attacks against federated learning in edge computing systems. IEEE Internet Things J. **8**(5), 3310–3322 (2020)

Part V
Applications and Case Studies

Chapter 19
Cyber Risk Assessment in IT/OT Convergence Using Machine Learning

Diego M. Mendez Mena and Maryam Karimi

19.1 Introduction

The demand to measure cyber risk and provide mitigation in any given environment has become more evident. The multiplication of global cybersecurity incidents over the last few years justifies more analysis and subsequent countermeasures. The growing number of available frameworks and "best practices" reflect this necessity. However, many are based on the perceptions of various types of experts and do not necessarily rely on statistical methods using large datasets. While expert opinion is a valid method for initiating cyber-risk assessments, such as defining the problem or classifying data, it often leads to estimations that do not produce effective risk-mitigating results when critical moments arise. As portrayed by [1], humans generally do not provide realistic results despite having access to more information, giving other stakeholders a false sense of confidence when making decisions based on such an analysis.

Based on such premise, the importance of implementing data-based approaches to make decisions on cyber risk has increased. It becomes even more relevant when the results are applied to the convergence of information technology (IT) and operational technology (OT). OT has taken the spotlight after being neglected for years regarding security. It is now more apparent after critical infrastructure became the target of many state-sponsored cyber threat actors that can use cyber attacks to create as much havoc as possible for adversaries.

This chapter extends the research presented in our earlier paper *"Data-Driven IT/OT Cyber Risk Evaluation and Mitigation Prioritization Using Machine Learning"* [2]. Our objective is providing a novel approach for risk assessment consider-

D. M. Mendez Mena · M. Karimi (✉)
Department of Automation, Digitalization and Cyber Research, Factory Mutual Insurance Company, Norwood, MA, USA
e-mail: maryam.karimi@fmglobal.com

E. Nowroozi et al. (eds.), *Adversarial Example Detection and Mitigation Using Machine Learning*, https://doi.org/10.1007/978-3-031-99447-0_19

ing two main aspects: Analyzing the correlation of attack methodologies grounded in factual data using Machine Learning (discussed in Sect. 19.4). Then, enabling a cyber defense strategy by calculating a probabilistic risk assessment on breach likelihood and financial impact (covered in Sect. 19.5). Using this approach, we can identify the hazards of IT/OT convergence, pinpoint risky attack techniques, and prioritize mitigation and detection strategies while reducing uncertainty. To our knowledge, this is the first research to use a data-driven approach and machine learning to determine the most risk-relevant ATT&CK techniques and their dependencies with mitigation and detection techniques.

19.2 Literature Review: Machine-Learning for Cyber Risk

This section reviews the literature on cyber risk assessment using Machine Learning (ML). Machine learning is mainly used in cybersecurity for threat detection, such as intrusion detection, malware detection, phishing detection, spam detection, and fraud detection. The common practice involves selecting relevant features and creating models like signature-based and anomaly-based ones [3]. However, the literature reviewed in this section, goes beyond the classic use of ML classification for threat detection.

In [4], authors analyzed 1000 instances from Twitter (X) using histograms and word clouds and developed neural network models to determine if these instances were linked to entries in the Common Vulnerabilities and Exposures (CVE) database. In [5], authors focused on tuning hyperparameters to enhance the effectiveness of ML and deep learning methods in mapping CVEs available in the National Institute of Standards and Technology (NIST) dataset to MITRE ATT&CK techniques. In [6], authors used Natural Language Processing (NLP) feature modeling to find correlations between MITRE ATT&CK Threats, Techniques, and Procedures (TTPs) and Common Weakness Enumerations (CWEs). In [7], authors provided a comprehensive statistical analysis of the MITRE ATT&CK framework, offering insights into relationships among various entities such as threat actors, software, campaigns, domains, platforms, tactics, and techniques and presenting recommendations to improve digital infrastructure security. In [8], authors investigated the relationship between operational context and vulnerability assessment effectiveness by comparing the context-agnostic CVSS v2 model with threat intelligence data from Alien Vault Open Threat Exchange and MITRE ATT&CK adversary tactics, techniques, and procedures. They derived a risk enumeration and mitigation priority model using unsupervised machine learning techniques to provide contextual priorities for mitigating cataloged vulnerabilities, such as CVEs. The study finds that incorporating threat and adversary tactics data with the CVSS "base score" offers better contextual risk assessment and prioritization for information security and technology professionals [8]. To date, no machine learning techniques have been employed to assess risk for attack techniques.

19.3 Background on Data and Cyber Security Frameworks

This section provides information on the data set and the cyber security frameworks we used for feature engineering.

19.3.1 Data

Obtaining reliable public cybersecurity data is challenging. Although sharing reliable Cyber Threat Intelligence (CTI) data is crucial for improving future cyber risk controls, there are challenges related to quality, legal, privacy, budgeting, and interoperability issues in CTI distribution [9]. After a thorough investigation of available data on breaches, we used VERIS (Vocabulary for Event Recording and Incident Sharing) Community Data Base (VCDB) [10], which is connected to Verizon's annual Data Breach Investigations Report. VERIS compiles structured postmortem information from cybersecurity incidents, allowing organizations to share detailed, anonymous reports. It provides incidents' root causes, demographics, and impacts. The database contained over 10,000 records of cybersecurity breaches encompassing 2671 attributes, including incident descriptions and victim information for each case from 2012 to 2023. These attributes are structured hierarchically under the major sections on Actors, Actions (with subsections of error, physical, social, malware, and hacking), Assets, Attributes (Impacts on confidentiality, integrity, and availability (CIA)), Value chain (required capabilities attacker must acquire before an attack), and Patterns (attack difficulty, and discovery method). It is important to stress that even though the database's strengths are evident, it also has limitations that might not provide a complete picture of the current cyber environment. However, more information is always helpful, especially when aiming to reduce uncertainty.

19.3.2 Cybersecurity Risk Frameworks

Essential security attack techniques and mitigations should be identified to determine the baseline for risk measurement. MITRE ATT&CK is a comprehensive framework for categorizing and understanding how attackers operate through cyber adversary tactics (high-level adversary objectives during an attack, e.g., gaining initial access, maintaining persistence,impact), techniques, and procedures (TTPs) [11]. The framework allows us to determine a structure that helps researchers model threats and how procedures should respond; as demonstrated by [12], the authors investigate the TTPs in the top 25 well-known ICS attacks and propose security controls for used techniques.

Other tools assist in determining severities and priorities for any existing risk. For example, NIST's Common Vulnerability Scoring System (CVSS) provides a qualitative computation of severity using base, supplemental, environmental, and threat metrics in its latest version (CVVS v4.0) [13]. CVSS v4.0 platform first creates a "CVSS vector" comprising essential criteria and then outputs a vulnerability Severity score based on these indicators. The CVSS v4.0 vector includes factors such as attack vector, attack complexity, privileges required, user interaction, and the impact on confidentiality, integrity, availability, and exploit maturity [14].

19.4 Machine Learning Modeling and Attack Techniques

We used VCDB and a novel method for feature engineering to derive ATT&CK techniques and label each attack with CVSS v4.0 scores and then preprocessed to achieve a clean and reliable data set. Following this, we developed a model to establish the correlation between attack vectors and corresponding scores for these attacks. Once the correlation was confirmed, we identified the most strongly correlated attack techniques. Then, we pinpointed the mitigation techniques associated with these highly correlated attack vectors. Finally, we identified frequent attack chains and created a model to predict the CVSS score for each attack chain. This section explained each of these steps in detail.

19.4.1 Feature Engineering and Preprocessing

We utilized the MITRE ATT&CK framework [11] and generated features that are attack techniques and derived the values from VERIS features. The MITRE ATT&CK framework provided a dictionary that maps VERIS features [15] to attack techniques, particularly those VERIS features that are subcategories of actions (categories of hacking, malware, and social) and value chain (categories of development, targeting, distribution, and non-distribution services). For example, the VERIS feature "action.hacking.variety.OS commandin" is mapped to the ATT&CK technique 'T1059':Command and Scripting Interpreter.

Our mapping process converted 160 features from the VERIS dataset into 594 MITRE ATT&CK techniques and sub-techniques. We then merged the sub-techniques and further refined the dataset by removing techniques with missing values and merging techniques that were referred to by the same VERIS features and covered the same category. This process resulted in 168 techniques in total.

Then, we employed the CVSS v4.0 score to label the data based on another set of features. We incorporated 167 features, including actions (categories of environmental, misuse, physical, and error), actor, asset (category of assets variety), attributes, plus (category of attack difficulty). For instance, if "action.misuse.vector.Web application" is True, the attack vector value in the CVSS v4.0 vector would be "Network." On the other hand, if "action.misuse.vector.LAN access" is True, the attack vector value would be "Adjacent." We ensured that the features deriving the CVSS v4.0 vector did not overlap with those used for MITRE ATT&CK techniques to avoid introducing dependencies between the features and the label in model training. To convert the CVSS v4.0 vector into a CVSS v4.0 score, we adapted the CVSS v4.0 Calculator [14] to generate a score ranging from 0 to 10. We then categorized these scores into high, medium, and low-risk levels based on thresholds of 4 and 7, as defined by the CVSS v4.0 standard [14]. We then constructed two sets of data frames:

- Original Features Database (OFD): This dataset includes the original 160 VERIS features labeled with the CVSS v4.0 score.
- ATT&CK Technique Database (ATD): This dataset contains the features mapped to ATT&CK techniques, labeled with the CVSS v4.0 score.

We aim to investigate the correlation between the features of these two datasets and the CVSS v4.0 score. We balanced and scaled the data and split it into a 80% training test with a 20% held out test set.

19.4.2 ML Modeling and Feature Selection

To evaluate the effectiveness of the feature sets in predicting risk scores, we developed several machine learning models using Scikit-learn and TensorFlow Python libraries [16–18], for both OFD and ATD. The models implemented include Random Forest (RF), Logistic Regression (LR), Support Vector Machines (SVM), and Neural Networks (NN) (initialized with sequential model and added three layers, 3 dense with ReLU and SoftMax activation functions and compiled using the Adam optimizer with a learning rate of 0.01) [17, 18]. We measured accuracy, precision, recall, and F1 score as performance metrics. The performance evaluations of these models on both datasets are presented in Table 19.1. The average F1 score for OFD is 0.62, and for ATD, it is 0.73, demonstrating a strong correlation between ATT&CK techniques and CVSS v4.0 scores.

Then, we employed several feature selection methods to identify the most related attack techniques to risk score. We subsequently evaluated these methods by building an RF model using the selected attack techniques. For the feature ranking methods that do not automatically select *n* features, we developed multiple

Table 19.1 Performance of models on original DB vs. MITRE techniques DB

	Original DB				MITRE techniques DB			
	RF	LR	SVM	NN	RF	LR	SVM	NN
F-score	0.624	0.613	0.627	0.620	0.733	0.720	0.734	0.730

Table 19.2 Performance metrics for different feature selection methods

	Pearson	Spearman	Thresholding	Top-k	RFE	L1 reg
F-score	0.540	0.558	0.727	0.679	0.732	**0.744**

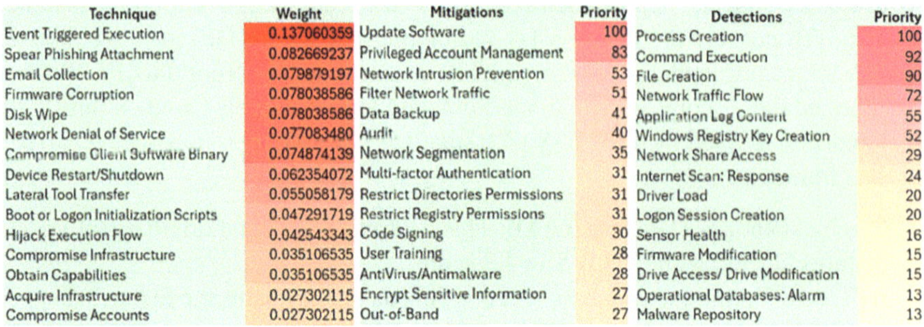

Technique	Weight	Mitigations	Priority	Detections	Priority
Event Triggered Execution	0.137060359	Update Software	100	Process Creation	100
Spear Phishing Attachment	0.082669237	Privileged Account Management	83	Command Execution	92
Email Collection	0.079879197	Network Intrusion Prevention	53	File Creation	90
Firmware Corruption	0.078038586	Filter Network Traffic	51	Network Traffic Flow	72
Disk Wipe	0.078038586	Data Backup	41	Application Log Content	55
Network Denial of Service	0.077083480	Audit	40	Windows Registry Key Creation	52
Compromise Client Software Binary	0.074874139	Network Segmentation	35	Network Share Access	29
Device Restart/Shutdown	0.062354072	Multi-factor Authentication	31	Internet Scan: Response	24
Lateral Tool Transfer	0.055058179	Restrict Directories Permissions	31	Driver Load	20
Boot or Logon Initialization Scripts	0.047291719	Restrict Registry Permissions	31	Logon Session Creation	20
Hijack Execution Flow	0.042543343	Code Signing	30	Sensor Health	16
Compromise Infrastructure	0.035106535	User Training	28	Firmware Modification	15
Obtain Capabilities	0.035106535	AntiVirus/Antimalware	28	Drive Access/ Drive Modification	15
Acquire Infrastructure	0.027302115	Encrypt Sensitive Information	27	Operational Databases: Alarm	13
Compromise Accounts	0.027302115	Out-of-Band	27	Malware Repository	13

Fig. 19.1 Top MITRE techniques, mitigation, and detections

models using the top n features for different ns. These methods included Pearson correlation, Spearman correlation, thresholding at 0.1, top-K selection, recursive feature elimination (RFE), and L1 regularization [16, 17]. As shown in Table 19.2 among these methods, L1 regularization with logistic regression proved to be the most effective in identifying the most critical ATT&CK techniques to predict score labels. It also resulted in a slight improvement in the model's performance.

Through this methodology, we developed an approach to predict risk-based scores and pinpoint the most effective and high-risk attack techniques based on historical data. Lasso regularization identified 28 attack techniques (18 for IT and 10 for OT) as the most prominent features. We created a RF model using only these 28 techniques to rank them by feature importance. The top 15 techniques and their importance (weight) are shown in the left table in Fig. 19.1. We cross-referenced these techniques with the mitigation and detections provided in the MITRE framework. 68 mitigation and 33 detection techniques can address all these attack techniques. The priorities for top mitigation and detection techniques in Fig. 19.1, are calculated by summing and normalizing the weights of the techniques addressed. Proposed method in Sect. 19.5 also considers dependencies.

19.4.3 Frequent Attack Chains and Sequential Learning

To complete our study, we analyzed 70 unique attack chains by sorting techniques based on their respective tactics to order them according to their position in the attack chain. We then utilized a Recurrent Neural Network (RNN) with a Long Short-Term Memory (LSTM) model to estimate the CVSS scores based on these attack patterns. Our model architecture included three layers: an embedding layer

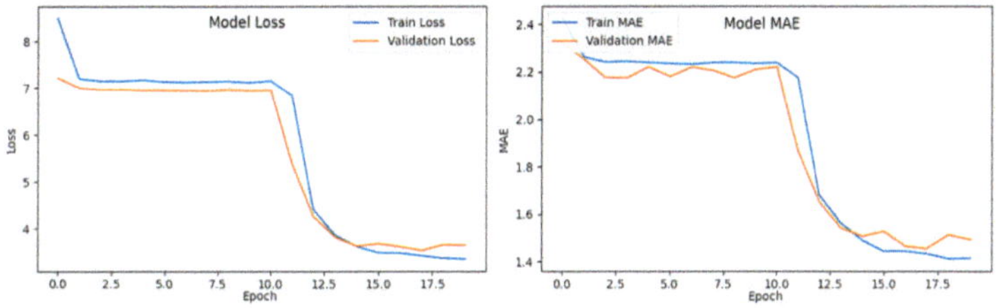

Fig. 19.2 RNN sequential model loss and MAE

with an input dimension corresponding to the vocabulary size and an output dimension of 128, an LSTM layer with 128 units, and a Dense layer with a linear activation function for regression tasks. The model was compiled using the mean squared error loss function and the Adam optimizer, with mean absolute error (MAE) as the evaluation metric as shown in Fig. 19.2 final model exhibited a test MAE of 1.3973, showcasing its effectiveness in predicting CVSS.

19.5 Probabilistic Analysis to Assess Risk

The probabilistic analysis comes after the results obtained from the machine learning correlation algorithm to reduce the uncertainty produced by the nature of cyber incidents. It also contributes to providing a conclusive risk assessment. It is based on the output (essential attack techniques) from the methods described in Sect. 19.4 to determine the probability density functions of the most persistent techniques and the associated risk in financial impact terms to determine Inherent Risk Curves (IRC). The resulting prominent Tactics and Techniques were probabilistically analyzed using a Bayesian approach, such as the one described by the NIST report number 8286A, where it is recommended that to improve the estimation of risk parameters, prior information can help make a more "objective determination" [19, p. 44].

The proposed function for analysis is the Beta distribution, which denotes random variables of finite length. Each record within the VCDB is considered a sample to determine likelihood. The probability density function (PDF) and cumulative distribution function are defined in [20, p. 176]. For this research project, the parameter α is considered the "hits" or successful specific attacks and defined during a period specified by $\alpha = Hits + 1$ (1), and beta can be calculated by $\beta = SampleSize - \alpha$ (2). The sample size is the number of companies monitored during several years, as calculated by [1, p. 193].

To construct the probability density function, some assumptions must be made to determine the sample universe. It is worth considering human/computing capabilities and limitations that may affect the comprehension of information, which may be determined by the resources needed for disclosing cybersecurity breaches. It is fair to assume that most organizations that use the framework to report an incident have

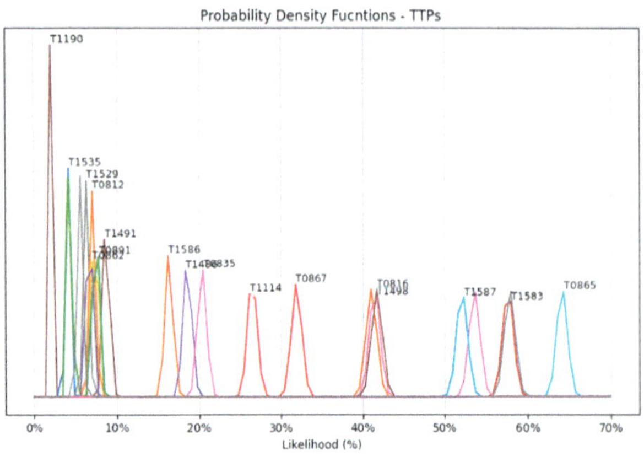

Fig. 19.3 PDFs based on prominent TTPs

data-collecting and incident response capabilities that pertain to conglomerates with higher funding and maturity levels. For this case, the researchers will assume that the sample universe is 500, which can be interpreted as the top 500 organizations with such capacity.

Each of the 28 Tactics and Techniques highlighted by the ML model was checked from the mappings attached to the original database, and total incident counts were performed. Each one of those successful quantities calculated was taken as "hits", and the unaccounted number of non-related incidents was taken as "misses". Figure 19.3 shows the result.

For example, T0865 was accounted for 3528 times, using equation (1), so $\alpha = 3529$. The sample size is 500 multiplied by 11, or 5502, since the VCDB shows breach data for the last 11 years. Therefore, by equation (2), $\beta = 1973$.

To generate the risk representation of the most influential Tactics and Techniques, it is necessary to determine the impact and show it in terms of each likelihood. Such representation takes place as "Inherent Risk Curves", which consider the probability of a loss event and its financial impact, like the ones plotted by [21] while quantifying the cyber risk of NASA missions. To determine the financial impact, IBM's "Cost of a Data Breach Report" for 2023 was considered, where a 95% confidence interval was used to determine lower and upper bounds for monetary estimates due to a successful cybersecurity breach. Figure 19.4 shows the calculation results. The inherent risk curves are calculated using the following: $P(X > x) = 1 - F(x)$ (3). The range of losses given by the data found in [22] were normalized for consistency.

A Python code was developed to calculate the Cumulative Density Function (CDF) of the given Beta function, using the "scipy.stats" library, where the function beta.cdf(range,alpha,beta) is available. One thousand points were used for this specific plotting, and the range was normalized based on this value. The function returns the CDF value defined in [20]. Equation (3) gives the graph's "y" value, and

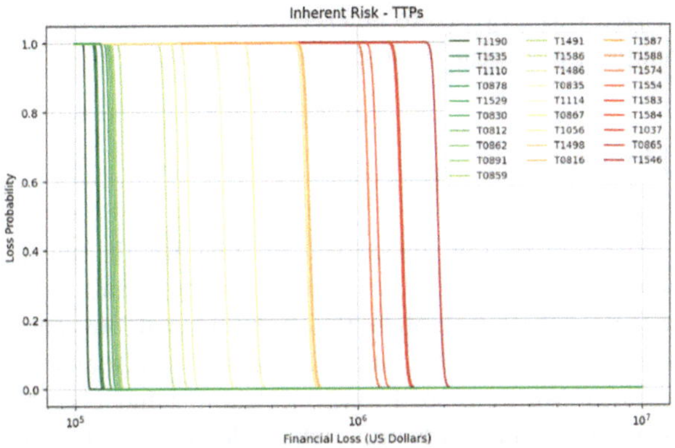

Fig. 19.4 IRCs based on prominent TTPs

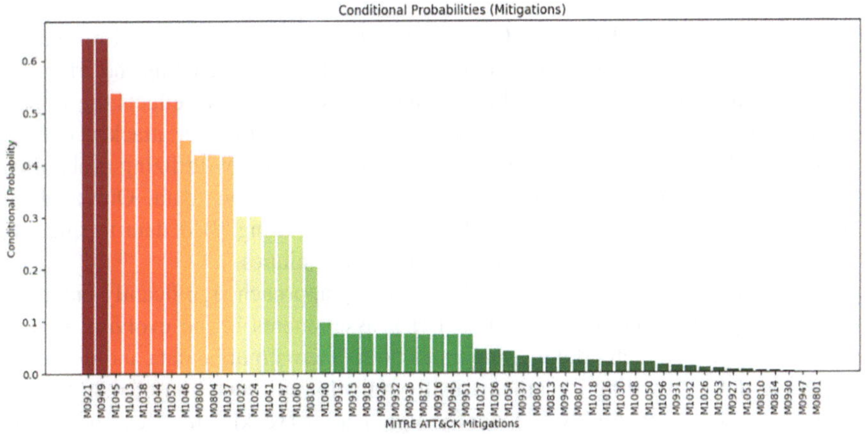

Fig. 19.5 MITRE ATT&CK mitigation probabilities based on technique dependencies

the financial loss data dictates the "x" values. This could be taken as a representation of the inherent risk caused by such threats.

Once inherent risk curves were determined, the next step was to find the linked mitigation and detection techniques that could shape the response strategy that prioritizes the most effective security controls against them. Given that mitigation and detection techniques often address more than one technique [23], simple convolution of PDFs would not relate to the premise as we cannot assume all variables are independent. Therefore, using the VCDB raw data, we calculated co-occurrences and conditional probabilities between TTPs, which provided the basis for the final calculation by using the chain rule for dependent probabilities [20, p. 202]. Figure 19.5 shows the joint probabilities that adjust to MITRE ATT&CK mitigation techniques derived from the VCDB dataset, which can be interpreted as the efficiency of such mitigations against their comprehended attack techniques.

The same approach was followed to determine the efficiency of MITRE ATT&CK detection techniques, which can be shared upon request.

19.6 Conclusion

The primary objective of this paper was to identify the most critical attack techniques with high risk using a data-driven approach and to suggest effective mitigation strategies. To accomplish this, we mapped our dataset to MITRE ATT&CK techniques to standardize and preprocess the data, and we derived CVSS v4.0 severity scores. We developed machine learning models, including random forest, regression, and neural networks, which demonstrated a strong correlation between ATT&CK techniques and CVSS v4.0 scores in our dataset. Using L1 regularization with logistic regression, we pinpointed the highest-risk ATT&CK techniques.

Leveraging the MITRE ATT&CK framework, we proposed essential mitigation and detection strategies for these high-risk techniques. Recognizing that risk is a combination of impact and probability, where the CVSS score reflects the impact severity of a vulnerability, we calculated the probability of each attack occurring along with the associated financial loss. This was achieved by utilizing probability density functions derived from the Beta distribution of our data. Our analysis enabled us to construct Inherent Risk Curves (IRC) using IBM data to quantify the financial impact of the most persistent attack techniques.

By focusing on these high-risk techniques, our approach significantly enhances the accuracy and effectiveness of cyber-risk assessments in the evolving IT and OT landscape. This methodology integrates both external/public data and internally generated data, providing a mathematically reliable representation of risk. Consequently, our results are highly relevant for organizations seeking to assess their cyber risk and develop informed strategies, ensuring their assessments are comprehensive and up-to-date.

Acknowledgments We thank FM's CEG and Cyber Research Team, especially Dr. Roland Schaefer, Tiago Dias, and Wade Chmielinski, for their support and invaluable discussions.

References

1. Hubbard, D., Seiersen, R.: How to Measure Anything in Cybersecurity Risk, 2nd edn. Wiley, London (2023)
2. Karimi, M., Mena, D.M.: Data-driven IT/OT cyber risk evaluation and mitigation prioritization using machine learning. In:2025 IEEE Conference on Artificial Intelligence (CAI), pp. 962–969. IEEE, Piscataway (2025, May 5)
3. Shaukat, K., Luo, S., Varadharajan, V., Hameed, I.A., Chen, S., Liu, D., Li, J.: Performance comparison and current challenges of using machine learning techniques in cybersecurity. Energies **13**(10), 2509 (2020)

4. Subroto, A., Apriyana, A.: Cyber risk prediction through social media big data analytics and statistical machine learning. J. Big Data **6**, 50 (2019)
5. el Jaouhari, S., Tamani, N., Jacob, R.I.: Improving ML-based solutions for linking of CVE to MITRE ATT &CK techniques. In: 48th Annual Computers, Software, and Applications Conference (COMPSAC), pp. 2442–2447. IEEE, Piscataway (2024)
6. Shenoy, N.: Correlation of vulnerabilities to MITRE ATT&CK using natural language processing feature-modeling. Ph.D. dissertation, Marymount University (2024)
7. Al-Sada, B., Sadighian, A., Oligeri, G.: Analysis and characterization of cyber threats leveraging the MITRE ATT&CK database. IEEE Access **12**, 1217 (2023)
8. Subbaratinam, S.: Machine learning based risk classification of vulnerabilities incorporating MITRE ATT&CK framework and threat intelligence. Ph.D. dissertation, Marymount University (2022)
9. MendezMena, D., Yang, B.: Decentralized actionable cyber threat intelligence for networks and the Internet of Things. IoT **2**, 1–16 (2020)
10. Verizon: Veris community database [Online]. Available verisframework.org
11. MITRE: Mitre ATT&CK®[Online]. Available https://attack.mitre.org/
12. Srinivasan, H., Karimi, M.: Threat-based security controls to protect industrial control systems (2025) [Online]. Available https://arxiv.org/abs/2501.13268
13. NIST: Nist cyber risk scoring [Online]. Available https://csrc.nist.gov
14. NVD: Vulnerability metrics, common vulnerability scoring system (CVSS) [Online]. Available https://nvd.nist.gov/vuln-metrics/cvss
15. Baker, J., Orchilles, J., Lim, N., Young, R.: Veris to ATT&CK mapping (2023). https://github.com/center-for-threat-informed-defense/attack_to_veris/tree/main/mappings/veris-1.3.7
16. Pedregosa, F., Varoquaux, G., et al.: Scikit-learn: machine learning in Python. J. Mach. Learn. Res. **12**, 2825–2830 (2011)
17. Raschka, S., Mirjalili, V., Python machine learning: machine learning and deep learning with Python, scikit-learn, and TensorFlow 2. Packt Publishing, Birmingham (2019)
18. Abadi, M., Barham, P., et al.: Tensorflow. In: 12th USENIX Symposium on Operating Systems Design and Implementation (OSDI 16), pp. 265–283 (2016)
19. Stine, K., Quinn, S., Ivy, N., Barrett, M., Witte, G., Feldman, L., Gardner, R.: Identifying and estimating cybersecurity risk for enterprise risk management. NIST Interagency/Internal Report (NISTIR) (2021)
20. Devore, J.L.: Probability and Statistics for Engineering and the Sciences. Cengage Learning, Belmont (2011)
21. Meshkat, L., Miller, R.L.: Quantifying cybersecurity risk for NASA missions. In: Reliability and Maintainability Symposium (RAMS), pp. 1–7. IEEE, Piscataway (2022)
22. IBM: Cost of a data breach 2023 | IBM (2024, Jan) [Online]. Available https://www.ibm.com/reports/data-breach
23. MITRE: Enterprise mitigations - MITRE ATT&CK (2025) [Online]. Available https://attack.mitre.org/mitigations/enterprise/. Accessed 23 Feb 2025

Chapter 20
Anomaly Detection Techniques in IoT Networks: Review and Comparative Analysis

Hamza Rhachi, Anas Bouayad, and Younes Balboul

20.1 Introduction

The term Internet of thing has evolved to encompass the entire ecosystem of interconnected objects. It begins in the physical world, with sensors that collect information. This information is then transmitted through the connection and integration of systems, and finally processed and stored for analysis and exploitation [1]. In an IoT network, even a non-sensitive device can be granted access to confidential data. Against the rapid growth of IoT systems, the proliferation of connected devices exposes companies to increased risk of cyberattacks. As a result, the security of the collected and transmitted data in the networks becomes a major issue in IoT.

Undoubtedly, IoT devices often operate in a dynamic environment, exposing them to various threats such as unauthorized access, data breaches, and denial-of-service attacks. These threats can be defined as anomalies. Anomalies in these networks, whether caused by cyber-attacks or faulty devices, can compromise system integrity, data confidentiality, and the security of critical applications. For this reason, anomaly detection is essential for identifying and mitigating threats by distinguishing normal behavior from unusual patterns, and reducing such risks.

Artificial intelligence (AI) is significantly changing the area of cybersecurity, providing powerful solutions to detect and respond immediately to threats. One of the major advantages of AI is its ability to identify previously unknown threats, also known as zero-day attacks. While traditional tools (IPS/IDS) rely on known signatures, AI can study network behaviors for unusual activity even before it's recognized as a threat. Along with artificial intelligence and machine learning, data

H. Rhachi (✉) · A. Bouayad · Y. Balboul
IASSE Laboratory, Sidi Mohamed Ben Abdellah University, Fez, Morocco
e-mail: hamza.rhachi@usmba.ac.ma; anas.bouayad@usmba.ac.ma; younes.balboul@usmba.ac.ma

E. Nowroozi et al. (eds.), *Adversarial Example Detection and Mitigation Using Machine Learning*, https://doi.org/10.1007/978-3-031-99447-0_20

processing can also advance much more, such as predicting issues and incidents or even developing predictive algorithms.

Furthermore, traditional methods are ineffective in finding new threats and require additional time for updates, which can be improved by using AI methods such as machine learning (ML) and deep learning (DL) techniques. These methods can apply advanced analytical approaches to handle the large and complicated data of IoT systems together to create a standard baseline for the network traffic of IoT devices. By analyzing this data, they can enable the detection of anomalies through learned relationships and patterns that differentiate normal from aberrant behavior [2]. A dataset must be created or generated from IoT devices, ideally consisting of real-time IoT network traffic, to train ML and DL algorithms. However, due to the complexity of IoT network data, datasets are created in advance by capturing different types of traffic on IoT networks and attack signatures. The effective utilization of these datasets highlights the importance of feature extraction, data cleaning, and preprocessing procedures. The dataset's accuracy compared to authentic data will lead to effective and trustworthy results from the AI algorithm's detection [3]. Frequently utilized datasets for anomaly detection in IoT networks include the IoT-23, NSL-KDD, BoT-IoT, and CICIoT2023.

Thus, examining the anomaly detection methods for sensing data in IoT networks is important, both scientifically and practically. This work offers a comprehensive review of past and present anomaly detection methods employed in IoT, especially in IoT data networks. This chapter is structured into several sections. Section 20.2 presents an overview of IoT networks, detailing their architecture, components, and the types of data generated. Section 20.3 provides the types of anomalies specific to IoT environments. Section 20.4 explores anomaly detection techniques, highlighting the use of ML, DL, and hybrid approaches. Section 20.5 offers a comparative analysis used to evaluate the strengths and weaknesses of various methods. Section 20.6 studies the challenges and limitations faced the anomaly detection methods in IoT networks. The last section is for the conclusion and future works.

20.2 IoT Architecture

The IoT is now a part of life; it can take the shape of smart homes, smart industries, connected cars, or connected cities. We can see that the number of connected objects and the data they produce is enormous. Therefore, an organized, effective IoT architecture is required to manage the increasing number of objects and data. The IoT architecture comprises several parts of connected IoT systems to ensure that data produced by sensors is gathered, stored, and processed in large data storage (data warehouses), and then the object actuators execute commands sent through a user application. However, there isn't one specific standard reference architecture for the IoT, as it encompasses a variety of technologies. The most basic and widely accepted format is the architecture with 4-Layers as presented in Fig. 20.1.

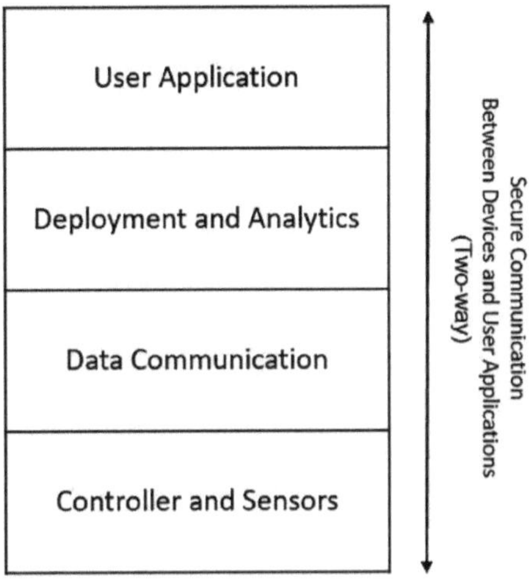

Fig. 20.1 Suggestion of IoT architecture with 4-layers

This means that no single model can be applied to all possible implementations. It must be sufficiently open with different protocols to support multiple network applications. In most cases, it consists of scalability, functionality, availability, and maintainability. Additionally, the larger volume of data produced by IoT sensors and devices has contributed to a major problem. This means that a strong IoT architecture should be able to handle the higher volume of data in the IoT system. The leading IoT system architectures are those that comprise the security of the data flowing over the various IoT networks. They are tasked with managing, observing, and examining a significant amount of information in IoT systems [4].

20.3 Data Generated in IoT

Sensors and actuators are essential in stage one of the IoT architecture. Sensors have the ability to detect data streams from the real world, they transmit this data, which can later be used for analysis [5]. One issue in applying machine learning to find anomalies in the IoT is the vast amount of data produced by IoT devices. This data might be noisy, high-dimensional, and miscellaneous, making it hard for machine learning models to identify anomalies precisely. Furthermore, the changing nature of the IoT systems leads to progress data patterns and associations that may evolve over time, necessitating frequent updates or retraining of the machine learning model [6].

Anomaly detection relies heavily on the variety of data produced in IoT networks, based on the application, device type, and motivation with anomaly detection. Here's a summary (Table 20.1) of the most types of data, transmitted in IoT Networks.

Anomaly detection methods, including machine learning, deep learning, and hybrid models, are used to find such irregularities, ensuring the reliability and

Table 20.1 Types of the data generated in IoT environments

IoT data type	Examples and their roles
Environmental	Such as temperature, air quality, and humidity, it help monitor natural disasters and guarantee sensor reliability
Hardware/device	Including battery levels and connection status, it identifies hardware faults and potential cyberattacks
Network	Such as packet logs, latency, and protocol usage, it supports and aids intrusion detection and prevents DDoS or injection attacks
IoT security and anomaly data	Including IDS alerts, authentication logs, and signatures of malicious activity, it identifies and detects new emerging threats, zero-day vulnerabilities, and cyberattacks of unauthorized access, such as multiple failed logins, and unauthorized IP access
Health and biometric	Like heart rate and activity logs, it offers early health warnings and ensures wearable dependability
Smart home	Such as appliance usage and security system logs, which improve home security, energy optimization, and unexpected door/window events

effectiveness of IoT networks. In addition, IoT networks typically generate data in the range of terabytes. Numerous sources, including smart sensors and smart types of equipment, produce this data. Both high-level generalized data and low-level raw data are possible in the IoT. Various data formats form the basis of data variation and diversity. For instance, the data in a "MAC Address" field may be separated by dashes (xx-xx-xx-xx-xx-xx), colons (xx:xx:xx:xx:xx:xx), or just 12 hexadecimal digits (xxxxxxxxxxxx). This represents a lack of consistency. Therefore, data uncertainty is caused by measurement inaccuracy and dynamic data change.

20.4 Types of Anomalies in IoT Environments

In the context of the IoT, a simple definition of an anomaly is: the observable effects of an unexpected change in a system's state that is outside of its local or global norm [7]. One crucial component of anomaly detection methods is the type of anomaly. Depending on the situation in which an irregularity is discovered, or on the way it happened, it can be abnormal or not. This factor can guide how the system will manage, comprehend, and detect anomalies, based on their type or cause.

Anomalies can be divided into two major categories: type-based anomalies and cause-based anomalies. Three subcategories make up the anomaly categorization depending on nature: Point anomalies, which are single data points that appear unusual compared to the other data, for example, a sudden and abnormal temperature spike from thermostats (e.g., a reading of 80 °C in a typical room) or abnormally large credit card purchases, the second is collective anomalies, i.e., a collection of data points that looks strange compared to the data set, although any point may be normal, these anomalies are more difficult to detect and often indicate complex problems, such as network attacks or system incompetence, the last is contextual anomalies; it happens when data looks normal in one situation but anomalous in another, as an example, a rise in network traffic might be usual during the day, but if it happens at night, it could suggest a security issue. On the other hand, cause-based anomaly classification focuses on the reasons behind the anomaly, taking into account whether its origin is malicious or non-malicious., which can include events caused by human or hardware errors, bugs, or individual users when using high traffic loads as an example. There are several types including spatial anomalies, operational anomalies, temporal anomalies, measurement anomalies, and network flow anomalies [8]. Network anomalies are a group of malicious actions aimed at disrupting, denying, or destroying some services of the IoT network system, thereby affecting its privacy, integrity, and accessibility [9].

20.5 IoT Anomaly Detection Techniques

Anomaly detection is the process of finding anomalies or outliers in a dataset, which makes it a crucial task in various sectors. The goal is to identify unusual patterns that may signify issues, fraud, or new attacks. Anomaly detection is being revolutionized by artificial intelligence (AI) through the use of sophisticated algorithms (machine learning and deep learning methods). With the ability to learn and adapt to new anomalies, AI methods can analyze and evaluate large datasets with surprising accuracy. This strong combination of AI and data analysis allows them to enhance operational efficiency and decision-making by identifying, managing, and preventing anomalies.

Anomalies can be detected with high accuracy using machine learning methods such as Isolation Forests (IF) and Support Vector Machines (SVMs). Additionally, deep learning approaches such as autoencoders (AE) and recurrent neural networks (RNN) can also be used on different datasets. It is important to mention that even when using the same model on two identical datasets, the results can vary depending on how the algorithm is used [10], and the performance can change based on how an algorithm is applied, like the features picked from the dataset or the importance assigned to each feature [11]. Below we explain the techniques used for anomaly detection.

- Machine learning techniques are essential for identifying anomalies in IoT networks. They can be classified into three primary types. First, supervised learning methods require labeled data (normal/abnormal), these methods are most effective when high-quality training data is available [12]. Typical methods include logistic regression (LR), SVMs (Support Vector Machines), decision trees (DT), and Random Forests (RF) [13], which are often used to model complex relationships while preventing overlearning. Secondly, unsupervised learning methods are suitable for situations where labeled data is limited or unavailable, we found common methods such as K-Means [14] segment data into groups, with anomalies often appearing in rare or isolated clusters, Isolation forest [15] is another efficient method for detecting anomalies by gradually isolating abnormal data points. Finally, there are semi-supervised learning methods, these merge labeled and unlabeled data to build more robust models. For example, methods based on label propagation use known data to enhance learning with unknown data [16], in this situation, hybrid methods that integrate clustering with a supervised approach, also offer a powerful alternative in this context.

– Deep learning techniques are especially effective at detecting anomalies in complex and voluminous data such as data gated from IoT networks. They mainly rely on deep neural networks that can learn complex representations [17]. Among these techniques, autoencoders are one of the most common, these unsupervised networks learn to reconstruct normal data and use the reconstruction errors to detect anomalies. Autoencoders come in several types, each type is adapted to a specific purpose (convolutional autoencoders (CAEs), variational autoencoders, sparse autoencoders, denoising autoencoders, etc.) [18]. Recurrent neural networks (RNNs) are particularly successful for processing sequential data, such as network traffic, they detect anomalies by recognizing unusual patterns in data streams. Furthermore, variant techniques like LSTM (Long Short-Term Memory) and GRU (Gated Recurrent Units) are favored for their capacity to capture long-term dependencies in the data. Additionally, Generative Adversarial Networks (GANs) stand out for their capability to generate data samples that resemble normal data, anomalies are identified as samples that the model cannot reproduce efficiently.

20.6 Comparative Analysis/Challenges and Limitations

In order to perform a detailed analysis of the various machine and deep learning techniques applied in different contexts, we have organized our findings in a table. This table provides an effective comparison of the contributions of the articles studied by highlighting several important aspects: the algorithms and approaches implemented in the research, such as neural networks and support vector machines. These techniques are included to evaluate their effectiveness and performance, along with the variety and quality of the datasets, and we have given the best technique for each article. Lastly, the challenges and limitations identified in each study, whether related to data treatments, computing resources, or model application. To gain a complete understanding of the challenges in the field of anomaly detection in IoT networks, Table 20.2 presents the best performing approaches for the period 2021–2024, while addressing the challenges and limitations encountered recently.

The analysis shown in the table highlights the strengths and weaknesses of the various machine and deep learning techniques used for AD in IoT network research. It is evident that some methods, such as deep neural networks, provide high performance because they can capture complex relationships in data. Still, they

Table 20.2 Comparison of techniques applied in different IoT contexts

Papers	Machine learning/deep learning techniques	Dataset(s) used	Best method	Challenges and limitations
E. Krzysztoń et al. [19]	This paper's writers compare anomaly detection techniques in IoT networks employing machine learning algorithms on popular datasets. They evaluate supervised (e.g., random forests, gradient boosting, SVM) and unsupervised (e.g., isolation forests, autoencoders) machine learning algorithms for identifying anomalies in IoT networks, utilizing datasets to test and evaluate the performance of the algorithms through common metrics like accuracy, precision, and recall	IoT-23 and CICIoT-2023	Random-forest and autoencoder methods	– New security issues are being created by the expansion of the Internet of Things and its incorporation into Industry 4.0 and 5.0. – Proper data pre-processing is essential for increasing the accuracy of anomaly detection techniques. – Limitations of centralized machine learning methods in distributed IoT networks.
Bhupal Naik D. S. et al. [20]	The authors examine machine learning techniques for Anomaly detection in IIoT datasets, discovering that Random Forest performs better than other techniques by comparing 5 machine learning methods: Random Forest, Logistic Regression, Light Gradient Boosting Machine, Decision Trees, and K Nearest Neighbors. They tested and examined these algorithms using reference datasets that contained abnormalities in IIoT data	Industrial IoT (IIoT) datasets contain information for predicting maintenance in filtration systems and identifying defects in semiconductor manufacturing. Additionally, Data for IoT Analytics supports modeling and resource management in settings. Finally, some data greatly supports energy optimization systems	The Random Forest (RF) method	Not mentioned in this study

(continued)

Table 20.2 (continued)

Papers	Machine learning/deep learning techniques	Dataset(s) used	Best method	Challenges and limitations
Shwe Sin Myat Than et al. [21]	This paper compared and analyzed CNN, LSTM, and autoencoder models for anomaly detection in IoT networks using the IoT-23 dataset. Measures such as accuracy, mean absolute error, precision, and AUC are used to assess the performance of the model	IoT-23	The CNN model	Not covered in this research
Maria Balega et al. [22]	This paper examines how machine learning models work well for finding anomalies in IoT networks by using various datasets. XGBoost, support vector machines (SVMs), and deep convolutional neural networks (DCNNs) are the supervised machine learning models that are employed. The authors reduced an objective function that includes a loss function and a regularization term for XGBoost and nonlinearly mapped the input to a high-dimensional feature space of the linear decision surface of the SVM model. All datasets were prepared for evaluation	IoT23, NSLKDD, and TONIoT datasets	XGBoost	The main challenges and limitations noted in this paper are: – The variety of IoT applications and changing network environments makes it difficult to identify optimal models for anomaly detection. – IoT systems in different environments can generate highly different data, necessitating models to be tested on several datasets.

(continued)

Table 20.2 (continued)

Papers	Machine learning/deep learning techniques	Dataset(s) used	Best method	Challenges and limitations
Marwa A. Elsayed et al. [23]	This research provides a comparison of methods and algorithms used to detect anomalies in IoT networks using publicly available datasets. The algorithms used are unsupervised learning algorithms: Isolation Forest (IF), Local Outlier Factor (LOF), One-Class Support Vector Machine (O-SVM), and Variational Autoencoder (VAE)	NF-UNSW-NB15	Autoencoder (AE) model	Not mentioned in this paper
Lerina Aversano et al. [24]	This paper introduces an effective deep learning method for detecting anomalies in IoT networks by using a large combined dataset that contains normal and abnormal IoT traffic. The researchers merged 5 distinct datasets containing anomalous IoT traffic. They perform hyper-parameter optimization and feature reduction using an autoencoder, along with assessing robustness to Gaussian noise	The combined dataset serves as a realistic benchmark for normal and abnormal IoT traffic, covering various attack types, including Mirai, Scanning, DDoS (TCP and UDP), DoS (HTTP, TCP, UDP), and Xbash	The autoencoder neural network model proves to be useful for anomaly detection in IoT network traffic, even with noise in the data	Not mentioned in this analysis
Imtiaz Ullah et al. [25]	The convolutional neural network (CNN) model, a deep learning-based anomaly detection model, is proposed in this research for IoT networks in 1D, 2D, and 3D configurations. The authors analyze the performance of the proposed model using metrics like accuracy, precision, recall, and F1 score. It is evaluated on multiple IoT intrusion detection datasets	The datasets used in the study were BoT-IoT, MQTT-IoT-IDS2020, and IoT-23	The convolutional neural networks model (CNN)	Not mentioned in this research

(continued)

Table 20.2 (continued)

Papers	Machine learning/deep learning techniques	Dataset(s) used	Best method	Challenges and limitations
Bauyrzhan Omarov et al. [26]	Convolutional neural networks and bidirectional long-term memory are combined in this study to offer a hybrid CNN-BiLSTM model for anomaly identification in IoT network data. The Data preprocessing procedures used for this research include data conversion into numerical format (Encoding method), normalization, and standardization. The authors implemented and applied the model in 3 steps: 1. Modeling the system for normal, faulty, and attack categories. 2. Creating datasets for normal, faulty, and attack behavior. 3. Assessing supervised machine learning algorithms on the datasets.	UNSW-NB15 dataset (contains 2,540,044 records)	The suggested hybrid model (CNN-BiLSTM) is the most effective. It reaches high accuracy compared to other evaluated machine learning techniques	– The challenge of differentiating malicious data from normal network traffic as attack assaults get more complex. – The difficulty of selecting the right (K) parameter in the K-nearest neighbor algorithm, as it influences processing time and classification accuracy.

frequently require large datasets and significant computing time. Conversely, more traditional methods, such as random forests or support vector machines, perform better when resources are limited, but they might not be accurate on complex datasets. Moreover, tab 1 emphasizes a marked dependence of results on the quality and diversity of the datasets used, confirming the importance of judicious data selection when designing a model. Finally, the recurring challenges and limitations identified present intriguing opportunities for future research addressing these issues.

20.7 Conclusion

In the IoT context, we are seeing a massive increase in the number of communicating devices deployed in different environments. This increase is driven by numerous applications that affect the traffic between connected objects. Most IoT networks are typically low complexity and therefore contain limited security mechanisms. In addition, the integration of artificial intelligence into IoT networks considerably enhances anomaly detection. It has an exceptional capability to detect anomalies and patterns in real time, allowing for proactive threat detection and thus reducing the risk of data breaches.

In summary, this chapter has examined the IoT architecture and the types of data they generate. It also conducted a comparative analysis of common machine learning and deep learning techniques used for anomaly detection, which highlighted their respective performances while also addressing the challenges and limitations of applying these techniques in various IoT environments.

In light of these challenges, several perspectives can be envisaged to improve the efficiency of anomaly detection. First, it is important to explore hybrid techniques that combine multiple detection algorithms to overcome the specific limitations of each method, obtain more robust results, and be able to capture complex IoT data. In addition, applying more sophisticated feature selection methods, especially those based on unsupervised machine learning methods could reduce computational costs.

References

1. Jot, E., Sharma, L.S.: Study of anomaly detection in IoT sensors. Int. J. Res. Appl. Sci. Eng. Technol. (IJRASET) **11**(8) (2023). https://doi.org/10.22214/ijraset.2023.55226
2. Rafique, S.H., Abdallah, A., Musa, N.S., Murugan, T.: Machine learning and deep learning techniques for internet of things network anomaly detection—current research trends. Sensors **24**(6), 1968 (2024). https://doi.org/10.3390/s24061968
3. Vigoya, L., Fernandez, D., Carneiro, V., Nóvoa, F.J.: IoT dataset validation using machine learning techniques for traffic anomaly detection. Electronics **10**(22), 2857 (2021). https://doi.org/10.3390/electronics10222857
4. Kumar, S., Tiwari, P., Zymbler, M.: Internet of Things is a revolutionary approach for future technology enhancement: a review. J. Big Data **6**, 111 (2019). https://doi.org/10.1186/s40537-019-0268-2
5. Masmali, F., Miah, S.: Different applications and technologies of Internet of Things (IoT) (2021). https://doi.org/10.48550/arXiv.2110.10452
6. Achiluzzi, E., Li, M., Georgy, M., Kashef, R.: Exploring the use of data-driven approaches for anomaly detection in the Internet of Things (IoT) environment (2022). https://doi.org/10.48550/arXiv.2301.00134
7. Cook, A.A., Mısırlı, G., Fan, Z.: Anomaly detection for IoT time-series data: a survey. IEEE Internet Things J. **7**(7), 6481–6494 (2020). https://doi.org/10.1109/JIOT.2019.2958185
8. Al-amri, R., Murugesan, R.K., Man, M., Abdulateef, A.F., Al-Sharafi, M.A., Alkahtani, A.A.: A review of machine learning and deep learning techniques for anomaly detection in IoT data. Appl. Sci. **11**(12), 5320 (2021). https://doi.org/10.3390/app11125320

9. Barford, P., Plonka, D.: Characteristics of network traffic flow anomalies. In: Proceedings of the ACM SIGCOMM Internet Measurement Workshop (2001). https://doi.org/10.1145/505202.505211
10. Rhachi, H., Bouayad, A., Balboul, Y., Aitmessaad, B.: Anomaly detection in IoT networks—classifications and analysis techniques. In: Farhaoui, Y., Hussain, A., Saba, T., Taherdoost, H., Verma, A. (eds.) Artificial Intelligence, Data Science and Applications. ICAISE 2023. Lecture Notes in Networks and Systems, vol. 838. Springer, Cham (2024). https://doi.org/10.1007/978-3-031-48573-2_67
11. Wang, S., Balarezo, J.F., Kandeepan, S., Al-Hourani, A., Chavez, K.G., Rubinstein, B.: Machine learning in network anomaly detection: a survey. IEEE Access **9**, 152379–152396 (2021). https://doi.org/10.1109/ACCESS.2021.3126834
12. Sarker, I.H., Kayes, A.S.M., Badsha, S., et al.: Cybersecurity data science: an overview from machine learning perspective. J. Big Data **7**, 41 (2020). https://doi.org/10.1186/s40537-020-00318-5
13. Osa, E., Oghenevbaire, O.E.: Comparative analysis of machine learning models in computer network intrusion detection. In: 2022 IEEE Nigeria 4th International Conference on Disruptive Technologies for Sustainable Development (NIGERCON), Lagos, Nigeria, pp. 1–5 (2022). https://doi.org/10.1109/NIGERCON54645.2022.9803175
14. Ahmed, M., Seraj, R., Islam, S.M.S.: The k-means algorithm: a comprehensive survey and performance evaluation. Electronics **9**(8), 1295 (2020). https://doi.org/10.3390/electronics9081295
15. Al Farizi, W.S., Hidayah, I., Rizal, M.N.: Isolation forest based anomaly detection: a systematic literature review. In: 2021 8th International Conference on Information Technology, Computer and Electrical Engineering (ICITACEE), Semarang, Indonesia, pp. 118–122 (2021). https://doi.org/10.1109/ICITACEE53184.2021.9617498
16. Villa-Pérez, M.E., Álvarez-Carmona, M.A., Loyola-González, O., Medina-Pérez, M.A., Velazco-Rossell, J.C., Choo, K.-K.R.: Semi-supervised anomaly detection algorithms: a comparative summary and future research directions. Knowl. Based Syst. **218**, 106878 (2021). https://doi.org/10.1016/j.knosys.2021.106878
17. Rafique, S.H., Abdallah, A., Musa, N.S., Murugan, T.: Machine learning and deep learning techniques for Internet of Things network anomaly detection—current research trends. Sensors **24**, 1968 (2024). https://doi.org/10.3390/s24061968
18. Nawaz, A., Khan, S.S., Ahmad, A.: Ensemble of autoencoders for anomaly detection in biomedical data: a narrative review. IEEE Access **12**, 17273–17289 (2024). https://doi.org/10.1109/ACCESS.2024.3360691
19. Krzysztoń, E., Rojek, I., Mikołajewski, D.: A comparative analysis of anomaly detection methods in IoT networks: an experimental study. Appl. Sci. **14**(24), 11545 (2024). https://doi.org/10.3390/app142411545
20. Bhupal Naik, D.S., Dondeti, V., Balakrishna, S.: Comparative analysis of machine learning-based algorithms for detection of anomalies in IIoT. Int. J. Inf. Retrieval Res. (IJIRR) **12**(1), 1–55 (2022). https://doi.org/10.4018/IJIRR.298647
21. Myat Than, S.S., Myint Soe, A., Maw, A.H.: A comparative analysis of CNN, LSTM, and autoencoder models of IoT intrusion detection. In: 2024 5th International Conference on Advanced Information Technologies (ICAIT), Yangon, Myanmar, pp. 1–6 (2024). https://doi.org/10.1109/ICAIT65209.2024.10754916
22. Balega, M., Farag, W., Wu, X.-W., Ezekiel, S., Good, Z.: Enhancing IoT security: optimizing anomaly detection through machine learning. Electronics **13**(11), 2148 (2024). https://doi.org/10.3390/electronics13112148
23. Elsayed, M., Russell, P., Nandy, B., Seddigh, N., Zincir-Heywood, A.: Anomaly detection for IoT networks: empirical study, pp. 432–437 (2023). https://doi.org/10.1109/CCECE58730.2023.10288813
24. Aversano, L., Bernardi, M.L., Cimitile, M., Pecori, R., Veltri, L.: Effective anomaly detection using deep learning in IoT systems. Wirel. Commun. Mob. Comput. **2021**, 9054336, 14 pp. (2021). https://doi.org/10.1155/2021/9054336

25. Ullah, I., Mahmoud, Q.H.: Design and development of a deep learning-based model for anomaly detection in IoT networks. IEEE Access **9**, 103906–103926 (2021). https://doi.org/10.1109/ACCESS.2021.3094024
26. Omarov, B., Auelbekov, O., Suliman, A., Zhaxanova, A.: CNN-BiLSTM hybrid model for network anomaly detection in Internet of Things. Int. J. Adv. Comput. Sci. Appl. (IJACSA) **14**(3) (2023). https://doi.org/10.14569/IJACSA.2023.0140349

Chapter 21
Bridging the Gap from Research to Reality: Methods for Fortifying Mitigation Measures Against Adversarial AI

Joel Brogan, Amir Sadovnik, David Bolme, Steven Young, Arka Daw, and Edmon Begoli

21.1 Introduction and Motivation

The research field of adversarial AI (AAI) and its mitigation (AAIM) is plagued by a gap between theory and reality. Some do not see AAI as physically realizable. Because it is bound to laboratory settings with poor efficacy in the wild, they might believe it poses no real threat, and therefore countermeasures against it unnecessary. But in fact, this could not be further from the truth. If we want our AI systems to stay hardened against attack, AAI countermeasures must stay one step ahead of the bad actors working to successfully deploy adversarial AI maliciously. But in practice, defense against AAI is difficult to achieve. In the words of Nicholas Carlini, one of the only effective defenses we know of is that of training ML models with adversarial examples. Therefore, this chapter focuses on strategies that support that methodology, along with strategies of preventing manipulation of AI models. Section 21.2 outlines the importance of hand-crafted adversarial examples for the purpose of fine-tuning models to mitigate their vulnerabilities, and introduces a framework and interface for producing these examples. Section 21.3 describes an approach to further bolster AI model robustness against attack via synthetic digital twin data. finally, Sect. 21.4 provides experiments with methods for preventing manipulation of AI models.

J. Brogan (✉) · A. Sadovnik · D. Bolme · S. Young · A. Daw · E. Begoli
Center for AI Security Research, Oak Ridge National Laboratory, Oak Ridge, TN, USA
e-mail: broganjr@ornl.gov; sadovnika@ornl.gov; bolmeds@ornl.gov; youngsr@ornl.gov; dawa@ornl.gov; begolie@ornl.gov
https://www.ornl.gov/caiser

E. Nowroozi et al. (eds.), *Adversarial Example Detection and Mitigation Using Machine Learning*, https://doi.org/10.1007/978-3-031-99447-0_21

21.2 The Case for Hand-Crafted Adversarial Examples

Adversarial AI mitigation measures require strong AI red-teaming efforts to develop datasets of adversarial examples with which to fine-tune models against. Often, large datasets of automatically-generated attacks fail to capture the true distribution of adversarial examples that might be seen by a model. Despite extensive research into automated attacks and defenses in academic settings, there is a lack of practical tools for designing *one-off, scenario-specific* attacks by practitioners or non-expert users. In realistic adversarial scenarios, an attacker often fine-tunes an attack on a *specific* input (or model) to achieve a desired outcome, rather than applying a generic attack blindly across a large dataset. The Lantern workbench is introduced to fill this gap. Lantern provides an interactive interface for rapidly crafting custom adversarial attacks on a case-by-case basis, allowing users to leverage domain expertise to hand-tune attack parameters and constraints for a given target sample. The goal is to "shine a light" on model vulnerabilities by enabling quick iteration and feedback, in contrast to purely automated large-scale attack evaluations.

21.2.1 Adversarial Evasion Attacks

Adversarial evasion attacks are a well-known vulnerability of deep neural networks. These attacks involve adding a small perturbation to the input (e.g. an image) that is generally imperceptible to humans but causes the model to misclassify the input. Since the discovery of this vulnerability in 2013 [18], a vast body of work has emerged exploring various attack algorithms and improvements. Key developments include one-step attacks like FGSM [8] and iterative attacks like PGD [11], among others. Researchers have improved attacks along multiple dimensions: increasing success rates (effectiveness in flipping model predictions), reducing the perceptibility of perturbations, and improving computational efficiency of generation. Comprehensive surveys (e.g., [19]) discuss the many challenges and techniques in this domain. However, much of this research evaluates attacks on entire datasets to report aggregate metrics, which does not necessarily reflect how an adversary would operate in a real scenario. In practice, an attacker often crafts a single adversarial example targeted to a specific input and goal, tweaking the method until it works for that case.

21.2.2 Adversarial Attack Tools and Workbenches

Given the importance of assessing model robustness, several libraries and tools have been developed to help evaluate adversarial attacks and defenses. For example, the *Adversarial Library* by Rony and Ayed [15] provides implementations of many evasion attacks and utilities for evaluation. This library, however, focuses on white-box attacks for classification and segmentation tasks and is built on PyTorch, limiting its use to PyTorch models. Another major toolkit is IBM's *Adversarial Robustness Toolbox (ART)* [12], which offers a wide range of attack algorithms (30+ evasion attacks) and supports models from all major deep learning frameworks (TensorFlow, Keras, PyTorch, MXNet, etc.). ART's modular design for swapping models and attacks makes it a strong foundation for building an interactive workbench.

Two Six Labs' Armory framework [16] is another tool aimed at evaluating AI model robustness, providing an environment to test models against standardized adversarial attacks. Armory focuses on reproducible, large-scale evaluations (often pairing with specific datasets and evaluation metrics). In contrast, Lantern is designed to enable custom, one-off attack generation in a user-friendly, interactive manner. Lantern leverages the extensive capabilities of ART under the hood, but augments it with a novel user interface and design principles to facilitate rapid experimentation on individual samples.

21.2.3 Lantern Workbench Design and Methodology

Lantern is an interactive adversarial AI workbench that allows users to visually and iteratively craft adversarial evasion attacks for specific inputs. The design strategy of Lantern centers on three key principles: (1) interactivity for rapid feedback, (2) modularity to easily extend to new attacks and models, and (3) introspection to automate the interface generation for attack/model parameters.

Interactive Experimentation Lantern provides a graphical user interface (GUI) (built with Streamlit [17]) that enables users to configure an attack and immediately see the results. Users can upload an input (e.g., an image), select a model, choose an attack method, and adjust parameters via slider controls or input fields. The effects of parameter changes can be observed in real time by running the attack and displaying the output (perturbed example and model prediction). This interactive loop allows for "hand-tuning" an attack—the user can iteratively refine parameters or apply constraints (like masking out certain image regions from perturbation) to meet their goals. This stands in contrast to scripting an attack offline and provides a more intuitive exploration of a model's weaknesses.

Modularity and Extensibility To support a wide range of attacks and ML frameworks, Lantern is designed to be highly modular. It builds on the inheritance structure of ART to allow new attacks or models to be added with minimal code changes. For example, developers can create a new attack class in Lantern that inherits from an existing ART attack class and a Lantern base class, automatically integrating default UI components. This multiple inheritance approach means that if an ART attack supports certain features (e.g., an adversarial patch or mask), Lantern can incorporate an interactive widget for it without additional implementation. The use of IBM ART as the backend ensures that any model/attack supported by ART can be plugged into Lantern. The workbench's code inspects available attack classes and model classes to populate the interface options dynamically.

Introspection for Automatic UI Generation Lantern exploits Python introspection to automate the creation of UI controls for attack and model parameters. When a new attack is integrated, Lantern uses the function signature (via the Python `inspect` module) to determine what parameters the attack requires and their default values. It then generates appropriate UI elements (sliders, text inputs, checkboxes, etc.) for those parameters in the Streamlit interface. This design dramatically reduces the effort needed to add new attacks or models: developers do not have to write separate GUI code for each parameter. The default parameter values and types are used to infer sensible ranges or options for the UI. This introspective approach ensures consistency and saves time, making the workbench easily extensible as new attack methods emerge.

Lantern's front-end is built with Streamlit (a lightweight web app framework for data science) and takes advantage of its extensive widget library for building the control panel. The combination of Streamlit for interactivity and ART for attack algorithms and model interfaces forms the core of Lantern's implementation.

21.2.4 Results and Example Scenarios

To demonstrate Lantern's capabilities, the paper presents two representative use-case scenarios highlighting how a user can craft and evaluate adversarial attacks through the interface. These examples show the advantage of hand-tuning attacks with Lantern's immediate feedback, compared to running pre-defined attacks blindly.

Attack on an Image Classification Model

The first example is an attack on an image classifier, targeting a misclassification on a specific image. The scenario involves an image of a *minivan*, which the adversary attempts to have the model misclassify as a *sports car* (a targeted attack). Using Lantern, the user selects a pretrained classifier and loads the minivan image. They then choose an attack algorithm (the example uses a gradient-based attack) and adjust parameters. A key feature demonstrated is the use of a *mask*: the user draws/defines a mask restricting perturbations to a specific region of the image (in this case, the vehicle's bumper area). This models a situation where an adversary might only be able to modify part of an image (or wants the perturbation confined to a certain area for stealth).

After configuration, the attack is run. Lantern then displays the original image, the adversarial image, and the difference (perturbation) image side by side for the

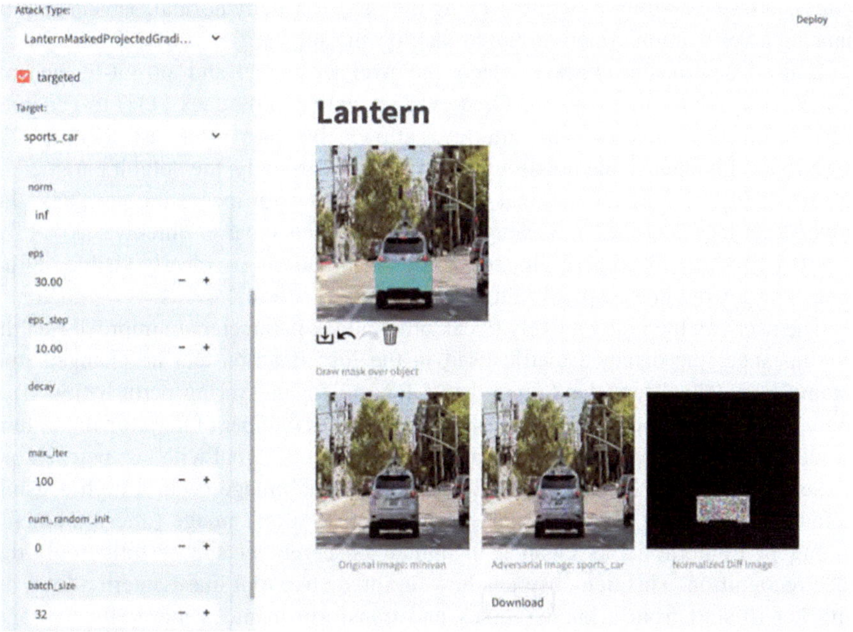

Fig. 21.1 Lantern example 1—Targeted misclassification attack on an image classifier. **Left**: Original input (image of a minivan). **Center**: Adversarial image with a perturbation applied (nearly indistinguishable from the original to the human eye). **Right**: Difference image (magnified perturbation), showing that the changes are concentrated on the minivan's bumper. In this attack, the model's prediction was flipped to incorrectly label the image as a sports car

user to examine. Figure 21.1 illustrates the outcome. In the example, the adversarial perturbation is indeed confined to the desired region (the bumper) and is almost imperceptible. Yet, the model's prediction for the altered image has changed to the target class ("sports car"), indicating a successful attack. This example highlights how Lantern enables fine-grained control: the attacker was able to achieve the misclassification with a very localized and subtle modification. Such a tailored attack might be more effective and realistic than an unconstrained perturbation over the whole image.

Attack on a Face Recognition System

The second example demonstrates using Lantern to attack a face recognition model. Unlike standard image classification, face recognition often uses a distance threshold or similarity score (not just discrete class labels). The target model in the paper is an implementation of the ArcFace face recognition model [5], which computes facial embeddings. The adversary's goal is to subtly alter a person's face image so that it no longer matches the same person's reference image in the database (thus evading recognition) while keeping the image visually normal. This is akin to causing a verification system to fail to identify the subject.

Using Lantern, the attacker selects the ArcFace model and an out-of-the-box attack from ART (the Projected Gradient Descent (PGD) attack [11]) to generate adversarial face images. The interface allows the user to adjust the attack's perturbation budget ϵ (the maximum size of noise). Initially, the default parameters are tried: Figure 5 of the original paper shows that with increasing ϵ values, the adversarial face becomes noticeably noisy. Lantern lets the user quickly see that for $\epsilon > 0.1$ (in normalized pixel intensity), the perturbations are clearly visible on the face, which would be suspicious in a real setting.

The user can then interactively tweak other attack parameters to improve stealth. For instance, the distance metric used in the loss function can be changed: the example switches from the L_2 norm to L_1 norm for measuring perturbation size. With such tuning (and possibly adjusting step sizes or number of iterations), Lantern produces a new adversarial image that achieves the attack goal with less perceivable noise. The result is shown in Fig. 21.2: the "before" image (with a high ϵ using default settings) has visible artifacts, whereas the "after" image (after parameter tuning in Lantern) looks clean to a human observer yet successfully fools the face recognition. This demonstrates how an interactive tool like Lantern allows an attacker to start from a known attack and transform it into a more effective and subtle version via guided trial-and-error. The rapid feedback loop is crucial—it enables converging on a working attack that balances misclassification success with imperceptibility of perturbations.

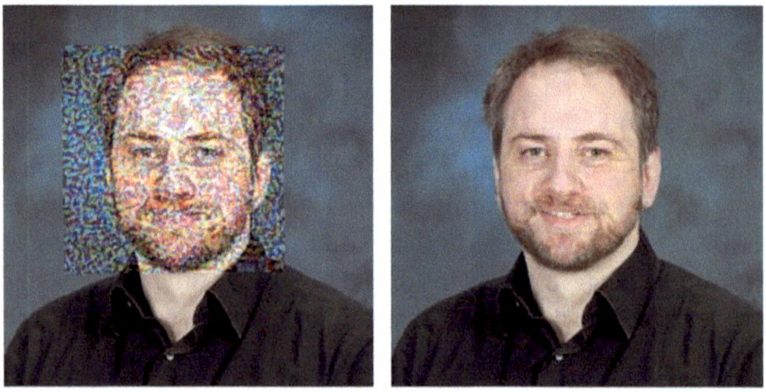

Fig. 21.2 Lantern example 2—Attacking a face recognition model. **Left**: Adversarial face image generated with a standard PGD attack at a relatively large perturbation level (ϵ), resulting in visible noise (making the attack impractical). **Right**: Adversarial face image after fine-tuning attack parameters in Lantern (using a different distance metric and optimized settings), which achieves misrecognition while also preserving natural image appearance. In a face verification system, the right image would evade matching with the original identity, while an ordinary viewer would not notice the tampering

21.2.5 Discussion: Hand-Crafted vs. Automated Adversarial Attacks

One of the key motivations behind Lantern is the recognition that *hand-crafted* adversarial attacks can be more effective in certain scenarios than fully automated, one-size-fits-all attacks. Automated attack frameworks typically run a fixed algorithm (or a set of algorithms) with preset parameters across many samples to evaluate overall robustness. While useful for benchmarking, this approach may miss targeted vulnerabilities in a specific input or may not optimize the attack for stealth on that particular input.

In contrast, manually fine-tuning an attack (with the aid of an interactive tool) allows an attacker or researcher to inject domain knowledge and intuition into the process. For example, a human can observe where an attack places perturbations and decide to constrain them to important regions (or conversely, to unimportant regions to avoid detection). They can try different attack strategies, adjust hyperparameters, or combine approaches in a way that an automated script might not. Hand-crafting attacks often reveals worst-case scenarios for a model. Indeed, an attacker in the real world only needs to fool the model on the one instance they care about—they are free to devote all their effort to optimizing that single attack, whereas automated evaluations stop once a generic attack succeeds or fails under given settings.

That said, crafting attacks by hand can be time-consuming and requires some expertise. Lantern strikes a balance by streamlining the manual process: it automates the low-level details of generating adversarial examples while giving the user high-level control. This way, even non-experts can experiment with different attack settings through a user-friendly interface, rather than coding and debugging algorithms. Another reason adversarial attacks are often manually tuned is to ensure they meet additional constraints not easily included in automated routines—such as visual imperceptibility under human inspection, physical realizability, or specific semantic objectives (e.g., "change this stop sign into a speed limit sign attack" might require trying a few approaches creatively). Automated attacks might fail under such custom constraints, or they might produce a solution that technically fools the model but looks obviously altered to a human. A human-in-the-loop can adjust the attack until these extra criteria are satisfied.

In summary, fully automated attacks provide breadth (testing many inputs quickly), whereas hand-crafted attacks provide depth on a particular instance. Lantern is designed to support the latter, giving researchers a sandbox to explore adversarial tactics beyond what is captured by standard automated benchmarks. By enabling rapid iteration with a human guiding the process, it helps uncover "weak spots" in models that automated attacks might overlook, and complements automated evaluations: together, they contribute to a more comprehensive understanding of model robustness.

21.2.6 Conclusions

Lantern represents an important step toward practical adversarial AI experimentation tools. It leverages and extends existing libraries (like ART) to allow machine learning engineers and researchers to craft adversarial examples in a flexible, interactive manner. With features such as modular integration of new attacks and models, and automatic UI generation for parameters, it greatly lowers the barrier to performing customized attack experiments. The case studies demonstrate that users can quickly create effective attacks that are tailored to specific inputs and constraints, something that would be difficult to achieve with purely automated attack pipelines.

By enabling users to go beyond generic attacks, Lantern helps illuminate the precise conditions under which a model can fail (hence the name *"Lantern"* for shining light on model vulnerabilities). This is valuable for defenders as well: insights from Lantern could guide the development of more robust models or targeted defenses. The authors suggest integrating Lantern with evaluation frameworks like Armory in the future, to combine interactive attack crafting with quantitative assessment on larger scales.

Overall, Lantern underscores the point that human-guided adversarial testing can complement automated methods in AI security. It provides an intuitive workbench for adversarial attack development, useful for both research and educational demonstrations. As adversarial machine learning continues to evolve, tools like Lantern can accelerate understanding and mitigation of real-world adversarial threats.

21.3 Using Digital Twins to Bootstrap Adversarial Training

We present a novel approach for generating synthetic datasets containing adversarial patches using a 3D digital twin system built in Unity. Our system simulates traffic cameras and CCTV environments, allowing us to apply adversarial AI patches to textures and run object detection and classification models in a controlled setting. By capturing synthetic data across various angles and poses, we construct robust adversarial fine-tuning datasets that enhance the resilience of computer vision models against AI-based attacks. Traditional datasets often fail to include adversarial perturbations, limiting their robustness. Our Unity-based 3D digital twin system capable of simulating real-world camera setups while embedding adversarial AI patches dynamically. This system generates large-scale synthetic datasets to strengthen models against adversarial manipulations (Fig. 21.3).

Fig. 21.3 A truck rendered in our 3D environment with a swappable adversarial patch. In this instance, it shows the "toaster" patch from [2]. Patches are interchanged from a either a dataset of pre-determined patches, or generated dynamically and applied within the environment to increase example sample size

21.3.1 Methodology

Our system is built in Unity and mimics real-world surveillance by simulating realistic environments with traffic cameras. It applies adversarial AI patches to object textures at various angles and positions, and runs object detection and classification models on the modified scenes. The adversarial patches are designed using established automated and hand-crafted attack methods, including those from Sect. 21.2, ensuring diversity in appearance and effectiveness. By varying object poses, lighting conditions, and camera perspectives, our dataset captures a wide range of adversarial scenarios.

21.3.2 Discussion

Providing fine-tuning datasets rich with adversarial examples is an imperative component when hardening ML models. Our method enhances the robustness of computer vision models against adversarial attacks by systematically varying the placement and presentation of adversarial patches. Future work will explore the integration of real-world adversarial data to further improve model generalization. Preliminary experiments show that models fine-tuned using our synthetic adversarial dataset demonstrate improved resilience to AI-based attacks. Detection models trained using these datasets exhibit a lower misclassification rate when exposed to real-world adversarial examples. Our approach offers scalability and better generalizability, enabling rapid dataset generation without significant manual effort.

21.4 Current Methods for Countering Model Manipulation

Another main mode of AAI is against the models themselves; either via weight extraction, model inversion, or even model weight manipulation, in which the model itself is discretely modified to behave incorrectly. In this section, we present current cryptologic and cryptographic methods for preventing these types of attacks, and experiment with their effectiveness and practicality.

This section presents a comparative analysis of the inference speed, memory usage, and proof generation time for a standard YOLOv8 model versus implementations using zk-SNARKs, zk-STARKs, and Fully Homomorphic Encryption (FHE). Experiments were performed on an NVIDIA A100-equivalent system, and results were visualized using whisker plots.

We experiment using YOLOv8 as it is a state-of-the-art object detection model known for real-time performance. For instance, a small YOLOv8n model can achieve 0.99 ms inference per image on an NVIDIA A100 GPU with TensorRT acceleration [9]. This speed makes YOLOv8 suitable for time-critical applications, but it operates on plaintext data, providing no inherent privacy or verifiability guarantees. We explore three main techniques that can enhance trust in ML inference and prevent weight manipulation or extraction:

- zk-SNARKs (Zero-Knowledge Succinct Non-Interactive Arguments of Knowledge): Cryptographic proofs that a computation (e.g., neural network inference) was carried out correctly, without revealing inputs or model. SNARKs have succinct proofs and fast verification, but proving can be computationally intensive, requiring conversion of the network to arithmetic circuits with millions of constraints [7]. zk-SNARKs can prevent weight manipulation, but not weight extraction.
- zk-STARKs (Zero-Knowledge Scalable Transparent ARKs): Similar goal as SNARKs but transparent (no trusted setup) and post-quantum secure. STARKs use arithmetic intermediate representations and rely on hash-based security, often leading to larger proof sizes and heavy computation. They trade off some efficiency for transparency [14]. zk-STARKs can prevent weight manipulation, but not weight extraction.
- Fully Homomorphic Encryption (FHE): Allows performing the inference on encrypted images directly, so the model never sees plaintext input. The result is also encrypted and can be decrypted by the client [1, 13]. FHE ensures data privacy but incurs significant computational cost (no proofs are generated in this approach). This method can prevent both weight manipulation and weight extraction, keeping the model private.

21.4.1 Methodology

The experiment involved running YOLOv8 in four configurations:

- Standard YOLOv8 using PyTorch on an NVIDIA A100.
- zk-SNARK-based YOLOv8 using Halo2 with an R1CS circuit representation.
- zk-STARK-based YOLOv8 using Risc0 ZKVM and libSTARK.
- FHE-based YOLOv8 using IBM HE layers and Microsoft SEAL with the CKKS scheme.

To ensure a fair performance analysis, all experiments were conducted under consistent conditions. Below, we detail the execution configurations for each approach below:

Hardware and Settings We used a 16-core CPU (Intel Xeon Silver @2.3GHz), 128GB RAM, and an NVIDIA Tesla V100 GPU (32GB). YOLOv8 [9] ran batch sizes of 1, 8, and 16 for standard GPU inference. For zk-SNARK/STARK and FHE, we used batch size 1 to manage proof-generation overhead and FHE's high computation cost. Each test image was 640×640 pixels.

Quantization for zk-SNARK/zk-STARK We quantized YOLOv8 to 8-bit fixed-point, replacing floating-point operations with integer arithmetic. ReLU became sign-bit checks, and a small number of convolution filters were pruned to reduce the constraint count. The final YOLOv8n model still had tens of millions of R1CS constraints, substantial but feasible for proof generation.

FHE Parallelization FHE inference employed CKKS with multithreading for large polynomial operations and ciphertext batching for parallelism. Parameters (e.g., polynomial degree 2^{15}, $\sim$40-bit modulus) were tuned for performance/precision trade-offs. We minimized bootstrapping by reordering network layers and reusing ciphertext objects. Despite optimizations, FHE was still far slower than GPU-based inference. All methods (SNARK, STARK, and FHE) were tested on 20 images, recording inference time, memory usage, and proof-generation time.

21.4.2 Experimental Setup

All experiments ran on an NVIDIA A100-equivalent system with 128GB RAM. YOLOv8 was quantized to 8-bit fixed-point arithmetic for zk-SNARK and zk-STARK. FHE inference leveraged multithreading and ciphertext batching. In our experiments, the libraries we used included PyTorch 1.13.1 with Ultralytics YOLOv8 framework, Halo2 and ezkl for zk-SNARK proving, Risc0 ZKVM v0.12 and libSTARK for zk-STARK proving, and the IBM HElayers (PyHELayers 1.3) and OpenFHE 1.0.0 for FHE.

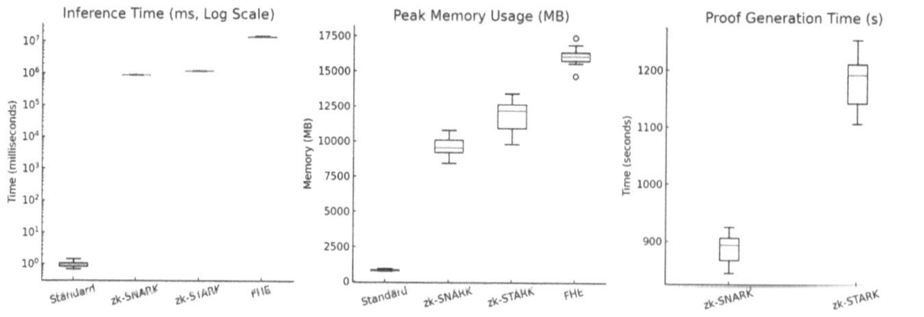

Fig. 21.4 Performance comparison of YOLOv8 implementations (10 runs each on random 640 × 640 noise images). Y-axis in left plot is log-scaled

21.4.3 Results and Analysis

A whisker plot was generated to visualize the comparison between different approaches. The performance metrics collected include inference time, memory usage, and proof generation time. The results are summarized in Fig. 21.4.

Figure 21.4 compares Inference Time, Memory Usage, and Proof Generation Time for Standard YOLOv8, zk-SNARK, zk-STARK, and FHE approaches, each tested on 10 runs of random 640 × 640 images. The left plot's Y-axis is log-scaled.

(1) **Inference Time.** Standard YOLOv8 on an A100 GPU achieves a median of about 1,ms per image. Moving to zk-SNARK inflates this to around 15 minutes due to the overhead of converting every operation into a large arithmetic circuit. Our result aligns with [10], which reported several hundred seconds for a MobileNet-based model under SNARK. STARK infers similarly slowly (around 20 minutes), partly because hash-based commitments and AIR (algebraic intermediate representation) add more steps. The big advantage is a trustless setup, not faster proofs. FHE is even more time-consuming, taking hours per image; homomorphic convolutions are notoriously expensive. Studies have shown multi-hour latencies for ImageNet-scale FHE [6], making real-time detection impractical.
(2) **Memory Usage.** Standard YOLOv8 needs minimal memory (under 1 GB GPU usage). In contrast, zk-SNARK's large R1CS constraints demand about 8–10 GB of RAM, while zk-STARK typically uses 10–12 GB. Both proving processes must handle huge intermediate data (witness values, polynomial commitments). FHE inflates memory further (16–18 GB) because encrypted data and homomorphic operations are inherently bulkier [4]. Although frameworks like Orion [6] and CryptoNets [3] have improved memory management, FHE still exceeds the other approaches in RAM usage.
(3) **Proof Generation Time.** Only the ZK methods produce proofs, and this step largely mirrors their extended inference times: about 15 min for SNARK, 20 min for STARK. Verification, however, is rapid (seconds or less [10]).

Standard YOLOv8 does no proof, and FHE relies on encrypted outputs rather than proof-based correctness. Future work aims to reduce these times via parallelization and more efficient protocols [14], but current overheads remain many orders of magnitude above real-time feasibility.

In summary, AAI tamper-proof YOLOv8 variants, and therefore any ML model variant aiming to incorporate these features that prevent weight extraction or manipulation, incur extreme slowdowns and much higher memory use. SNARK/STARK approaches add proof generation overhead in the tens of minutes, while FHE can take hours—trading speed for confidentiality or verifiability.

In particular, our key findings are the following:

- Standard YOLOv8 achieves real-time inference (<1 ms), while zk-SNARK and zk-STARK take several minutes per image.
- FHE inference is significantly slower, taking several hours per image.
- zk-SNARKs have lower proof generation time than zk-STARKs, but require a trusted setup.
- Memory usage is highest for FHE due to encrypted computations.

While the field of zkml is still young and has provided promising progress, there are still catastrophic performance trade-offs when using privacy-enhancing techniques for deep learning inference. zk-SNARKs and zk-STARKs provide verifiability at the cost of extreme latency, while FHE ensures privacy but is impractically slow for real-time object detection. Future work should focus on optimizing these cryptographic methods to improve feasibility in real-world applications.

Acknowledgments This manuscript has been authored by UT-Battelle, LLC, under contract DE-AC05-00OR22725 with the US Department of Energy (DOE). The US government retains and the publisher, by accepting the article for publication, acknowledges that the US government retains a nonexclusive, paid-up, irrevocable, worldwide license to publish or reproduce the published form of this manuscript, or allow others to do so, for US government purposes. DOE will provide public access to these results of federally sponsored research in accordance with the DOE Public Access Plan (http://energy.gov/downloads/doe-public-access-plan).

References

1. Brogan, J., Barber, N., Cornett, D., Bolme, D.: VDiSC: an open source framework for distributed smart city vision and biometric surveillance networks. In: Proceedings of the IEEE/CVF Winter Conference on Applications of Computer Vision, pp. 148–154 (2023)
2. Brown, T.B., Mané, D., Roy, A., Abadi, M., Gilmer, J.: Adversarial patch. Preprint. arXiv:1712.09665 (2017)
3. Brutzkus, A., Elisha, O., Gilad-Bachrach, R., Navon, A.: Low latency privacy preserving inference. In: Proceedings of Machine Learning Research (PMLR), vol. 97, pp. 812–821 (2019)
4. Crypto Stack Exchange User: ZK-SNARKs vs. homomorphic encryption. Crypto Stack Exchange (2020). Accessed 04 Mar 2025

5. Deng, J., Guo, J., Xue, N., Zafeiriou, S.: Arcface: additive angular margin loss for deep face recognition. In: IEEE/CVF Conference on Computer Vision and Pattern Recognition (CVPR), pp. 4690–4699 (2019)
6. Ebel, A., Garimella, K., Reagen, B.: Orion: a fully homomorphic encryption framework for deep learning. arXiv e-prints, pages arXiv–2311 (2023)
7. Feng, B., Qin, L., Zhang, Z., Ding, Y., Chu, S.: Zen: An optimizing compiler for verifiable, zero-knowledge neural network inferences. Cryptology ePrint Archive (2021)
8. Goodfellow, I.J., Shlens, J., Szegedy, C.: Explaining and harnessing adversarial examples. Preprint. arXiv:1412.6572 (2014)
9. Jocher, G., Chaurasia, A., Qiu, J.: Ultralytics YOLOv8: advanced YOLO for object detection (2023). Accessed 01 Dec 2023
10. Kang, D., Hashimoto, T., Stoica, I., Sun, Y.: Scaling up trustless DNN inference with zero-knowledge proofs. Preprint. arXiv:2210.08674 (2022)
11. Madry, A., Makelov, A., Schmidt, L., Tsipras, D., Vladu, A.: Towards deep learning models resistant to adversarial attacks. In: International Conference on Learning Representations (ICLR) (2018)
12. Nicolae, M.I., Sinn, M., Tran, M.N., Buesser, B., Rawat, A., Wistuba, M., Zantedeschi, V., Baracaldo, N., Chen, B., Ludwig, H., Molloy, I., Edwards, B.: Adversarial robustness toolbox v1.2.0. CoRR, abs/1807.01069 (2018)
13. Park, J., Kim, D., Kim, J., Kim, S., Jung, W., Cheon, J.H., Ahn, J.H.: Toward practical privacy-preserving convolutional neural networks exploiting fully homomorphic encryption. Preprint. arXiv:2310.16530 (2023)
14. Peng, Z., Wang, T., Zhao, C., Liao, G., Lin, Z., Liu, Y., Cao, B., Shi, L., Yang, Q., Zhang, S.: A survey of zero-knowledge proof based verifiable machine learning. Preprint. arXiv:2502.18535 (2025)
15. Rony, J., Ayed, I.B.: Adversarial Library (computer software). Zenodo (2024, Feb). https://doi.org/10.5281/zenodo.5815063
16. Slater, D., Cadalzo, L.: Two six GARD armory (Version 0.16.4). Zenodo (2023). https://doi.org/10.5281/zenodo.7561756
17. Streamlit: Streamlit: an open-source framework for data app creation (2024). https://github.com/streamlit/streamlit
18. Szegedy, C., Zaremba, W., Sutskever, I., Bruna, J., Erhan, D., Goodfellow, I., Fergus, R.: Intriguing properties of neural networks. Preprint. arXiv:1312.6199 (2013)
19. Zhang, J., Li, C.: Adversarial examples: opportunities and challenges. IEEE Trans. Neural Netw. Learn. Syst. **31**(7), 2578–2593 (2020)

Index

E. Nowroozi et al. (eds.), *Adversarial Example Detection and Mitigation Using Machine Learning*, https://doi.org/10.1007/978-3-031-99447-0

MIX
Papier aus verantwortungsvollen Quellen
Paper from responsible sources
FSC® C105338

If you have any concerns about our products,
you can contact us on
ProductSafety@springernature.com

In case Publisher is established outside the EU,
the EU authorized representative is:
Springer Nature Customer Service Center GmbH
Europaplatz 3, 69115 Heidelberg, Germany

Printed by Libri Plureos GmbH
in Hamburg, Germany